The Nature of Sustainability

Seeking the Physics and Patterns
that Underlie Sustainability

Steve A. Thomas

Chapbook Press

Grand Rapids, Michigan, USA

Chapbook Press

Schuler Books
2660 28th Street SE
Grand Rapids, MI 49512
(616) 942-7330
www.schulerbooks.com

Copyright © 2016 Steve A. Thomas. All rights reserved. No part of this publication may be reproduced, distributed, stored in a retrieval system, or transmitted in any form or by any means, including recording, scanning, photocopying, or other electronic or mechanical methods without the prior written permission of the publisher, except in the case of brief quotations in critical reviews and certain noncommercial uses permitted by copyright law.

ISBN: 9781943359394

Library of Congress Control Number: 2016944185

Recommended citation:

Thomas, Steve A. 2016. The Nature of Sustainability. Chapbook Press. Grand Rapids, Michigan.

Cover art by Liz Bradford. Editing assistance by Julianne Geiger.

Acknowledgements

I want to thank Dave Sollenberger, Laurel Ross, John Balaban, Jane Balaban, Steve Packard, Jerry Wilhelm, and Ralph Thornton. Each of you helped me to see nature in a new way, and I continue to learn from you every day.

I want to thank *everyone* in my family. Without you, I wouldn't be here. But more than that, I wouldn't know *who* I am.

Prologue

Toward the end of the fifteenth century, the Europeans came upon what they saw as a new land, and for many of them, what would become a new home. As they explored North America, they encountered tremendous natural bounty and hundreds of civilizations. They sailed to this new land by the hundreds, and then the thousands. They came with ideas, hopes, values, attitudes, ambition, love, threads of wisdom, and technological power. They came not only for personal aspirations, but also with a full intention of establishing a better home and a better place—more freedom, more justice, and a better economy.

With these goals, they established. They subjugated. They toiled. They fought. They had children. They loved. They prayed. And many of them died for principle. With all the wisdom they had, they set into motion something beautiful and creative. They did this for themselves, for their children, for their children's children, and for all of us here today.

Years, decades, and now centuries have passed. Each of us has a different family story—some reading this may be direct immigrants, while others may be native with ancestry stretching hundreds of generations back. For many families that settled near the middle of this country, it has now been about six generations since the long sea voyage.

Native Iroquois wisdom encourages people to guard the welfare of their people not just in the moment, but across a span of seven generations. As we shift through time as a nation—many of us moving into our seventh generation on this continent—it is a good time to reflect upon our nation and our lives. It is time for us to ask what has been achieved and where we might be headed. This honors that first generation (whether ours is recent or distant), our grandparents, our parents, and our children and grandchildren. Clearly, this nation is on the cusp of changes. These changes might turn out well, but only if we look at our lives and nation with clarity and great honesty.

More than 200 years after this nation was born, would the founders be proud of how things are turning out? Would they feel that this nation, as it stands today, was worth fighting for and dying for? Have we established a nation we are satisfied with? Proud of? Are we honored to pass it forward to the next generation? Are we living up to our potential? Are we even close to living up to our potential? Are governments and society operating as we need? As we would expect? As planned? As we wish they would?

What about the other institutions and systems that have been established over decades? Is our infrastructure working properly? Are our corporations working in favor of the nation? If not, toward what purpose do they work? Is our environment the setting we'd want, if there were a choice? Are the schools working as needed to best promote the welfare of the nation? What about our families? Our neighborhoods?

From broad to narrow scale, and from the impersonal down to our very own bodies, it is an important time to ask if this nation is producing a present and a future of high value and desirability. Or are we watching our potential fade and our dreams decay? Even if there is no single right answer, it is time to ask these questions…

This book is one person's view regarding the state of our livelihoods within this nation, and the state of our nation. This view will include an effort to understand what has led to that state, and what can possibly be done to improve that state, if we desire. There will be an effort to look at how things work, including why they work, and why they don't work. These are the views and ideas of one person from about the sixth generation.

Table of Contents

PART I NOTES FROM THE SIXTH GENERATION 1
   Introduction 3
   Chapter 1 Our New Lives 5
   Chapter 2 Why is this Happening? 8
   Chapter 3 Sources of the Problems 11
   Chapter 4 Solving the Problems 13

PART II THE PHYSICS OF SUSTAINABILITY 15
   Chapter 5 Introducing Functionality and Sustainability 17
      5.1 A Superball Example 19
      5.2 A Glacier Example 20
      5.3 Definitions of Functionality and Sustainability 22
      5.4 More with the Superball 24
      5.5 Feedback Loops 28
      5.6 Independent Forces 30
      5.7 A Watermill Example 31
      5.8 Introducing Complex Systems 45
      5.9 Compatibility 46
      5.10 Compatibility of the Waterwheel 47
      5.11 A Village School Example 50
      5.12 The Village School Example, continued 59
      5.13 Discussion: Designing a Village School and Complex System Analysis 72

PART III NATURAL SYSTEM SUSTAINABILITY 77
   Chapter 6 Natural Systems 79
      6.1 Translation to Complex Human Systems 84
   Chapter 7 Dynamics and Hierarchy 90
   Chapter 8 Dependence and Interdependence 103
   Chapter 9 Income, Wealth, and Waste 112
   Chapter 10 Competition, Cooperation, and Succession 123
   Chapter 11 Time 137
   Chapter 12 Group Decision Making 158

Chapter 13 Expression, Protections, and Group Formation 170
Chapter 14 Government 196
Chapter 15 Resilience 205
Chapter 16 Diversity 244
Chapter 17 Innovation and Management 259
PART IV APPLICATION 281
Chapter 18 Return to the Village School 283
   18.1 Pattern Review 285
   18.2 Sustainability Profiles for the Alternatives 300
   18.3 Sustainability Pattern Application Summary 319

PART I

NOTES FROM THE SIXTH GENERATION

Introduction

The values and desires of this country's first European settlers and institutional architects were not all in perfect alignment. They each drew from different perspectives. Many had witnessed immense human suffering and social and political disasters they sought to avoid repeating. Many of them had strong Christian values. Others leaned toward personal morals, ethics, convictions, or theories. There were many personalities, subcultures, and differences of opinion present prior to and during the birth of this nation. However, we might be able to ascribe one common value to all of these people—*the idea that this nation could and should live up to its potential, to a potential that so many other governments and societies fell short of.*

That idea that this nation could and should live up to a higher potential was a founding theme. In other words, there was an overall belief that from the local level all the way to the national level, the newly established and empowered institutions that sustain livelihoods would reach their potential value, and would help (or at least not prevent) other institutions and organizations from reaching their value. The concept of achieving a higher potential would carry into areas of the economy, the government, educational institutions, civil society, the landscape and environment, families, and individual wellbeing.

For example, the federal government would operate only where fundamentally necessary in order to support but not interfere with a strong economy, a livable landscape, a safe and just civilization, coherent families and communities, and productive educational institutions. Similarly, a healthy civil society would operate freely while providing enough guidance and support for a unified and just government, a healthy economy, a safe environment, intact families and communities, individual prosperity, and relevant educational institutions.

Families could grow safely, and raise children who could eventually help the nation conduct its business. The environment would provide plenty of resources for a prosperous nation: wood, productive farmland, fish, game, and space for free enterprise, among other things. By accomplishing important goals without interfering with other aspects of people's livelihoods, the institutions, organizations, systems, and structures of this new blessed nation would live up to a high potential. The Founding Fathers attempted to establish this over long debates, negotiations, and a number of extremely innovative institutional provisions.

More than two hundred years have now passed since our national journey began. Are these aspirations being born out? Have institutions such as the government and civil society developed as the nation's founders expected or wished? Is our government operating cleanly and carefully, preserving personal liberty, establishing fairness, enabling a prosperous economy, enabling strong and healthy families, allowing for stable and functional communities, and promoting a livable landscape? Does our economy support our wants and needs? Does it provide rewards for good work and allow improvements to move forward? Does it provide a surplus for a healthy government and leave behind a place worth living? Does it do all this without consuming the time and space needed for happiness and fulfillment?

Is this a nation full of people that feel a sense of achievement in their lives? Are these people active in keeping their government and communities on a straight path? Do they have the tools and leverage to leave things in good condition for the future?

Do we have communities and families that share a common purpose and are strong enough to give back more than they need to take? Do people feel a sense of belonging and a satisfaction in the fact that they share a common fate? Do we have a landscape that has what we need in the deepest sense, such as productive land, clean air, clean water, manageable weather, and the living natural backdrop that reminds us that we're home?

This is not to ask whether things are perfect, but whether most of the time, in the most important ways, these systems and structures accomplish their most important goals. Do they?

Chapter 1 Our New Lives

We have a massive federal government that often seems too large or complex to act sensibly and quickly, and while recently there have been no violent revolutions, divisions are so severe that paralysis is standard. Despite this paralysis, government spending grows rapidly, and a crushing debt looms. We have educational institutions that are criticized for creating average or unqualified students at high cost. Schools are often embattled between teacher's interests, administrators, parents, and politicians, and educational efforts seem pulled in so many directions that they seem schizophrenic.

Our economy is as easy to move ahead as a mule and as easy to anticipate as a bucking bronco, and in the past few decades many people have been completely thrown off and had to learn to carry their own packs. The plain reality for many people is that they can barely afford a rent payment, home ownership, a car, furniture, new clothes for the kids, daycare, college tuition, the electric bill, quality food, a short vacation, or tickets to a ball game or a movie.

The "growth economy" seems unwilling to grow lots of what we want it to, such as jobs for all the willing workers, workers for all the open jobs (ironic but true), a somewhat level playing field, good income for hard work, and services that don't require us to skip our next credit card payment.

Each year the environment seems to act less like a friend and a little more like a sparring opponent. Weather swings back and forth, trouncing old temperature and storm records, and the idea of "normal weather" is becoming an oxymoron. Each year more people are finding out the hard way that they live in harsh climates where fires, hurricanes, tornadoes, drought, ice storms, blizzards, rising seas, or flooding are a costly reality or continual danger.

If we're *lucky*—and we may not be—crop productivity will continue to rise as our population grows, but that is possibly only due to the use of more chemicals and irrigation, which raise the cost of producing food. Most of the plants and animals that were abundant when Washington, Lincoln, and Roosevelt walked the landscape are no longer encountered during a typical day—now they're often relegated to widely scattered remnant areas or little corners of land behind a preservation sign. And even there, they continue to decline. At the same time, our landscape is so full of irritating weeds and pest animals (such as starlings, quagga mussels, feral pigs, Asian carp, pigeons, raccoons, ragweed, poison ivy, knapweed, cheatgrass, rush skeletonweed, giant phragmites, Australian paperbark, and kudzu) that we hardly know what nature means anymore. In many places drawing inspiration from our contemporary landscape can require mental effort.

It would be nice through all of this if everyone could say, "At least I have my health," but so many of us don't. Many of our bodies, and even our children's bodies, seem like they aren't doing so great. We are often tired, heavy, forgetful, and with endless types of aches, pains, and maladies. Whether it's some type of skin, sinus, eye, stomach, bone, joint, headache, hair, hormone, weight, sleep, anxiety, energy, attitude, viral, fungal, bacterial, or mood problem, one might wonder if we'll someday run out of names for all of our syndromes and sicknesses, and run out of letter combinations for all the drugs that we'll be taking to treat them.

Through all of these stresses, our families and communities should be supportive. If you are fortunate, your family is dependable, understanding, and respectful, and your community feels like part of your home—full of places and people you can count on. Whenever you have a real problem, someone is there (literally there) to help. Maybe you and a neighbor carpool the kids to sports to reduce your trips. Maybe another neighbor lends their pickup truck when you have to haul bricks from the garden center. Perhaps you don't feel strange chasing your dog through neighborhood yards, because these are people you know.

Or, maybe these scenes seem like part of some old children's story that is no longer realistic. Maybe like many people nowadays, the family and community you assumed would be there have slipped away. Maybe you have landed in a place where the family unit and neighborhood seem absent, broken, or only make matters worse.

Do you spend a quarter of your life running your kids around to activities, buying them things, driving to or from their mom or dad, following a divorce? Do you come home to find your kids sitting in front of a blaring TV, watching something you hardly approve of? Are you glad they are indoors, because if they were out, they might be at the neighbors', whose kids have gotten into drugs? Maybe your parents or other family members are divorced with troubles of their own, and are unable to support one another. Have you become a caretaker for the sick in your family, spread so thin that you have nothing left for those that are healthy? Is your neighborhood too unsafe or unknown to take a walk at night? Do you have trouble finding time for your best friend, or the most important people in your life?

If this is your life, you might wonder how long you can continue. If some of this sounds like your life, you aren't alone, and despite how it feels you are less alone with each year.

Perhaps many of us assumed the governments and institutions that predated our birth would simply work correctly, or that if they did have problems, they could be solved. But attempts to solve anything at the national level seem to lead to endless debate and antagonism. At the individual level, many of us assumed we could do our schoolwork, make friends, eat what mom and dad told us to, get married, have children, get a good job, work hard, have a satisfying life at work and at home, and have time and money to spare. Perhaps we put in our best effort, working hard and with good intentions, but ended up without a very good job, without a good income, with no home, without a sense of belonging in our neighborhood, few real friends, divorced, short on time (or very short on time), feeling powerless, feeling hopeless, without much energy or good health, and without any indications that all of this could possibly get better. The life many people expected from their efforts is MIA, and they just can't figure out how to get back on track.

And so, on behalf of our nation's founders, ourselves, our children, and their children, *is this nation living up to its potential*? Are the institutions, systems, and structures that frame our lives—from the largest to the most local—operating such that they are providing their highest potential value? Are they helping (or at least not preventing) other sectors of our lives from reaching their potential value? We might be able to close our eyes and force out an optimistic "yes," but a critical inspection suggests that at best, the answer would be mixed.

At least in many ways, our institutions and organizations are not especially working as intended. Worse though is that the trend line seems pointed away from the goals originally envisioned, and new fault lines seem to be appearing regularly. Many people have an increasing skepticism that this ship is pointed in the right direction, or that it can be righted with the tools at hand.

In light of these growing problems, antagonism, and disconnections, perhaps the most realistic answer is "*No,*" *despite the intelligence, innovation, and best intentions of the nation's architects and many who have since followed, from many fundamental perspectives, many of our most critical systems are not meeting their potentials, and they are missing the mark by enough that we have large problems*. More troubling, the trend lines are mainly negative, such that what looked like mere difficulties 30 years ago seem to be seeping in as existential threats that look ready to swallow us. These are the findings of one person from the sixth generation.

Maybe you accept the argument that our collective condition is now somewhat troubled and in decline. But that doesn't explain *why* it is happening. If we don't know why these downtrends are occurring, we probably won't be able to fix them. Some possible causes of these trends are the focus of the next chapter.

Chapter 2 Why is this Happening?

If you happen to agree that our systems, structures, and institutions are not working well, then you probably agree that they ought to be fixed. What are the odds that we are going to fix them? The odds are probably zero unless we can first understand *why* they aren't working. How likely is it that you can fix your car by replacing a randomly selected part? Pretty low. Thus, we need to ask, "*Why* are our systems, structures, and institutions failing us in significant ways, and not living up to their potentials?" Let's consider four possibilities:

Explanation #1

Maybe it is physically impossible for our systems, structures, and institutions to prosper/succeed. For example, perhaps physics dictates that employment levels cannot exceed 90%, that no form of large government can run without incurring a large debt, that no planet can continue too long before the climate is destabilized and water is poisoned, and that no communities can exist without major alienation and strife. In other words, maybe there are laws or limitation of physics for systems, structures, and institutions, almost like the laws that dictate the speed of light and the forces of magnetism.

While this argument might seem to have some potential validity, further inspection indicates that it is faulty. For one thing, deterministic physics does not necessarily operate at the level of higher-order systems or complex systems such as governments, communities, and corporations. In fact, complex systems exhibit unpredictable behavior, and if you believe that humans have free will, then that also means a given complex system (such as a government) probably doesn't have a predetermined fate.

As additional evidence against the "laws of physics" argument, there are many cases where nations are not beset with the bourgeoning problems of ours. In fact, our own nation in prior years is an example of relative success, making this explanation seem weak.

Explanation #2

Our systems, structures, and institutions could prosper and succeed, but unqualified people with poor attitudes and selfish behavior are in charge of operating them. If people are continually bungling and incompetent or looking out only for themselves, they will display poor judgment and will do simple things incorrectly, like not following installation instructions, taxiing to the wrong runway, or making mathematical errors. This is causing our systems, structures, and institutions to fail. Is it possible then that most people are simply so self-centered or careless that they fumble through work, planting the seeds of destruction?

Like the prior argument, further consideration suggests this simply does not explain the fix we are now in. Professional mistakes occur, but those usually stand out because they are not the standard. There are exceptions, and everyone is imperfect. But on the whole, this nation is filled with people who are attentive, well intended, and sufficiently qualified for their occupations—construction workers, teachers, surveyors, marketers, nurses, mathematicians, policemen, mayors, truck drivers, welders, human resource managers, journalists, administrators, engineers, computer technicians, air traffic controllers, accountants, etc. This list could include roles such as mothers and fathers as well.

Even if you still believe in an explanation of personal incompetence multiplied, there isn't strong supporting evidence for this, such as perhaps a high proportion of the ordinary citizenry in jail, a high proportion of firings from skilled jobs, or a lack of motivation or work ethic, which is instilled in all school children or job trainees. An explanation of selfishness or incompetence can explain some individual situations, but it cannot explain the broad pattern of failure and difficulty that seems ubiquitous now. Finally, to believe that we are all this incompetent is also to forego hope of long-term achievements.

Explanation #3

Collectively, we have been unconsciously seeking our own demise. To suit a somewhat hidden psychological drive, we have been designing systems, structures, and institutions to deliver negative outcomes.

It may seem illogical that people would put so much effort into things that won't satisfy them. But if you have ever watched someone defiantly forge ahead with destructive choices, despite pleas or warning from those around them, then you might agree that this can occur. This cannot explain all the malfunction we witness, but it is a critical piece. This explanation is discussed further below.

Explanation #4

A fourth potential explanation for the bind we find ourselves in is that under current conditions, and as currently designed, our systems, structures, and institutions are simply incapable of producing the results that we want. This isn't just saying that "something isn't working." It is going a step further, and saying "something *cannot* work," at least not as intended.

What does this mean? If you saw a huge sailboat with a main sail the size of a bed sheet, or a car with really big tires on the left side and really small tires on the right side, or someone trying to carve a pumpkin with a door key, you'd know immediately that you were looking at things that *cannot* work well, even with the best of efforts. Is it possible that our systems, structures, and institutions are in a similar situation?

It is easy to imagine a failed ship powered by a tiny sail, a failed car with lopsided tires, and a failed pumpkin carved with a little key. But when it comes to complicated systems, structures, and institutions, failing designs aren't so obvious. What would these failing designs look like? What if the local town simply isn't laid out very well for current conditions, and the layout is interfering with the formation of a healthy community? What if the university is teaching too few subjects and is thereby stifling the culture? What if the average workday is too short and is creating economic hardships for families? Maybe food prices are too high, and it's impacting the availability of labor.

But how would anyone know how a town should be laid out? One person might like the layout, but another might not. Who can gauge the number of subjects that the university should teach? Isn't that just a matter of preference for university administrators, and unrelated to culture? And the same for workday length and food prices. Some people might find them suitable and others might not, but what does that have to do with family needs or the availability of workers?

The argument that will be focused upon herein is based upon the fourth explanation—that indeed our systems, structure, and institutions are failing in their objectives because they aren't well designed for existing conditions. Further, a hypothesis will be presented. The hypothesis is that these complicated systems, structures, and institutions have real functions and processes that are patterned. This means their functions and processes are not arbitrary, and not random; nor are they based upon whimsical opinions, or upon what any of us happens to want.

Thus, the town's layout might be interfering with the community, even if everyone in town "is used to" the layout. The university might be stifling the culture without intending to do so at all. The workday length might be causing financial stresses, even if everyone has agreed to the same schedules. And yes, perhaps food prices are affecting worker availability, even if nobody is aware of the connection.

More than anything, this book is an attempt to help people make these connections, and to help them understand why their systems, structures, and institutions are producing the results that they are.

Chapter 3 Sources of the Problems

The view presented herein is that we, along with our systems, structures, and institutions, are not sufficiently living up to our potentials, and that we have inadvertently been causing this to occur through two avenues. First, we have often set our sights on circumstances or events that will not satisfy us; actually, they will *hinder* us from reaching our potentials and satisfaction. Second, we have set up many of our systems, structures, and institutions such that they are, quite simply, incapable of providing their intended result; thus they cannot fully support us as we try to reach our potential, no matter how hard we try, no matter how angry we get, and no matter how long we stick it out with them.

It probably seems crazy that many of our problems are born from our strong attraction to things that reduce our level of fulfillment, but we see similar behavior in children. They refuse the bitter medicine that they need to get better. They whack another child and are stunned when they receive a revengeful conk on the head. They want to play with chemicals, matches, or fireworks that injure them. They choose the mindless video game over learning something. The list goes on and on. Clearly, the limited level of awareness that children have influences them to make choices that don't really benefit them in the medium term or longer term.

While it's easy to point out this unaware behavior in our children, it can be harder to see in our adult neighbors, and even harder to see in ourselves or in close associates. All of us, even adults, have developed a strong tendency to crave and work for short-term substitutes, rather than for what fulfills us, or for what our well-being apparently requires. We pursue things that are immediately exciting or pleasing but aren't beneficial in the longer run. With this "candy mentality," we crave surface over substance.

To fulfill our craving for eye "candy," we choose a shiny appearance over quality. For our ears, we often choose continual talking and noise or synthesized stimulus over silence, natural sounds, or personally moving music. For our noses, we drape our bodies and clothing in manufactured chemical scents, rather than create an environment with its own good smells.

For our social side, we follow the lives of celebrities and dozens of keyboard friends, rather than spend time with the most important people in our lives. For activities, we spend weekend days watching other people play sports on TV. For occupation, we work for higher pay and pass up the chance to work on something we love. When it comes to romance, many people choose sex over intimacy. Wanting to find their souls, many take drugs rather than trying to see and understand the divine already in and around them. And of course, many people find that the candy in their mouths tastes so much better than the real food their bodies need to be healthy.

There's nothing intrinsically wrong with good looks, background music, making friends on the internet, getting paid well, watching TV, and jelly beans. And we all need time to rest and periodically escape from our regular lives. But when we have a "candy mentality" we take it a step further and actually begin to strive for these superficial things.

We all struggle with this mindset from time to time. Unfortunately, once the candy has become the goal instead of the occasional treat, then no matter how much we get, it doesn't connect us to our happiness, lasting success, or continued fulfillment.

But that's just one problem. There's a second problem—our systems, structures, and institutions contain design flaws or are mismatched with conditions. If they are poorly designed, it means that well-intentioned and smart people have been creating and working within systems that include incorrect assumption about how things work.

If all this is true, then we should just be able to admit the truth and get everyone on board to make any changes that are necessary. But unfortunately, the notion of changing things that don't work often meets great resistance from people. Why?

No one likes the idea of being wrong. What's so uncomfortable about being wrong? Maybe it is frightening for people to think that goodwill and hard work can *still* result in utter failure to reach a goal (Wasn't the War on Poverty well intended? How could it fail?). Maybe it's just arrogance, and people would prefer to believe that humans are too sharp to err in this manner, or that the Founding Fathers were aware of every national pitfall, and that the Constitution essentially set us on a path of permanent success, or that this nation is blessed with a benevolent cadre of government officials that are too perfect to make significant missteps.

Maybe the culture has convinced people that "winning" is too important. We have become used to seeing people finger-point and refuse to accept blame for any problems. We are all affected by the desperation to win. However, while we hold onto winning and being right, we are also incapable of making improvements.

In summary our nation is moving toward the brink of destruction because we are seeking superficial satisfaction at the expense of long-term gain, and our systems, structures, and institutions are simply incapable of delivering what we want, at least under contemporary conditions. Next we will look into dealing with these unproductive conditions.

Chapter 4 Solving the Problems

Our craving for excitement and all types of candy—the taste, the sound, the lifestyle—can cause us to choose short-term pleasure at the cost of our greater livelihood. We all face the same periodic dilemma: we crave something in the short term that will not benefit us much in the long term. In the face of that dilemma, sometimes we choose short-term benefits; other times we choose long-term benefits.

We all have moments of strength and weakness, and we all need rest, relief from toil, and some indulgence at times. But what makes the difference for each of us individually and then in mass, is whether we have come to believe that our greater fulfillment might eventually be found *in the candy*. Once we collectively adopt this belief, we start giving up whole aspects of our potential for one temporary pleasure or impulse after another. In contrast, when we know our happiness doesn't really dwell inside the candy, then we use our energies to work toward a better future.

If we want to turn the corner and make progress with the problems that are growing before us, we will also need to challenge our attitude about the value of being right. We will need to shift our priority *away from being right,* and toward *getting it right*. We will need to observe attentively, instead of cherry picking the information that complements our own point of view. We will need to pay attention to new information and research findings that defy our expectations.

We will need to question the groups and organizations that make us comfortable, and that we normally agree with based upon our political leaning, our race, education level, religion, personal preferences, etc. We will need to stop making kneejerk assumptions about how everything works. This means asking hard questions. It means questioning information that is convenient and comfortable. It means continuing to inquire until we have arrived at something solid.

And finally, to address the problems at hand *we will have to become better at understanding how things work, and predicting whether specific choices will lead to outcomes we want.* This will require some creativity, and is the focus of the remainder of this book. From the personal to the national, the changes we will need to make may be difficult. They may be frightening. On the other hand, if you look at where things are headed now, where do you think we'll end up if we don't make some fundamental changes? Plus, is there anything more worthwhile than creating a much better future for yourself, your family, and your nation?

Next we begin to look very fundamentally at how things work.

PART II

THE PHYSICS OF SUSTAINABILITY

Chapter 5 Introducing Functionality and Sustainability

We will need to do three main things to regain our direction. First, we will have to stop seeking immediate satisfaction by way of habit and lifestyle. Second, we will need to step back and observe instead of emphasizing winning and being right. Third, we will have to become better at forming systems, structures, and institutions that produce more successes and fewer failures.

If our systems, structures, and institutions (from here onward, "systems, structures, and institutions" may be referred to simply as "systems") are not working the way we'd like, then they should be fixed. But how can they be fixed? Where do we begin? This probably seems like a daunting task.

Let's consider an analogous situation. If your bike wasn't working properly, you would want to repair it. If you had to do the repair yourself, but didn't really know how, you might first read the owner's manual that came with the bike. The manual could help direct you to the source of the problem, and explain how to fix it. If the manual was missing or unhelpful, you might seek somebody else's assistance—someone who knew enough to help fix a bike problem.

But if you couldn't find a skilled helper, then as a last resort, you might try to tinker with the bike. You could start watching each moving part, and then you could start gradually taking the bike apart. As you deconstruct the bike, you could carefully investigate each part. By doing this you would begin to understand just what each part was designed to accomplish.

If all goes well, you will eventually come to a part or connection that looks broken, warped, misaligned, etc. If you were careful and kept track of each bike part along the way, you would be able to carefully reconstruct the bike with any necessary fix or adjustment. If successful, your thorough analysis will have led you to the problem, the solution, and a working bike.

Perhaps we should be able to go through the same steps with our broken systems. Things such as economies, educational systems, and land-use patterns don't come with owner manuals, so perhaps our first inclination would be to ask an "expert" how these complicated things should be fixed. However, if we consult the experts, we will find a wide array of opinions, claims, and counterclaims.

Perhaps more troublesome would be the fact that many of the experts actually appear to be overseeing many of the systems that are producing unwanted results. Because experts often disagree with each other, relying upon them in many cases would require us to choose sides, and would simultaneously place trust in practices that appear to be at least associated with negative results. Odds are, by using expert guidance, we'll do no better than the status quo.

So here, we will try the third option—looking at these systems, bit by bit, in an attempt to understand all the parts. If we can understand all the fundamental units of something such as a regional economy, then perhaps we can theoretically figure out how to "disassemble" the troubled economy and reassemble one that works better.

We may find it difficult as we attempt to understand how our systems work (or should work) by looking at their individual constituents and interactions. And our quest should be *all* the systems, structures, and institutions (not just some of them). Therefore, our analysis will begin in the most fundamental, simple manner possible. Our process will have well-defined rules, but it will need to remain general enough to pertain to just about anything.

Our process will begin by answering this simple question: Why do some things *work*, while others don't? In other words, what properties allow one system to accomplish what we want it to, but cause another system to fail?

Our discussion of why things work or don't work (i.e., fail) will be aided by two closely intertwined concepts: *functionality* and *sustainability*. These will be formally defined later. But for now, a useful concept for *functionality* is whether something works, or how well it works.

A useful concept for *sustainability* is whether something can continue to work.

Why are some things functional or sustainable? Why are other things nonfunctional or non-sustainable? Here is an answer:

All objects, groups of objects, contraptions, systems, and organizations of any kind have this in common: they are able to function for—or toward—a given *objective* if and only if they have some amount of these two things:

1. Internal designs, conditions, and forces that support the objective.
2. External designs, conditions, and forces that support the objective.

The objective can mean anything—for a bike it might mean carrying a person, or for a door handle it might mean holding the door closed periodically. In this case the word *internal* refers to that which occurs within the thing or *of* the thing we'd like to have function. *External* refers to outside of that thing.

The idea is fairly simple. It means that when there are conditions, processes, or energies that support objects, systems, contraptions, and organizations in regard to a particular objective, then those objects, systems, contraptions, and organizations can function toward that objective.

Next, we can apply these concepts of internal and external support to a simple hypothetical example.

5.1 A Superball Example

We want to look at a simple case of how internal and external support can lead to functionality and sustainability. First we need a *thing*, and we need to know just what we want it to *do* (i.e., its *objective*), or what we expect it to accomplish. So let's start with a simple thing.

Let's say the *thing* is a *superball*, and let's say that its *objective*, or what we expect it to do, is *bounce up and down*. We've probably all had fun with that at some point in our lives.

According to criteria #1 in our expression of functionality above, for this superball to bounce up and down, it must have internal designs, conditions, and forces that support the bouncing-up-and-down objective. What does that mean for a superball? Without taking too long to become a toy products expert, we could surmise that to be successful at bouncing up and down, the ball will have to be liftable and throwable (by a human being), be able to bounce up after striking a surface, be able to bounce repeatedly, bounce with some predictability of direction, and to be reusable.

Similarly, according to criteria #2 above, for this superball to bounce up and down, it must also have external designs, conditions, and forces that support bouncing up and down. What does that mean for a typical superball? Without overanalyzing the external environment and getting ourselves confused, we could suggest that for the ball to behave successfully, there will need to be a hard flat surface, someone to periodically throw or drop the ball, and something to pull the ball back down after bouncing, like gravity.

Whew! That might seem a little awkward, but this type of analysis isn't so hard to do. We made it through the first step of a simple and silly example of understanding the criteria for a thing to work. This exercise may seem very foreign at first, but there is a greater purpose to this.

Okay, we figured out how a simple thing can work. But to have a stronger understanding, we also have to be able to see the other side of the coin. Why do things fail to work? Or more simply put, why do some things not work? It isn't very complicated, because it is just the *inverse of what does work*. We could say that all objects, groups of objects, systems, contraptions, and organizations of any kind are *unable* to achieve or work well toward a given *objective* if they have a prevalence of these two things:

1. Internal designs, conditions, or forces that limit, weaken, or fail to support the objective.
2. External designs, conditions, or forces that limit, weaken, or fail to support the objective.

To see a case example of this, let's stick with the bouncing superball. We said that to work well with regard to its internal design/conditions/forces, the ball would have to be liftable and throwable (by a human being), be able to bounce up after striking a surface, be able to bounce repeatedly with some predictability of direction, and to be reusable.

Now let's imagine taking away liftability and throwability. Will the superball work if it is too large or too heavy to lift and throw? No. Will it work if you pick it up and it sticks to your hand like superglue? No. Will it work if the ball is made out of soft, fluffy material? No. Will it work if it is shaped like a cube with sharp corners? No. Will it work if it cracks open after one or two bounces? Of course not. What about external conditions?

We said that to work well with regard to its external design/conditions/forces, it needs a hard flat surface, someone to periodically throw or drop the ball, and gravity. If we take these things away, will the superball work? Will it work if it is used against a pile of dirt or the surface of a pond? No. Will it work if nobody drops or bounces it? No, it will just sit there. Will it work without gravity? No, it will fly into space and never return after its first bounce (unless you keep it in a confined room).

It's really that simple. Any *thing*, any *object*, any *machine*, any *system*, any *organization*, any *institution*, or any *program* that we create needs internal and external designs, conditions, and forces that support their objective or intended function. Without those, they do not work optimally, and may be confined to working poorly, barely working, producing unwanted results, or not working at all. (You may notice that technically, the requirement for functionality can be met with only internal support, and lacking external support; or only external support, and lacking internal support. However, such situations require extreme conditions that don't provide much insight on our particular quest.)

Interestingly, the same requirements for functionality hold true for things that we didn't create. Next we'll look at another fairly simple example of something that no person built.

5.2 A Glacier Example

We can do the same type of functionality analysis on natural phenomena such as a stream, a cloud, sunshine, or a glacier. Let's try it on a *glacier*. First we have to define our objective or function of interest for the glacier (you have to know a little bit about glaciers to do this, but it's easy to follow along if you don't). Let's assume that our objective really just reflects typical glacial behavior: to continually grow or gain new snow and ice on the upper parts while the lower parts continually move downslope and degenerate into meltwater.

According to #1 of our requirements for functionality, for this glacier to continually grow on one side and melt or slough off on the other, it must have internal designs, conditions, and forces that support this objective. For a glacier, these would be an ability to transform snow to packed ice, and an ability for that ice to melt at warmer temperatures, and essentially to flow under pressure. The external design/conditions/forces (second requirement for functionality) that enable this often are a mountainside with cold winters, cool summers, and ample snowfall. That's it. We're done with what is required for our hypothetical glacier to behave like a typical glacier.

Now to broaden our understanding, we want to also determine things that will *prevent* the glacier from achieving its objective. In this case, this means to prevent it from behaving like a typical glacier. Recall that all objects, groups of objects, systems, contraptions, and organizations of any kind are unable to function well for or toward a given objective or purpose if they have a prevalence of these two things:

1. Internal designs, conditions, or forces that limit, weaken, or fail to support the objective.
2. External designs, conditions, or forces that limit, weaken, or fail to support the objective.

We said that to work well with regard to its internal design/conditions/forces, the glacier would need the ability to transform snow to ice, and the ability for that ice to melt at warmer temperatures, and to essentially flow under pressure. Now we have to turn this around in the logical sense. Very simply, if water and ice had different physical properties, the whole glacier thing might not work.

If, for example, water had the physical properties of sand, our accumulating sand piles would never compress into rock (like snow compresses into ice chunks), and if they did solidify, the rock wouldn't tend to flow downslope (the way ice does under pressure). Another way to imagine a failed glacier would be one in which salt was spread on the surface after every snowfall, causing the snow to melt and preventing thick ice from ever forming.

We said that to work well with regard to its *external* design/conditions/forces, the glacier would need to be on a mountainside with cold winters, cool summers, and ample snowfall. What if we take any of these things away? Take away cold winters, and there is no period when ice accumulates. If summers are too hot, there is too much melting relative to ice accumulation. If there isn't enough snowfall, the glacier will not form, or will evaporate and sublimate rather than accumulate. Put the glacier on a flat area, and it could still exist, but it might need more snow to survive a seasonal onslaught of sunshine, and it will flow more like a flattening pancake, rather than the downslope movement that is more typical.

Chapter 5 Introducing Functionality and Sustainability | 21

It might seem strange or logically circular to describe the causes for functionality of something like a glacier when it already works by forces of nature—forces which we have nothing to do with. True, it wasn't a human idea that caused snow to pack into dense ice, and for that ice to alter its shape under high pressure. And people didn't build the cold mountainside. And none of us established the thermal and physical properties of water.

Yet, without these things, the glacier would not be what we know as a glacier. And, does it matter who or what forces created something when it comes to assessing its functionality, or, as we'll soon consider, its sustainability? If we think about it further, no person actually causes the rubber in a superball to have the physical properties it does, either. Rather, somebody looked for or created materials that had those properties, and caused them to set in the shape of a sphere.

Even if we hadn't made the materials, those materials had the physical capability awaiting, even prior to our discoveries. And does that potentially mean that many more systems, structures, and institutions that we have not discovered, created, or attempted, and of which we do not yet know, have properties that might be very valuable? Yes, it does.

5.3 Definitions of Functionality and Sustainability

Now that we have dabbled in understanding what separates things that work from things that don't, let's present more precise definitions of functionality and sustainability.

First, here is our more precise definition of functionality:

> *Functionality* (noun) is the degree to which a particular system, structure, or institution is able to accomplish a particular objective under a given set of external conditions.

A closely related term that follows on its heels is sustainability:

> *Sustainability* (noun) is the degree to which a particular system, structure, or institution is able to begin and continue to accomplish a particular objective under a given set of external conditions.

Notice that the definitions are almost the exact same, except that functionality is a snapshot view while sustainability introduces an element of continuation across time.

A simple visualization of *functionality* is sort of like a measurement or determination of how well a light bulb is capable of giving light. We would say the light bulb has "high" or "strong" functionality if, under a certain set of external conditions (such as receiving electricity), it gives the amount of light we want. We would say the light bulb has "low" or "poor" functionality if, under a certain set of external conditions (such as receiving electricity), it gives no light or the wrong amount of light.

A simple visualization of *sustainability* could be a measurement or determination of how well a light bulb is capable of not just giving light, but giving it over the course of time. We would say the light bulb has "high" or "strong" sustainability if, under a certain set of external conditions (such as receiving electricity), it can give the amount of light we want for a long time. We would say the light bulb has "low" or "poor" sustainability if, under a certain set of external conditions (such as receiving electricity), it does not give the amount of light we want for long.

While functionality is like asking if the light bulb gives light at all, sustainability is like asking if the light bulb will burn out quickly or if it will last a long time. With these definitions, any system, structure, or institution that is not functional per an objective is also not sustainable per that objective (e.g., the light bulb that won't turn on at all certainly can't stay on). However, the reverse cannot be said. Just because a system, structure, or institution has low sustainability does not mean that it can't function for a limited amount of time.

Functionality happens to be influenced most strongly by cause-and-effect relationships. So for functionality, if we want to produce a particular effect, then all we need to do is find a cause—something which causes that effect. If we want heat, we simply have to supply something that causes heat—having caused the heat, we have achieved functionality.

When we misunderstand cause-and-effect relationships, we tend to develop nonfunctional systems or solutions. An example would be taking antibiotics to kill bacteria in order to cure the flu. The flu (influenza) is not caused by bacteria, so taking antibiotics is a nonfunctional solution, or cure, for the flu.

Understanding cause-and-effect relationships is something we all begin to do naturally as children. We quickly learn that things fall when they are dropped, that very hot things cause pain, and that dogs can lick or bite. We continue to learn about these relationships as we age, but some of these relationships are complex and remain obscured and indiscernible, even if we are smart.

The importance of cause-and-effect relationships and their control over functionality is one reason it is so important that we shift our emphasis from promoting our viewpoints to listening, science, and discovery.

While functionality is strongly tied to cause-and-effect relationships, sustainability is strongly influenced by feedback mechanisms and interplay between systems, structures, or institutions. Understanding sustainability often requires a greater degree of awareness—above the most obvious cause-and-effect relationships.

So to create something that is sustainable, we have to be skilled at understanding how different parts within systems and between systems affect each other, and how they change through time. This is not a simple task, and it is seemingly a realm in which we are experiencing many collective failures. For now, though, we should continue to see if we can understand how various principles apply to fairly simple systems.

5.4 More with the Superball

Let's see if we can distinguish between functionality and sustainability for the superball example. Earlier we said that anything can function for, or toward, a given *objective* if—and only if—it has some amount of these two things:

1. Internal designs, conditions, and forces that support the objective.
2. External designs, conditions, and forces that support the objective.

This is true for both functionality and sustainability. But whereas functionality simply requires these things, sustainability requires them to keep occurring. To be very specific then, functionality requires:

1. Internal designs, conditions, and forces that support the objective *at a given point in time*.

and/or

2. External designs, conditions, and forces that support the objective *at a given point in time*.

Whereas sustainability requires:

1. Internal designs, conditions, and forces that support the objective *through the course of time*.

and/or

2. External designs, conditions, and forces that support the objective *through the course of time*.

We said that for the superball to work as intended, it would have to have internal designs, conditions, and forces that support the objective of bouncing up and down. We found that the ball needed to be liftable and throwable, able to bounce up after striking a surface, able to bounce repeatedly, able to bounce with some predictability of direction, and reusable. We can now conduct an analysis to predict both functionality *and* sustainability by looking at each criteria.

Internal properties of the superball:

#1: *To be liftable and throwable.*

Functionality: If we make the ball small enough and light enough, it will be liftable and throwable (most people can lift something weighing 30 pounds).

Sustainability: Is there any physical event we can imagine that will change the ball's liftability or throwability over time? Will it somehow become too heavy to throw after we start using it? Under normal conditions, nothing would cause the ball to become heavier. But if we had made the ball very heavy in the first place (say over 5 pounds), somebody might tire quickly of using it, which would reduce its sustainability according to our criteria above (i.e., we would have reduced internal designs, conditions, and forces that support the objective over the course of time). Thus, our only criteria for sustainability is to make the ball easy to lift and throw over and over.

#2: *To be able to bounce up after striking a surface.*

Functionality: If we make the ball out of flexible material that has a rapid rate of recovery after deformation (such as hard, springy rubber), it will tend to bounce back with high speed off of a hard surface, reaching a substantial rebound height.

Sustainability: Is there any physical event we can imagine that will change the ball's ability to bounce after being used previously? If our material fractured after a few impacts, eventually shattering, this would greatly diminish its sustainability. If we can't think of any others, then our only criteria for sustainability in this regard is to make the ball out of a material that will hold together at its projected impact speed—let's say about 100 mph or less.

#3: *To be able to bounce repeatedly.*

Functionality: If our ball does not alter its shape or qualities much after each impact, it will also continue bouncing, instead of rolling away or crumbling.

Sustainability: Can we imagine any event that will change the ball's ability to bounce again and again? Similarly to our prior finding, if the ball is made out of a long-lasting material that doesn't undergo physical changes at normal temperatures, and under normal use, it will tend to be sustainable.

#4: *To bounce with some predictability of direction.*

Functionality: If we make the ball in the shape of a perfect sphere, its bounce angle should be close to the impact angle (not including spin).

Sustainability: For sustainability this property needs to hold true while the ball is used over and over. Thus, as long as the ball does not change shape by being used (by denting or chipping), it should also be sustainable in this regard.

#5: *To be reusable.*

Functionality: For the ball to be reusable, it should not break or change shape, or lose elasticity after striking surfaces. That takes care of functionality.

Sustainability: This requires that reusability remain after using the ball over and over. Thus, as long as the ball is resilient amid repeated physical stress (i.e., it can strike a surface again and again without changing its shape or properties), it should also be sustainable in this regard.

That covers internal design requirements for functionality and sustainability of the superball. Functionality required a weight limit, construction out of flexible material that has an extremely rapid rate of recovery after deformation, an ability to hold shape and material properties after striking a surface, a spherical shape, and non-deterioration from usage. Sustainability required an even lower weight limit, durability of shape (no breakage or wear) through use and high impact speed, and durability/stability of material properties.

This exercise, while perhaps tedious for a simple toy, is still illuminating. It shows that even starting with almost no expertise on a structure or system, we can develop at least some understanding of what is required internally for that structure or system to have functionality and sustainability. It is true that this superball analysis might leave out physical factors, but the point is that the analysis is possible.

Note that there can be overlap between functionality and sustainability. That is, we may find that a certain design attribute is required to provide both functionality and sustainability, and in some cases we might not see a sharp cutoff when functionality ends and sustainability begins. In practice it can be difficult to distinguish between an instantaneous result (functionality) and one spread over time (sustainability). Nothing in our lives actually occurs instantaneously anyway, which means that the separation between functionality and sustainability is somewhat arbitrary. However, don't let this lead you into thinking there is no point in distinguishing between them, because understanding the difference between functionality and sustainability will enable a much higher level of awareness when it comes to complex systems.

You might have noticed that we only completed half of the analysis for the superball. To be thorough, we need to go through the same steps for external conditions. Without that, we really don't know if our internal design (the ball coming out of the factory) matches the external design or conditions (use by a person on planet Earth).

We said that to work as intended, the superball should have external designs, conditions, and forces that support the objective of bouncing up and down. We determined that for this it needs a hard flat surface, someone to periodically throw or drop the ball, and gravity. We will go through each criteria and look at whether we have both functionality and sustainability.

External properties of the superball:

#1: *A hard flat surface.*

Functionality: The hard flat surface would be required for functionality. Hard floors are ubiquitous in the Westernized world. We could use a sidewalk, a paved street, a tile floor, etc.

Sustainability: This would require that our hard and flat surface remain into the future. First, we must be sure our surface will not disappear soon. Is our tile floor about to be replaced by carpeting? Is our paved road beginning to deteriorate? Then we must ask, "Will our ball tend to destroy the surface it bounces upon?" We have to make sure that our hard surface is even harder than whatever material we made the ball out of. Or perhaps the surface is not harder, but it is shatter-proof or crack-proof under normal conditions. So, for sustainability of our hard and flat surface, we need to make sure that surface is likely to stay around and will not be damaged by using the ball.

#2: Someone to periodically throw or drop the ball.

Functionality: This is anybody capable of doing the job. As long as we could do this ourselves or find someone to do it, we should have achieved functionality.

Sustainability: This would require that someone is there to bounce the ball again and again. This doesn't sound difficult, but most adults will get quickly bored from superball bouncing. So we might not be able to achieve sustainability unless a child is doing the throwing and dropping.

#3: Gravity.

Functionality: If we have gravity, we have functionality. Gravity can be found at a reasonable strength in the vicinity of any enormous object, such as a planet or a star, so that would address functionality.

Sustainability: Gravity is permanent as long as you stay near the object, so staying, say, on Earth, would help to make the superball scenario sustainable.

It really is silly to spend this much energy analyzing how to have a bouncing superball be functional and sustainable, isn't it? Silly, yes; but crazy, no.

Remember that our goal at this point is to learn how to theoretically deconstruct and reconstruct any system so that it can work better. So, figuring out the superball is actually a valuable start to something much larger.

You might even be surprised at how much this silly effort with the superball produces. If you asked a regular person, "How do you get a bouncing superball to work well?" They might reply, "Well, I guess it just takes a ball that bounces really well." Or someone who looks at things more intellectually might say, "Hmmm, this can be achieved with a hard rubber ball that you can afford." And of course, in loose terms they are correct.

But by taking it apart, and putting together the picture one bit at a time, we have found that *to work well* means to be *functional and sustainable*, and that this apparently requires: a ball that is small enough to pick up, less than 5 pounds, made out of flexible material that has a rapid rate of recovery after deformation, made of material that has an ability to regain its shape following repeated impacts at 100 mph or less, is perfectly spherical, has material properties that are resilient to stress that is repeated many times over, a hard flat surface that we can depend upon to remain, a hard flat surface that is either crack proof or harder than the material used for the superball, a child, and proximity to a planet or star, with no plans to leave. Now, who would have thought of all that?

It is true that we could have made different findings with the superball. Maybe we forgot about something, like how it would be important for the ball to not have a poison that soaks into the user's skin that may incapacitate them. We could think of many additional qualities that the ball or the external environment should have in order to be functional and sustainable. But that isn't really the main point. The main point now is to learn a process for viewing the functionality and sustainability of anything (yes, hopefully, the ability of *anything* to work correctly).

If something works in the moment it is functional, but if it cannot continue to work, it is not sustainable. How is it that something can be functional but not sustainable? The reason is simply that things often change over the course of time. We might set out with one set of conditions and wind up with a totally different set. This may happen in less than one second, in hours, over months, or over centuries.

These changes can cause something that was perfectly functional to become completely nonfunctional (i.e., unsustainable). It happens all the time in our lives. We turn on the light…the bulb gives light, and then overheats and melts—no more light. We turn on the light…one of the kids tosses a football and the bulb breaks—no more light. We turn on the light…someone else turns on a hair dryer and a fuse blows—no more light. We turn on the light…eleven months later the filament burns out—no more light.

Why are things changing?

We will see that changes in conditions can come from within or without. Changes can occur due to the activity of our system, structure, or institution, or they can occur independently of its activity. These changes can lead to system improvement, or demise. Our next point of discussion will be categorizing the changes in conditions that arise and come to affect sustainability trajectories.

5.5 Feedback Loops

In trying to uncover the keys to the sustainability of a bouncing superball, you might have noticed that we had to address several potential problems that could arise while using the superball over and over. For example, we said that functionality required that the ball be light enough to lift and throw or drop—30 pounds or less. But when it came to sustainability, we considered whether the ball would be light enough to lift over and over. Because of that, we had to be stricter on the weight limit—5 pounds.

Why the switch in maximum weight? Because conditions change *while* the superball is being used and *because* it is being used. This illustrates one of the common mechanisms that affect sustainability—feedback loops.

Feedback loops are things that occur *because of* the operation of any system, structure, or institution. Virtually every event or action in the world changes the world in some way, and feedback loops are merely that subset of changes that affect the ongoing event or action itself. Feedback loops can be positive (maybe the superball's material gets bouncier after warming up with a few bounces).

Feedback loops can also be negative (perhaps the child's arm becomes fatigued after using the superball for 15 minutes). Feedback loops can occur inside the system, structure, or institution (say the ball cracks after being used), or outside the system, structure, or institution (say the floor gets chipped and uneven in numerous places after being struck by the bouncing ball).

Feedback loops can even include interactions between the inside and the outside system, structure, or institution. For example, maybe the ball bounces so well that for even more excitement, the child wonders what will happen if he slams it into a brick wall with his tennis racquet. He does this, which cracks the face of the ball, causing the child to decide to discard the ball.

Feedback loops can be direct or indirect; that is, the change incurred by the system, structure, or institution can act immediately back upon the system, structure, or institution. Or, like dominoes there can be a chain or series of interactions that eventually come back to affect the system, structure, or institution.

Here is a formal definition of feedback loop:

> *Feedback Loops* (noun) are internal or external conditions or forces that arise as a result of a system, structure, or institution's normal activity or use. These conditions or forces either augment or support the initiation, continuation, or expansion of the system, structure, or institution and its function of interest; or they limit or weaken the initiation, continuation, or expansion of the system, structure, or institution and its function of interest.

Feedback loops have a lot to do with the sustainability of many systems, structures, and institutions. People are familiar with them but probably just don't think of them so analytically. Here's an example: You go out for a jog one day. You might start out feeling a little creaky—not warmed up, tired, or dragging yourself along. But as your blood starts to circulate and your muscles get used to the work, you start to feel a groove, and feel like you can move along well. If this has ever happened to you, then you've experienced a positive (internal) feedback loop. The act of jogging made jogging get easier.

But let's continue this example further down the road: You aren't a marathon runner, and after about four miles, your muscles are tiring. You're feeling a little dehydrated, and decide that you should end the run as soon as you get back to your home block. If this has ever happened to you, then you've experienced a negative feedback loop. The act of jogging made jogging get harder.

It may be surprising that the exact same activity initiates a positive feedback loop in one case, and a negative feedback loop in another case. How can that be?

It happens because our body was changing during the jog, and due to the jog. The jogging was initially supported by the body it altered, but it was eventually unsupported by the body it altered even further.

This example helps to illustrate one reason there can be such disagreement and confusion over many topics—a particular action can result in one outcome sometimes, and in the opposite outcome other times. If we aren't fully aware of where these separate, we end up fighting over which outcome is true and which is false, when indeed both arguments may be simultaneously true, and simultaneously false. We will come back and include feedback loop considerations in our analysis later.

5.6 Independent Forces

In addition to feedback loops, there are a class of conditions and forces that don't relate as directly to our system, structure, or institution of interest. Someone learning about feedback loops might respond by saying, "I get it, but most things that happen seem random. If I had a superball in the house, my dog would chew it and swallow it. If I plan an outdoor party it's probably going to rain, but that can't have anything to do with me, can it? And somebody at my office won several thousand dollars in the lottery last year, and took a month-long vacation with the money."

It is true. There are successful designs, failed designs, positive feedback loops, and negative feedback loops; and then there are other things that seem to come out of nowhere. There are events that arise independently of anything we're doing.

We can categorize these events as "independent forces." Independent forces commonly arise, or do not arise, based upon unrelated factors and regardless of activity of the system, structure, or institution we are focused upon.

Here is a formal definition of independent force:

> *Independent Forces* (noun) are internal or external conditions or forces that occur during a system, structure, or institution's normal activity or use, but they do not occur because of its use. These conditions or forces either augment or support the initiation, continuation, or expansion of the system, structure, or institution and its function of interest; or they limit or weaken the initiation, continuation, or expansion of the system, structure, or institution and its function of interest.

If we stay with the jogging analogy a little bit longer, maybe you go for your jog and a thundercloud that seemed distant at first begins to unload 15 minutes into your run, and you turn back home. Or perhaps you go for a jog, and just as you get down the block, you remember that there's an important youth sports league meeting for your son in 30 minutes, which prompts you to run half your route, stop back home, and change clothes before the meeting. Maybe you go for a jog and the nature trail that has been closed and under repair for three months is now open, letting you take a longer and more scenic route. Or your plans may be stifled when twenty minutes before your jog, you feel a migraine headache coming on, and decide to lie down instead.

These are all examples of independent forces that affect your jog—how easy it will be, how far you will go, whether you go at all. They can arise internally (a headache) or externally (the open nature trail). Independent forces are sometimes predictable (you know that temperatures will be higher in July than in April), sometimes partially predictable (the high face on the mountain has a predicted 50/50 chance of an avalanche in the next month), and sometimes hardly predictable at all (a local deluge during the regional dry season).

Feedback loops and independent forces are simply mechanisms that affect how sustainable something is, and how much support it has to carry forth through time. But they don't necessarily dictate sustainability. Oftentimes, they contribute pressure in the direction of sustainability or against it, and only help to determine sustainability in balance with other forces. In the jogging examples, the migraine headache would make the run difficult, but it wouldn't necessarily prevent your run if you were strong willed and determined to make it happen.

Feedback loops and independent forces can be produced by an essentially mechanical force, like the weather that causes a storm. But they can also be produced by free-will choices, like the automobile purchases that people make.

Similarly, impacts of feedback loops and independent forces can be essentially mechanical, or they can be attenuated by free will. Consider the glacier that shifts from a stable year-to-year mass to a declining mass as annual snowfall decreases due to warmer winters. Or consider the ball that begins to develop small cracks at high-impact speeds. The glacier and the ball have no real free will to control their responses to these forces, yet their sustainability is strongly influenced by them. However, the jogging examples tend to involve an element of free will (on the part of the jogger).

5.7 A Watermill Example

We've come a long way now from the general question of why things don't seem to be working well anymore. Like trying to fix a bike without training or an owner's manual, we set off on a quest to see if our systems can be understood by carefully looking at all their parts and relationships. By understanding all these parts and relationships, theoretically, we can fix our broken systems.

We can see that for something to work like we want, it needs to be both functional and sustainable. To be functional and sustainable in turn requires support to come from within and/or without. To have support from within requires internal designs and conditions that support an objective, and to have external support requires external designs and conditions that support an objective. The internal designs and conditions are controlled by design choices, feedback loops, and independent forces. The external designs and conditions are also controlled by design choices, feedback loops, and independent forces (although not all internal and external designs are "choices"; some are just existing conditions).

This understanding should help us begin to answer the original question, "Why don't things seem to be working well anymore?" Things that "work" have adequately supportive internal and external designs, feedback loops, and independent forces affecting them. But things that "don't work" have one or more unsupportive internal and external design aspects, feedback loops, or independent forces acting upon them. This means that our important collective systems aren't working well because of some combination of unsupportive designs, unsupportive feedback loops, and/or unsupportive independent forces.

Before moving into an analysis of very complex systems, let's ground all of these theories with a look at something just slightly more complex—something beautiful—a watermill. Watermills were common in Europe and in early America. They were constructed along river edges to power machinery, grindstones, sawmills, etc.

There were several different types of watermills. The most simple and maybe most common type was the undershot watermill, which has a wheel (the waterwheel) spinning on a horizontal axis, which makes the waterwheel appear to stand on end like a Ferris wheel.

The bottom end of the waterwheel is in a stream or river and is pushed by the current, thus providing power to the axis of the wheel. The axis of the wheel can then accomplish heavy work inside a millhouse.

Let's imagine that we work on a project design team. Suppose that due to our work on a superball project, word has gotten around that we are proficient in determining system functionality and sustainability. So, our supervisor hands over a watermill design project to our team.

The watermill is to be designed for the small foreign village of Flodskov. We are given little information about Flodskov or the nation it is within—all we know is that the region's culture and technologies are different than ours, and the watermill will be used to cut lumber. Rather than designing the entire sawmill system, we are asked to narrow our focus to the spinning wheel parts of the system (waterwheel) instead of the broader view of what happens in the millhouse.

Our analysis for the watermill will be lengthier than what we did for the superball, and it could become complicated. To avoid getting mixed up, we can separate the functionality analysis from the sustainability analysis and number each step.

Let's number the functionality analysis steps *1* through *10*. These steps may seem confusing at first, so reviewing the steps more than once might help them seem more intuitive. Just remember that the main goal of all this is simply to develop a list of the factors that lead to or detract from functionality and sustainability of any particular system.

Here are the ten steps for the functionality analysis:

1. Name: Name of the system that you want to learn more about: **A waterwheel.**
2. Goal: The thing that the system will hopefully accomplish or the function that you are focused upon. This requires something like an action or activity, so a verb must be part of the *goal*. **The waterwheel turns with significant force or torque around its axis.**
3. Boundary: An imaginary box or imaginary shrink-wrap around the system. Inside the box or shrink-wrap is *internal*, and outside the box or shrink-wrap is *external.* **The outer surfaces of the waterwheel, an axis (shaft) upon which it spins, and the support structures holding up the waterwheel and shaft.**
4. Internal Design and Process: A basic description of the system's internal components and how, in theory, they would act together to help attain the *goal*. **The waterwheel can be fitted with paddles that are pushed by the river. Spokes can connect the wheel to a shaft. The shaft upon which the waterwheel rotates can be supported on both ends—one bearing in the millhouse, and one on a stone abutment set on the river bottom.**
5. Positive Indicators: Signs or evidence that would indicate to you that the system is achieving its *goal*. (Hint: If your system (1) is able to function (2) according to design (4), these are some of the desired or appropriate things you would observe.) **The waterwheel turns at a relatively steady rate, while transmitting significant force (torque) and energy through the shaft.**
6. Negative Indicators: Signs or evidence that would indicate to you that the system is *not* achieving its *goal*. (Hint: If your system (1) is unable to function (2) according to design (4), these are some of the unwanted or counter things you would observe.) **The waterwheel stops or has unreliable motion. It spins weakly, only turning if work requirements are removed from the shaft.**
7. Supporting Internal Conditions: Internal conditions under which the system is predicted or expected to be able to attain its desired *goal*, given its *internal design and process*. Allow yourself to aim for the *positive indicators* above. (Hint: These are the internal aspects of your draft design (4) which will theoretically lead to the positive indicators above (5).) **The paddles must be numerous enough that they are continually in the river. The paddles must be large enough and faced to create substantial pressure that spins the waterwheel. The paddles must be held tightly enough to withstand pressure from the river. The wheel must be large enough (have a large enough diameter) to produce significant torque upon the shaft. The spokes that hold the waterwheel rims to the shaft must be strong enough to withstand stress and slight deformation without snapping. The shaft must be strong enough to withstand great torque without cracking or snapping.**

The waterwheel on the whole is balanced about its axis, such that it will rotate at a uniform speed and not preferentially come to rest on one side. The contact faces between the shaft and the bearings in the millhouse and abutment must have low friction.

8. Supporting External Conditions: External conditions under which the system is predicted or expected to be able to attain its desired *goal*, given its *internal design and process*. Allow yourself to aim for the *positive indicators* above. (Hint: These are the external aspects pertaining to your draft design (4), which will theoretically lead to the positive indicators above (5).) **There is a flowing river contacting the bottom of the waterwheel. There is a stable support or ground upon which the bearing in the millhouse and the stone abutment rest.**

9. Non-Supporting Internal Conditions: Internal conditions under which the system is predicted or expected to be *unable* to attain its desired *goal*, given its *internal design and process*. Ask yourself what would help to create the *negative indicators* above. (Hint: These are the internal aspects of your draft design (4), which will theoretically lead to the negative indicators above (6).) **The paddles are sparsely distributed, causing the waterwheel to slow or stop. The paddles are small or angled incorrectly, failing to transmit significant power through the shaft. The paddles are attached poorly, possibly leading to detachment. The wheel has a small diameter, failing to transmit significant power through the shaft. The spokes are weak, leading to warping and breakage. The shaft is not strong enough, and cracks. The waterwheel is asymmetrical or unbalanced, causing the shaft speed to vary slightly or even causing the waterwheel to stop. The contact faces between the shaft and the bearings in the millhouse and abutment have high friction providing less energy to perform work.**

10. Non-Supporting External Conditions: External conditions under which the system is predicted or expected to be unable to attain its desired *goal*, given its *internal design and process*. Ask yourself what would help to create the *negative indicators* above. (Hint: These are the external aspects pertaining to your draft design (4), which will theoretically lead to the negative indicators above (6).) **There is no flowing river or only weak current contacting the bottom of the waterwheel. The support or ground upon which the bearing in the millhouse and the stone abutment rest is unstable, causing misalignment to grow. The workload placed upon the shaft is too high, forcing the wheel to a halt. The waterwheel paddles are blocked by an object, forcing the wheel to a halt.**

The primary "positive" results for functionality are summarized in #7 and #8 above, while primary "negative" results for functionality are summarized in #9 and #10. With this, we can see that given our overall design and understanding of how this device should work, *functionality* requires the following:

PF1. The paddles must be numerous enough that they are continually in the river.

PF2. The paddles must be large enough and faced to create substantial pressure that spins the waterwheel.

PF3. The paddles must be held tightly enough to withstand pressure from the river.

PF4. The wheel must be large enough (have a large enough diameter) to produce significant torque upon the shaft.

PF5. The spokes that hold the waterwheel rims to the shaft must be strong enough to withstand stress and slight deformation without snapping.

PF6. The shaft must be strong enough to withstand great torque without cracking or snapping.

PF7. The waterwheel on the whole is balanced about its axis, such that it will rotate at a uniform speed and not preferentially come to rest on one side.

PF8. The contact faces between the shaft and the bearings in the millhouse and abutment must have low friction.

PF9. There is a flowing river contacting the bottom of the waterwheel.

PF10. There is a stable support or ground upon which the bearing in the millhouse and the stone abutment rest.

In contrast, lack of functionality would tend to be brought on by the following:

NF1. The paddles are sparsely distributed, causing the waterwheel to slow or stop.

NF2. The paddles are small or angled incorrectly, failing to transmit significant power through the shaft.

NF3. The paddles are attached poorly, possibly leading to detachment.

NF4. The wheel has a small diameter, failing to transmit significant power through the shaft.

NF5. The spokes are weak, leading to warping and breakage.

NF6. The shaft is not strong enough, and cracks.

NF7. The waterwheel is asymmetrical or unbalanced, causing the shaft speed to vary slightly or even causing the waterwheel to stop.

NF8. The contact faces between the shaft and the bearings in the millhouse and abutment have high friction, providing less energy to perform work.

NF9. There is no flowing river or only weak current contacting the bottom of the waterwheel.

NF10. The support or ground upon which the bearing in the millhouse and the stone abutment rest is unstable, causing misalignment to grow.

NF11. The workload placed upon the shaft is too high, forcing the wheel to a halt.

NF12. The waterwheel paddles are blocked by an object, forcing the wheel to a halt.

Certainly, anyone who is an expert in this technology will probably have a much more developed and correct understanding of what makes a good waterwheel and watermill. But it is interesting that not even being experts, we can go from knowing almost nothing to understanding some basic working principles of this elegant machine.

Now to understand the sustainability of this system, we simply have to take these functionality concepts and draw the waterwheel through time to see what could arise. We will assume for the sake of this exercise that our functionality analysis above was reasonably accurate (though in this hypothetical case we don't know just how accurate). To move forward, sustainability simply requires the extension or improvement of the positive functionality results through time, and little or no development of the negative functionality results through time.

We've already learned why conditions change through time—feedback loops and independent forces. Typically we cannot anticipate all the possible feedback loops and independent forces that might arise to affect a system, but by keeping our analysis carefully regimented to cause-and-effect relations, we can pick out many of them—hopefully including those that are more likely.

We will first look for positive feedback loops that could arise while our system operates. Positive feedback loops are conditions or forces that will change because of system operation, in favor of its continued operation. Remember that feedback loops can be complex and include multiple cause-and-effect relationships or cause-and-effect chains. They can affect internal or external conditions to the system.

While we will keep positive and negative feedback loops separate in our review, we won't try to separate feedbacks further into internal or external categories, because that could become overly complicated. Also, as we explore feedback loops, we will include some hypothetical scenarios that are merely for illustration purposes and not to suggest that these scenarios are highly likely.

To determine sustainability, we can imagine we have built the waterwheel. It operates a sawmill inside the millhouse. If we did a good job of developing our list, we can rely on our supporting internal conditions and supporting external conditions for functionality to entertain positive feedback loops (PF1 through PF10 above). As the watermill runs, does the waterwheel have any tendency to improve in terms of any of the following positive internal properties?

PF1. The paddles must be numerous enough that they are continually in the river. No, the number of paddles won't normally improve due to the turning of the waterwheel, and they won't tend to stay in the river any more than if the wheel came to a stop. No positive feedback loop.

PF2. The paddles must be large enough and faced to create substantial pressure that spins the waterwheel. No, the turning waterwheel won't grow or improve the pressure on the paddles over time. No positive feedback loop.

PF3. The paddles must be held tightly enough to withstand pressure from the river. In general, if the paddles were not secured tightly in the first place, they would be unlikely to become more secure during operation. But wood can expand when it gets wet. If the paddles are wooden, if the wheel rim is wooden, and/or if they are secured with wooden pegs, these could swell. If planned well, the attached parts could tighten with each other once the waterwheel began operation. So there is a potential positive feedback loop, depending upon the materials and sizing of the parts.

PF4. The wheel must be large enough (have a large enough diameter) to produce significant torque upon the shaft. No, the turning waterwheel will not alter the size of the wheel in order to increase torque over time. No positive feedback loop.

PF5. The spokes that hold the waterwheel rims to the shaft must be strong enough to withstand stress and slight deformation without snapping. No, because the waterwheel is turning, the spokes will not tend to strengthen over time. No positive feedback loop.

PF6. The shaft must be strong enough to withstand great torque without cracking or snapping. As the waterwheel turns, the shaft will not tend to strengthen or withstand higher torque over time, unless perhaps it is made of metal (some metals harden through use or slight deformation). A metal shaft could be very expensive, or could be extremely heavy if it were not hollow. Let's say here that there is a potential small positive feedback loop if the shaft material includes metal.

PF7. The waterwheel on the whole is balanced about its axis, such that it will rotate at a uniform speed and not preferentially come to rest on one side. No, the weight balance of the waterwheel is unlikely to improve as a result of turning. No positive feedback loop.

PF8. The contact faces between the shaft and the bearings in the millhouse and abutment must have low friction. As the wheel turns, it is likely to wear smooth along the bearings. This could result in less friction. As long as we don't operate it poorly at first such that the bearings initially overheat or scar, friction could tend to improve slightly through use. Let's say here that there is a probable small positive feedback loop that would occur over the first few months of operation.

PF9. There is a flowing river contacting the bottom of the waterwheel. In general, the river will function on its own accord, independent of the watermill and waterwheel. However, we can imagine that just maybe, if the stream sediment (the bottom of the stream) is composed of small materials such as sand or pebbles, the stream water will run rapidly under the paddles, gouging the stream bottom below the paddles. Will this increase the adjacent stream flow rate and make the waterwheel spin faster? Or will it tend to do the opposite—depositing sediment in that vicinity, and slowing near the waterwheel? Whether this happens might have to do with sediment size, stream flow rate, slope, depth, width, bank stability, and waterwheel width. This all seems like complex physics and stream morphology that has its own feedback loops. We might not be able to solve this one quickly just by thinking about it. Regarding a positive feedback loop, let's just say maybe, maybe not.

Another scenario could be that the watermill's sawmill operation is very profitable, enriching the owner. This could help to give the mill owner lots of political influence. Perhaps villages upstream of the mill have dammed the river to create reservoirs, and traditionally they have relied upon that water during dry periods. If a very dry period comes one year, the mill owner could use his political influence to require the villages to release some of their reservoir water to help keep river levels more stable. This would be an indirect feedback loop, but nevertheless a positive one.

PF10. There is a stable support or ground upon which the bearing in the millhouse and the stone abutment rest. The movement/turning of the waterwheel could cause some settling of weight and material below the abutment and the support in the millhouse. If this settling is small, it could be harmless or even stabilizing and helpful (but if there's too much settling, it could throw the system out of alignment). So, there might be a moderate positive feedback loop, depending upon the amount of settling that occurs.

Next we look for *negative* feedback loops that could arise within and around our system while it operates. These are things that will change antagonistically toward the system's continued operation, because of its operation. If we did a good job of developing them, we can rely on our non-supporting internal conditions and our non-supporting external conditions (NF1 through NF12 above) to entertain negative feedback loops.

As the watermill runs, does it cause conditions to shift in the direction of any of the following?

NF1. The paddles are sparsely distributed, causing the waterwheel to slow or stop. The waterwheel's spinning motion and normal operation are unlikely to alter the number of paddles once it is built, unless the paddles were attached in an extremely flimsy manner. Thus there is a possibility of a negative feedback loop depending upon the quality of our work.

NF2. The paddles are small or angled incorrectly, failing to transmit significant power through the shaft. This negative feedback loop could occur quickly if poorly attached paddles shift. But even if they are well designed and attached, over time we can expect the slow wear of contact parts such as paddles. Let's assume at least a small negative feedback loop could be present in this way.

NF3. The paddles are attached poorly, possibly leading to detachment. If the paddles are not attached securely, then during operation it is possible that they would slowly loosen and become dislodged. This would constitute a negative feedback loop.

NF4. The wheel has a small diameter, failing to transmit significant power through the shaft. The diameter of the waterwheel will not shift just from being used, so no negative feedback loop.

NF5. The spokes are weak, leading to warping and breakage. Whether or not the spokes were made strong enough in the first place, they aren't likely to grow weaker just because of the turning waterwheel, so no negative feedback loop.

NF6. The shaft is not strong enough, and cracks. The operation of the waterwheel will tend to stress the shaft in at least two directions (torque and shear). It is possible that the shaft, if not strong at first, would bend and then weaken through operation. There is a potential negative feedback loop, but only if the shaft was not strong in the first place.

NF7. The waterwheel is asymmetrical or unbalanced, causing the shaft speed to vary slightly or even causing the waterwheel to stop. The waterwheel is unlikely to become imbalanced from use, unless as above the poorly attached paddles fall off one side, or a weak shaft bends. There is some potential for a negative feedback loop under these scenarios.

NF8. The contact faces between the shaft and the bearings in the millhouse and abutment have high friction, providing less energy to perform work. This problem could occur if we operate the waterwheel at high speed (creating heat and burning surfaces), or if we operate it carelessly without smoothed surfaces to begin with, or if the shaft or bearing materials have a tendency to wear at uneven rates, producing bumpy surfaces. However, even if the system is functioning as designed, we might expect frictional surfaces to wear slowly. This is an area of potential negative feedback loops.

NF9. There is no flowing river, or only a weak current contacting the bottom of the waterwheel. The river might function on its own accord, regardless of the watermill and waterwheel. However, we can imagine that maybe if the stream flow is somewhat slow and steady, and the stream carries a high load of fine sediment, the water could slow in the vicinity of the paddles, causing it to drop sediment in that location, and reducing water flow near the paddles. In this imaginary scenario, *yes*, operation of the water wheel would cause a negative feedback loop. However, without becoming experts in stream morphology, we will not be able to determine this with any certainty. Regarding a negative feedback loop, let's just say maybe, maybe not.

Another potential scenario is that since our waterwheel will operate a sawmill, there must be an economic incentive for producing lumber. Maybe the sawmill is supporting a booming city or a shipbuilding industry. Nearby landowners might be able to reap a large profit by cutting their forests. If this happens over the course of several years in the valleys that supply the river with runoff, there could be much more flooding than normal when spring rains come (due to excessive runoff from cleared forestland). This could cause upstream dams to burst, sending a torrent of water through the river. Such a torrent could easily rework the river channel and shift larger boulders such that they severely limit river flow near the waterwheel. This hypothetical scenario would be a negative feedback loop.

NF10. The support or ground upon which the bearing in the millhouse and the stone abutment rest is unstable, causing misalignment to grow. As the waterwheel turns, we would expect subtle vibrations to be transferred into attached structures and the ground. If the bearing or stone abutment were constructed upon weak ground, like peat or unpacked dirt, they would shift during watermill operation, leading to inefficiency or misalignment and wear between various parts. This would be another negative feedback loop.

NF11. The workload placed upon the shaft is too high, forcing the wheel to a halt. The spinning waterwheel turns a shaft that powers a sawmill. As the sawblade cuts more and more logs, it will dull, and the force required to saw logs will increase. This could cause the waterwheel to be slowed (a negative feedback loop) unless we periodically sharpen or replace the sawblade.

NF12. The waterwheel paddles are blocked by an object, forcing the wheel to a halt. Perhaps in the scenario above where increased logging leads to deforestation and a flood event, the strong current could also pull logs and downed trees into the channel. These could lodge obliquely into the base of the waterwheel, jamming it against the river bottom and cracking paddles. This is a negative feedback loop.

We could certainly imagine more feedback loops, but this is already a pretty long list to work with.

You might be wondering why we need to look at these negative feedback loops. If we discover the positive attributes for functionality and sustainability, isn't that all we need? Unfortunately, we can only develop a partial understanding of a system's sustainability by focusing on its supporting mechanisms. Life is full of changes, and determining only the supporting mechanisms doesn't help us steer clear of stifling conditions. As an analogy, if you were to explain football to someone by merely describing the need to reach the end zone, they'll end up out of bounds way before they ever score a touchdown. So, this deeper analysis allows us to look ahead at some of the conditions that could arise to diminish the sustainability of our waterwheel.

We are slowly building a long list of design requirements for the functionality and sustainability of the waterwheel. Notice that several negative feedback loops are associated with things we have control over, like how well we attach the paddles. This gives us the ability to purposely create our system more sustainably. But we aren't done yet, and we don't fully know what we will learn until we have really finished the analysis. So to continue…

Now we need to look at *independent forces*. Remember that independent forces are conditions or forces that arise independently of (and not due to) a system's normal function or use. So with few exceptions, an independent force pertaining to our waterwheel would have the same odds of arising whether or not the waterwheel is performing its function of interest (turning). Stated another way, an independent force is one that arises to affect the turning waterwheel, but does *not* arise *because* of the turning waterwheel.

To make it easier for us to envision independent forces, we can separate them into internal, external, positive, and negative categories. We will look at <u>positive internal</u> independent forces first. These would arise from within the waterwheel to affect anything in the waterwheel or its relevant environment. Recall first that these forces would positively affect any of the following:

PF1. **The paddles must be numerous enough that they are continually in the river.**

PF2. **The paddles must be large enough and faced to create substantial pressure that spins the waterwheel.**

PF3. **The paddles must be held tightly enough to withstand pressure from the river.**

PF4. **The wheel must be large enough (have a large enough diameter) to produce significant torque upon the shaft.**

PF5. **The spokes that hold the waterwheel rims to the shaft must be strong enough to withstand stress and slight deformation without snapping.**

PF6. **The shaft must be strong enough to withstand great torque without cracking or snapping.**

PF7. **The waterwheel on the whole is balanced about its axis, such that it will rotate at a uniform speed and not preferentially come to rest on one side.**

PF8. **The contact faces between the shaft and the bearings in the millhouse and abutment must have low friction.**

PF9. **There is a flowing river contacting the bottom of the waterwheel.**

PF10. **There is a stable support or ground upon which the bearing in the millhouse and the stone abutment rest.**

Is there anything that will happen within the waterwheel or its bearings that will improve its turning or its prospects for turning? One easy way to imagine an independent force is to pretend the waterwheel is just standing still, not rotating. While it stands there, is it likely to experience transformations that will improve its prospects or its performance, or is it likely to generate an improved environment for itself? As a static object, the waterwheel probably won't generate much favorable change. So no positive internal independent forces are likely in this case.

Then what about positive external independent forces? Is there anything that would come from the outside to positively affect the waterwheel's prospects, either by affecting the watermill or the river? Such a force would have to positively impact the positive functionality traits we just listed above.

Without more information, we don't know what scenarios are realistic, though many independent forces are theoretically possible. We could imagine that the kingdom or state that contains the watermill is poised for warfare and relies heavily upon its navy. Because of this, local bureaucrats could be ordered to hasten lumber production for shipbuilding. The local governor could declare that anything that slows lumber production, such as the diversion of river water for irrigation, must be halted until further notice. This would ensure minimal flows for the watermill into the foreseeable future.

Next we can consider negative internal independent forces. Is there anything that could happen within the waterwheel or bearings that will diminish its turning prospects? We can refer back to our list of negative functionality conditions again:

NF1. **The paddles are sparsely distributed, causing the waterwheel to slow or stop.**

NF2. **The paddles are small or angled incorrectly, failing to transmit significant power through the shaft.**

NF3. **The paddles are attached poorly, possibly leading to detachment.**

NF4. **The wheel has a small diameter, failing to transmit significant power through the shaft.**

NF5. **The spokes are weak, leading to warping and breakage.**

NF6. **The shaft is not strong enough, and it cracks.**

NF7. **The waterwheel is asymmetrical or unbalanced, causing the shaft speed to vary slightly or even causing the waterwheel to stop.**

NF8. **The contact faces between the shaft and the bearings in the millhouse and abutment have high friction, providing less energy to perform work.**

NF9. **There is no flowing river, or only weak current, contacting the bottom of the waterwheel.**

NF10. **The support or ground upon which the bearing in the millhouse and the stone abutment rest is unstable, causing misalignment to grow.**

NF11. **The workload placed upon the shaft is too high, forcing the wheel to a halt.**

NF12. **The waterwheel paddles are blocked by an object, forcing the wheel to a halt.**

The negative internal forces we're looking for would have occurred on their own accord, even if the waterwheel were just standing still. As it stands idly, is the waterwheel likely to experience transformations that will lessen its prospects or its performance? Yes, this could happen. Some materials do undergo chemical or physical deterioration over time, whether or not they are being used.

If the waterwheel or bearings contain iron components, those could rust though regular atmospheric exposure. Large wood components won't normally change much over time unless they are subject to very high humidity or insects. So we could surmise some possibility for negative internal independent forces if there are iron parts on the waterwheel or if some of the wood that was used contains boring insects.

For negative external independent forces, is there anything that could come from the outside to negatively affect the waterwheel's prospects, either by affecting the watermill or the river? To consider this, refer again to the negative functionality conditions above. Many such forces are theoretically possible.

Perhaps the following could occur, hypothetically:

The villages in the valley upslope of the watermill might be growing in population, and as a result, become more dependent upon intensive agriculture. As this happens, they might begin to divert river water each summer to increase crop yields. This would lower the river levels when they would already be at their lowest, causing weak waterwheel rotation and forcing the sawmill to shut down during summer.

In another scenario, beavers upstream of the watermill could release log segments downriver. If a log wedges between a paddle and the abutment, it could break a paddle. The resulting imbalance might cause uneven torque and additional wear upon the shaft until it is fixed.

In a third scenario, lightening could start a forest fire that races through the valley, burning trees and any wooden structures, including the millhouse and part of the shaft. Even if the waterwheel survives, after the fire the shaft could be damaged and paddles could be broken by woody debris in the river.

Regarding the watermill and waterwheel example, we have reached the limit of our current model for understanding and predicting functionality and sustainability.

To recap how this was done, we analyzed functionality by carefully naming and defining our system/structure/institution, its goal, and its internal and external designs or context. This led us to list its internal and external conditions for functionality and for non-functionality. We then shifted the focus to sustainability by looking at this list according to feedback loops and independent forces.

While we may not be experts in watermill construction, and while we may lack the complete picture of local conditions and historical context, our analysis still results in a surprisingly in-depth list of needs for functionality and sustainability. We found that initial functionality would hinge upon attaining ten things (PF1 through PF10) and avoiding twelve things (NF1 through NF12). In short, we need to exercise care in the materials and design of every part of the waterwheel, and also in the established relationship between the waterwheel and the river, the ground, and the millhouse workload.

If we hadn't been studying the idea of sustainability first, we might have wrongly assumed that if we could get this waterwheel going, it would keep going. But our review of feedback loops and independent forces suggests it isn't so simple. If we managed to get everything right and functionality is achieved, we might also see forces work in favor of sustainability because of:

- A slight sturdiness benefit if correctly sized wood attaches the paddles to the wheel rim.
- A slight benefit if the shaft is made of certain types of metal.
- A reduction in friction due to safe initial operation of the watermill.
- A potential but very uncertain increase in the water flow rate caused by channel changes after the waterwheel starts operating.
- The wealth and political influence of the watermill's owner.
- A small improvement in material stability and footing after the waterwheel starts operating, depending upon local surface geology or construction methods.
- The importance of the watermill's services to national defense.

But forces might work against sustainability if:

- The paddles were attached in an extremely flimsy manner.
- We aren't prepared for slow wear of the paddles.
- The paddles are made of the flimsiest material.
- The shaft is not strong in the direction of torque or shear stress.
- We operate the waterwheel at high speed.
- We operate the waterwheel without smoothed contact (rubbing) surfaces to begin with.
- The shaft or bearings are made of materials that have a tendency to wear at uneven rates.
- We aren't prepared for slow wear of the shaft-bearings and contact faces.
- The stream slows down once the waterwheel starts, essentially bypassing the waterwheel.
- The valley is deforested.
- There are large dams upstream of the waterwheel.
- We build upon soft or shifting material.
- Nobody sharpens or periodically replaces the sawblade.
- The waterwheel is not protected from free-moving logs in the river channel.
- There are large or growing populations upstream who want to use the river water.
- Beavers are active nearby.
- The site is subject to forest fires or other natural disasters.
- Waterwheel parts are made of iron.

- We don't check our wood building materials for defects such as insect infestations.
- It operates under conditions of high humidity.

These are interesting lists. They suggest sustainability gains regarding seven things and sustainability losses regarding twenty things. And this is *after* we determined how to obtain initial functionality. Of course we don't really know details about the local economy, politics, environment, stream morphology, and human populations. But running through this thorough deconstruction/reconstruction process illustrates how complex the sustainability requirements can be for a relatively "simple" system like a watermill.

5.8 Introducing Complex Systems

Why refer to the watermill as a relatively "simple" system? Let's take a moment to distinguish between *simple systems* and *complex systems*.

Generally, *simple systems* have fewer parts or components, and those parts and components are more likely to remain as part of the simple system. The parts that make up simple systems generally behave according to principles of math, physics, chemistry, and material science. Simple systems often have fairly well-defined boundaries. Simple systems are determinate. The outcomes of simple systems may vary slightly in practice, but that variation is very small in comparison to the consistency of the outcome. The better we are with our math and science, the better we do with simple systems. Some examples of simple systems are rotating or moving objects, motors, robots, chemical reactions, computers, clocks, and musical instruments.

Complex systems tend to have larger numbers of parts or components. Their parts and components may enter and exit the complex system with high frequency. The components that make up complex systems are not only controlled by known cause-and-effect relationships, but also by indirect relationships, uncertain relationships, varying relationships, and by transforming or evolving conditions.

Complex systems often have boundaries that are shifting or blurred. Because of these factors, we generally *cannot* predict complex system behaviors and outcomes with exact precision. *Complex systems are indeterminate*. The behavior of complex systems is probabilistic, but even so, those exact numerical probabilities are elusive.

Skill with math, physics, chemistry, and material science are helpful, but these sciences don't allow us to completely predict or control complex systems. With complex systems, it is often helpful to understand motivation, strategy, evolution, momentum, and balance. Examples of complex systems are human behavior, families, governments, corporations, sports, economies, weather, and ecosystems.

We began our analysis looking at superballs, which are simple systems. But we aren't doing all this analysis because there are broken superballs scattered everywhere. It is the apparent failure of our complex systems, structures, and institutions that brought us here. We began with simple systems, but we have been gradually building up toward complex systems.

So, we figured out that the waterwheel might be part of a complex system. What can we do now? Whether a system is simple or complex, we can look over our list of supporting and non-supporting factors for functionality and sustainability. Then, we can try to make smart plans to provide the supporting factors and limit the non-supporting factors.

To make sure the parts are well selected and crafted, maybe we'll hire a millwright instead of getting Uncle Joe to manage the job. Instead of leaving a lot of river behavior up to chance, maybe we'll recommend damming a section of the river and installing a log baffle and chute for the water to reach the waterwheel at an even rate. Instead of building it downstream of the growing villages, maybe we'll get on the upstream end. Instead of taking chances on political protection for water flow, maybe we'll send a lobbyist to the prince or governor to make our case. Instead of waiting for the beavers to send a big log downstream, maybe we'll pay the local kids 20 cents for every beaver tail they turn in.

5.9 Compatibility

We are on the verge of working with complex systems. But we have one more thing to consider so that we have the correct tools and language to interpret complex system behavior. We started out wanting to know how things work, and we recognized that to work means first to be functional. We defined functionality (the degree to which a particular system, structure, or institution will accomplish a particular objective under a given set of external conditions) and sustainability (the degree to which a particular system, structure, or institution will be able to begin and continue to accomplish a particular objective under a given set of external conditions).

Functionality and sustainability are very closely intertwined—really just separated by the passage of time. Then, especially with complex systems, there is a third property that is intertwined along with them. This last piece, *compatibility*, has to do with interaction between different systems, structures, and institutions:

> *Compatibility* (noun) is the degree to which one system, structure, or institution supports the functionality or sustainability of another system, structure, or institution.

Compatibility is a new term herein, but technically it doesn't contain any new things. It is a measure of support for sustainability between different systems, and it results from nothing other than feedback loops and independent forces between those different systems. Compatibility works both ways. Presumably, any pool of interacting systems have the ability to create positive or negative compatibility conditions for each other.

Why should we care about compatibility? If we are going through all the trouble to determine what will make a system sustainable, then we probably want to understand what types of things it will affect, or be affected by. What good is a well-designed watermill if you live in a lawless land and can't protect your property? Or if your nation is about to be ransacked by barbarians? If the technology you are banking on is being replaced by the internal combustion engine? If you will cause the loss of resources on which the region depends? Or if disease is moving through the land and a labor shortfall is about to ensue?

Without an understanding of compatibility, we are left confused with well-intentioned piecemeal endeavors and solutions that are trumped, turned around, or reshaped by other systems. If we don't take a look at compatibility interactions in a deconstruction analytical process, we will be short of understanding why, in the big picture, our complex of systems, structures, and institutions might be malfunctioning.

5.10 Compatibility of the Waterwheel

To create a fully sustainable watermill, we should conduct compatibility analyses both *to* and *from* the watermill, relative to all the systems it interacts with. So, we could conduct mutual compatibility analyses between the watermill and the local economy, and the watermill and the ecosystem, etc. However, we don't have enough information on these other systems to really determine how they would impact the watermill. Nevertheless, we can at least investigate which impacts the watermill might have upon other nearby systems and see what can be learned in the process.

Up until now, we have focused almost selfishly upon our own system. We paid no regard to other systems unless we thought feedback loops or independent forces were going to affect our system objective. This may seem confusing, but to understand the compatibility of our system for other systems, we have to turn the other way and look at how anything on the outside will be changed because of our system.

There is no math or computer formula to do this for us. But as before, we will continue to rely upon the application of logic or simple cause-and-effect relationships. We can start the compatibility analysis with two lists. First we list all materials, energy, and activities that have to be created, redirected, or initiated in order to create and sustain our system.

Second, we list any material, energy, or conditions that results as an intentional or unintentional consequence of creating or sustaining the system. The first two steps in a watermill compatibility analysis follow:

1. Inputs: List all materials, energy, and activities that have to be redirected, created, or initiated in order to create and sustain the system. This can include anything that is directly required, or likely to be indirectly required for the system.

We can refer to our findings from the waterwheel's functionality and sustainability analyses (supporting internal/external conditions and non-supporting internal/external conditions from the functionality analysis; and feedback loops, independent forces, and our potential responses to those from the sustainability analysis). As we review those findings, here are possible results:

   a. We will need many **wood parts to make the waterwheel**.
   b. Many of the wood parts are specialized and have small error tolerances, so we'll have to **hire a millwright**.
   c. Some of the specialized parts might be made of metal, and big sawblades have to stay sharpened. We might need to **hire a metal smith or blacksmith**.
   d. To control river behavior to some degree, and keep out interfering objects, we will probably want to **build a dam, a baffle, and a millrace** (waterwheel chute).
   e. We will probably have to **hire laborers** to construct and hoist the waterwheel, and build the dam, baffle, and millrace.
   f. To prevent river uses that limit flow, we might **pay someone to influence** nearby **government officials**.
   g. To avoid damage to the watermill and channel, we will **pay people for beaver skins**.
   h. We might have to **buy a plot of solid riverside ground**.

2. Outputs: List all material, energy, or conditions that result as an intentional or unintentional consequence of creating or sustaining the system. (This can include anything that is a direct product of the waterwheel, the purpose for creating the watermill, or the actions we conduct to make or operate it. Again, it helps to refer back to our results for functionality and sustainability.)

   i. The waterwheel will **power a sawmill**.
   j. **Trees will have to be cut down to make** some of **the watermill** parts.
   k. **Rock and sand will have to be quarried** for the cobble and cement that create the dam, baffle, millrace, and abutment.
   l. We will have to **replace an area of forest along the riverside with a mill**.
   m. We will **replace the noise of the gurgling river with a swooshing or splashing noise** of the waterwheel.
   n. The **river flow will change** somewhat in response to the dam and millrace.

The third and final step in the compatibility analysis is to infer which systems will be impacted by the inputs and outputs.

3. Compatibility: With inputs and outputs *a* through *n*, we then list any systems (excluding our own), or conditions that (1)—would tend to be directly

strengthened or gain support because of our system; or (2)—would tend to be directly weakened or lose support because of our system. This is probably easiest to do in tabular format.

Inputs and Outputs	**Supported** Systems, Structures, Institutions, or Conditions	**UnSupported** Systems, Structures, Institutions, or Conditions
a. wood parts to make the waterwheel	None in particular	None in particular (assuming material required to build the watermill and sawmill are not too large)
b. hire a millwright	Skilled tradesmen	None in particular
c. hire a metal smith or blacksmith	Skilled tradesmen	None in particular
d. build a dam, a baffle, and a millrace	Fish or ducks that might congregate in ponds	Migratory or cold-water fish; natural channel morphology
e. hire laborers	Carpenters or construction workers	None in particular
f. pay someone to influence government officials	A system of decision-making based on influence peddling	Conditions experienced by farmers or other interest groups
g. pay people for beaver skins	Bounty hunting	Beavers and the wetlands that they make
h. buy a plot of solid riverside ground	None in particular (we'd only need one riverside parcel)	None in particular
i. power a sawmill	Lumberjacks and the construction industry; sawmill workers	Tree and forests, especially tall and straight trees and dense forests on level ground
j. trees will have to be cut down to make the watermill	None in particular (only a few trees may need to be removed)	None in particular
k. rock and sand will have to be quarried	Laborers	One probable riverside area of natural habitat where the mining occurs
l. replace an area of forest along the riverside with a mill	A tree cutting crew (but only a few acres may need to be removed); people that derive a sense of calm from the spinning of the waterwheel	A river with a continuous stretch of natural habitat; people and animals that value or need undisturbed natural expanses

Inputs and Outputs	<u>Supported</u> Systems, Structures, Institutions, or Conditions	<u>UnSupported</u> Systems, Structures, Institutions, or Conditions
m. replace the noise of gurgling river with a swooshing or splashing noise	None in particular	People that value tranquility if the watermill makes noise in a location that was formerly quiet; none in particular if the river had been a loud, fast river in the first place
n. river flow will change	None in particular	Fish or mussel habitat, especially for any sensitive species

Table 1. Watermill compatibility

You might be wondering if anything is missing from this compatibility analysis. In addition to anything we simply forgot to consider, aren't the resulting sawmill, lumber, and employed sawmill workers going to have their own economic and community impacts? And what about the impacts of those impacts? Shouldn't we consider those issues too?

We could, but in this exercise we are trying to address direct outfalls of the waterwheel building and operation, while ignoring further repercussions. Additional issues could always be addressed by running a compatibility analysis upon the sawmill, the employed workers, etc.

To sum it up, the compatibility analysis suggests our simple watermill is having lots of impacts on the world right around it. Interestingly, it seems to have multiple positive impacts on the local economy, and negative impacts on the local environment.

Of course, we lack details about the watermill's location, local culture, and economy. So we can't be sure our findings are accurate. But running through this methodical process still provides lots of information, potentially allowing for a better understanding of what could lay ahead.

5.11 A Village School Example

We began with the goal of fixing broken systems by developing a process to essentially deconstruct them, understand them, and reconstruct them to work better. We saw that to work means to be functional, and to continue working means to be sustainable. We have seen that functionality and sustainability require internal and/or external support. Finally we have an analysis process to help determine areas of support and lack thereof, based upon internal designs, external conditions, feedback loops, and independent forces. Then a compatibility analysis process allows us to learn even more about how a given system might enhance or disrupt other systems.

We finally have all the pieces we need to analyze any system, structure, or institution. We are now ready to use this review process for a more complex system.

Let's say that our watermill analysis was viewed favorably by our supervisor. Soon after our sustainable watermill design was completed, we were assigned a new project—a school.

The school is planned in a rural village named Landesby, which is upstream of the watermill. We learn that Landesby has several thousand inhabitants, including several hundred children who are raised in very traditional patterns. Their way of life is characterized by living among extended families, and learning by doing. This includes learning the village customs and religion, and usually as they mature, learning a craft or vocation.

It happens that a group of people believe Landesby's children should have a more intellectually rigorous upbringing, so they want to create a school. The school proponents believe that this education (in reading, math, history, science, etc.) will make the children more capable of solving problems and making good decisions. They believe this will be good for the children, and by extension good for all of Landesby.

But there are also some school opponents who believe the children learn enough to get by already, and they don't want to see resources directed to something that has never before been necessary.

The disagreement has led the school proponents to seek our assistance. They want us to verify that indeed, a village school is a sustainable idea. Further, they want us to help plan the school's establishment or design in ways that will be sustainable. After listening to the claims of some school proponents, most members of our design team agree that a school sounds like a good idea, and they are eager to begin investigating the issue.

If we do our job well, a sustainability analysis should even yield helpful information on specific school designs or elements that are more sustainable, and it should highlight issues or designs that are problematic or less sustainable.

Even though we will be working for the school proponents, and even though we might have an initial tendency to agree with them, we recognize that our job is to carefully analyze the situation—remaining as open-minded and neutral as possible. Our design team moves forward, hoping for an accurate analysis that will improve the final decision and diminish disagreements.

Verbally, we are given some brief anecdotal information about Landesby's setting, including local prosperity levels, religion, skirmishes over land control, and cultural concerns. Even though we lack a lot of specific information that would make an analysis more exacting, we decide to run through the exercise and see what happens based on what little we do know.

Let's start by defining the system. Let's call it the *school*. Let's also say the school's *objective* is to cause Landesby's children to be more educated in ways that will benefit them. We can make the boundary of our system around the school building, including any person or thing inside of it that is intended to contribute to the school's objective, such as the teachers, desks, staff, books, walls, toys, etc.

Let's go back to the ten-step functionality analysis. Recall that hints to each step accompany the watermill analysis:

1. Name: **A school.**
2. Goal: **The school conveys more advanced and valuable information to children than they otherwise would have had access to, and in such a way that is understandable for them.**
3. Boundary: **A school building, including any teacher, administrator, book, activity, or other instrument of learning.**
4. Internal Design and Process: **Adults (teachers) teach the children inside of the building. They utilize teaching aids in many cases, such as books, chalkboards, paper, and game-like situations intended to increase learning by the children.**
5. Positive Indicators: **The children attend school, and afterward they are consistently better able to understand relevant science, math, history or social information about their environment or world.**
6. Negative Indicators: **There may be poor school attendance. Many of the children do not seem to be absorbing the intended information, or are unable to understand its applications. The children are learning information or developing perspectives on the world that are contrary to the school's teachings.**
7. Supporting Internal Conditions: **There is a secure and comfortable building in which the school can operate. The teachers are trained or experienced in the areas of math, science, history, social studies, language, and other relevant topics. The teachers coordinate their curriculum with each other and between different classes and different grades. The teachers know how to translate their adult-level knowledge into child-level understanding. The teachers have books and exercises directed toward children at various age levels that are interesting for children of that age.**
8. Supporting External Conditions: **The teachers are paid to teach to prevent them from being distracted by another job. The children attend the school frequently enough to learn and retain the information. The children take the schooling seriously. The children arrive at school without extreme distractions such as hunger or fatigue.**
9. Non-Supporting Internal Conditions: **The teaching location is forced to continually change, or is unsafe or distractingly uncomfortable. The teachers are not significantly knowledgeable in the areas of math, science, history, social studies, language, and other relevant topics. The teachers fail to coordinate their curriculum with each other and between different classes and different grades. The teachers are not able to translate or not**

aware of the need to translate their adult-level knowledge into child-level understanding. The teachers lack tools and techniques that engage the children.

10. **Non-Supporting External Conditions:** **The teachers are not paid or materially supported for their work, and have to make ends meet through some other avenue. The children do not attend the school frequently or consistently enough to learn and retain the information. The children do not take the schooling seriously. The children often arrive at school with extreme distractions such as hunger, fatigue, or fear.**

The primary positive results are summarized in #7 and #8 above, while primary negative results are summarized in #9 and #10 above. With this, given our overall design and understanding of how this institution works, functionality requires that:

PF1. **There is a secure and comfortable building in which the school can operate.**

PF2. **The teachers are trained or experienced in the areas of math, science, history, social studies, language, and other relevant topics.**

PF3. **The teachers coordinate their curriculum with each other and between different classes and different grades.**

PF4. **The teachers know how to translate their adult-level knowledge into child-level understanding.**

PF5. **The teachers have books and exercises directed toward children at various age levels that are interesting for children of that age.**

PF6. **The teachers are paid to teach to prevent them from being distracted by another job.**

PF7. **The children attend the school frequently enough to learn and retain the information.**

PF8. **The children take the schooling seriously.**

PF9. **The children arrive at school without extreme distractions such as hunger or fatigue.**

And in contrast, lack of functionality would tend to be brought on by the following:

NF1. **The teaching location is forced to continually change, or is unsafe or distractingly uncomfortable.**

NF2. **The teachers are not significantly knowledgeable in the areas of math, science, history, social studies, language, and other relevant topics.**

NF3. The teachers fail to coordinate their curriculum with each other and between different classes and different grades.

NF4. The teachers are not able to translate or not aware of the need to translate their adult-level knowledge into child-level understanding.

NF5. The teachers lack tools and techniques to engage the children.

NF6. The teachers are not paid or materially supported for their work and have to make ends meet through some other avenue.

NF7. The children do not attend the school frequently or consistently enough to learn and retain the information.

NF8. The children do not take the schooling seriously.

NF9. The children often arrive at school with extreme distractions such as hunger, fatigue, or fear.

That completes a simple functionality analysis. Just like we did with the watermill, we can use these lists as a baseline for a sustainability analysis of the school. Sustainability is functionality through the course of time. Forces or factors that enhance the positive (or limit the negative) functionality items are likely to enhance sustainability, and any forces or factors which enhance the negative (or limit the positive) functionality items are likely to diminish sustainability. Let's start by looking for positive feedback loops.

PF1. There is a secure and comfortable building in which the school can operate. Like all buildings, unless intervention is taken, whatever building the school uses is likely to experience some deterioration over time, so we would not expect a positive feedback loop here.

PF2. The teachers are trained or experienced in the areas of math, science, history, social studies, language, and other relevant topics. Although the teachers are expected to be trained in the first place, they might gain subject-matter knowledge by teaching those subjects. So let's say yes, there is a small positive feedback loop.

PF3. The teachers coordinate their curriculum with each other and between different classes and different grades. It is conceivable that the longer the teachers work together, the more refined they will become at coordinating with each other. Again let's say yes, this is a small positive feedback.

PF4. The teachers know how to translate their adult-level knowledge into child-level understanding. It is possible that teachers will adjust their teaching over time in ways that improve the children's learning. This could be another small positive feedback loop, especially if teachers are personally motivated in what they do.

PF5. The teachers have books and exercises directed toward children at various age levels that are interesting for children of that age. It seems reasonable that over time the teachers will accumulate more books or better books, equipment, and tools, but only if there is adequate funding. This would be a positive feedback loop.

PF6. The teachers are paid to teach to prevent them from being distracted by another job. This aspect would improve for the better during school operation only if the villagers are very pleased with the results of the school, or if the village tends to become more prosperous as a result of the school—something that may or may not happen. So, a positive feedback loop in this regard would depend on how well the school is perceived and how it affects the economy.

PF7. The children attend the school frequently enough to learn and retain the information. The prospect of attendance improvements does exist, but probably only if the parents and villagers are very happy with the school. A positive feedback loop in this regard would depend on how well the school is perceived.

PF8. The children take the schooling seriously. This outcome could be supported if the villagers are quite happy with the school's effect on their children. Similarly, if the local spiritual leader is pleased because the school is teaching the holy text in their curriculum, he might use his authority to encourage schooling as a religious duty. Positive feedbacks here are again contingent.

PF9. The children arrive at school without extreme distractions such as hunger or fatigue. This outcome would not tend to be supported by the school's actions. Extreme distractions such as hunger or warfare would not tend to diminish because of the school. However in the very long term, if the school helped to create students who became financially successful, it is possible they would use their wealth for the benefit of the village over time. There is a potential long-term positive feedback loop here.

Now we consider whether the school is likely to be the source of any *negative* feedback loops. Remember, these are forces or conditions caused by the school's operation that result in diminished support for the school's educational objective.

NF1. The teaching location is forced to continually change, or is unsafe or distractingly uncomfortable. A negative feedback could occur if the school building was not maintained.

NF2. The teachers are not significantly knowledgeable in the areas of math, science, history, social studies, language, and other relevant topics. This is not likely to occur as a result of the school's operation, so no negative feedback loop.

NF3. The teachers fail to coordinate their curriculum with each other and between different classes and different grades. This, too, is unlikely to be a growing issue resulting from the school's operation. However, this might happen if the teachers harbored resentment toward the school or toward each other. These attitudes could arise if teachers feel unappreciated, under-compensated, or if some teachers feel that the other teachers don't value their teaching style or goals.

NF4. The teachers are not able to or not aware of the need to translate their adult-level knowledge into child-level understanding. It seems very unlikely that the teachers would become less adept at this translation because the school was in-progress.

NF5. The teachers lack tools and techniques to engage the children. A negative feedback could occur if the school had no ongoing funding and books and other useful objects were progressively worn out.

NF6. The teachers are not paid or materially supported for their work and have to make ends meet through some other avenue. A negative feedback loop could occur if the village was not extremely prosperous in the first place, and critical resources were diverted to send the children to school and to fund the school, leading to progressively less ability to pay the teachers.

NF7. The children do not attend the school frequently or consistently enough to learn and retain the information. If the village is not financially improved by the operation of the school, or if the parents don't value the schooling, or if the children dislike school and have other alternatives to school available, then yes, this outcome could be caused by a negative feedback loop.

NF8. The children do not take the schooling seriously. A negative feedback loop in this regard is possible if the parents and villagers don't value what the school teaches, or if the school is perceived as contributing to local financial or situational hardship.

NF9. The children often arrive at school with extreme distractions such as hunger, fatigue, or fear. If the village already operated near the margin of survival, then it is possible that the additional expenditures on the school could actually heighten stress levels on the children. Also, if the village exists in a dangerous part of the world where tribal or village battles for resources are prevalent, then the diversion of boys from warfare activities (weapon making or battle practice) could make the village more vulnerable to attacks.

The next step is trying to predict independent forces that could affect the school. First we will look at positive independent forces by referencing our positive conditions for functionality. Independent forces are often unpredictable. So similar to the analysis with the waterwheel, we can add hypothetical scenarios in our analysis in order to help demonstrate situations and designs that could impact the school's sustainability.

PF1. There is a secure and comfortable building in which the school can operate. In one hypothetical scenario, a national police building in the village is deemed no longer necessary by the authorities, leaving the village with a well-built structure for a school. In another possible scenario, a neighboring village opens a brick factory, making bricks much easier and cheaper to obtain for building the school.

PF2. The teachers are trained or experienced in the areas of math, science, history, social studies, language, and other relevant topics. Perhaps a university is established in the national capital, making it possible to become well trained in math, science, etc.

PF3. The teachers coordinate their curriculum with each other and between different classes and different grades. A grant could become available to build a small housing complex for the teachers. This lets them live and work in close proximity to each other, making it easy to coordinate their curriculum.

PF4. The teachers know how to translate their adult-level knowledge into child-level understanding. A teacher's college could be established at the national university, making it possible to receive professional training in effective teaching methods.

PF5. The teachers have books and exercises directed toward children at various age levels that are interesting for children of that age. A wealthy patron could purchase and donate dozens of schoolbooks in the village's native language. Or, a trade blockade might end, allowing large amounts of paper to be imported into the kingdom for the first time in a decade.

PF6. The teachers are paid to teach to prevent them from being distracted by another job. Maybe several villagers discover gold near a rock outcrop a few miles from the village. This would be lucrative for the villagers, who would become wealthier, enabling them to raise teacher pay. In another scenario, all teachers are enrolled by law in a new public service union, which guarantees relatively high wages and generous benefits.

PF7. The children attend the school frequently enough to learn and retain the information. A new national law could declare that education is compulsory through age 12, which would cause attendance to increase. Or, the invention or introduction of gasoline motors could accomplish the agricultural work that had previously been accomplished by human and animal labor. Children would not be needed for farm work as often, and school attendance would improve.

PF8. The children take the schooling seriously. The prime minister, wanting more trained engineers for his military, might offer a significant reward to the parents of students who pass a national science exam at age 16, prompting many parents to urge their children to take school seriously.

PF9. The children arrive at school without extreme distractions such as hunger or fatigue. A new well could be dug for the village, replacing river water, which along with population growth was becoming a source of frequent illness and stress for families. Or, the village could make an agreement with a neighboring mountain village to share extra food whenever a poor harvest occurs for one but not the other—a frequent occurrence since weather patterns each year usually support ample crop production in either the valleys or the mountains, but not both. Or, shelling and skirmish attacks that have occurred along a nearby border for two decades might stop when a peace treaty is made, and the frightening blasts would begin to fade into memory.

Continuing now with *negative* independent forces, we reference our negative conditions for functionality:

NF1. The teaching location is forced to continually change, or is unsafe or distractingly uncomfortable. In one possible scenario, the only available building lacks heating and air conditioning and has a long and narrow shape. Along with the lack of windows along one side, it is very difficult to keep temperatures even.

Or, maybe the school could set up in an unused building, but a government building inspector closes the school repeatedly for code violations—insufficient exits, a broken floorboard, an uninspected electrical hookup, and then no earthquake-proofing. These closures would be disruptive and cause the school to temporarily shift back and forth to some other warehouse building or barn.

In another scenario, swamp drainage projects upstream could gradually cause part of the village (including the school building) to be in the local river's floodplain. The building could flood each spring. This would damage furniture, stain the floors, and cause mold and dank air. This would lower the overall sense of comfort and might force school closures.

NF2. The teachers are not significantly knowledgeable in the areas of math, science, history, social studies, language, and other relevant topics. Budget cuts might force a reduction in the number and breadth of classes offered at a national teacher prep school.

NF3. The teachers fail to coordinate their curriculum with each other and between different classes and different grades. A school principal could discourage collaboration among the teachers and instead prefer the teachers clear plans through him, complicating teacher efforts to work together seamlessly. In another scenario, anger between teachers over unrelated extended family issues could cause them to work together inconsistently.

NF4. The teachers are not able to or not aware of the need to translate their adult-level knowledge into child-level understanding. The villagers might be a cultural minority that the government is trying to assimilate. With the threat of job loss, the government could forbid teachers to teach in the village's native language.

NF5. The teachers lack the tools and techniques to engage the children. The village could be taxed at a high rate by the national government, limiting its ability to pay for basic necessities, such as books or furniture. Or, if the village language is spoken within a restricted geographic area, the books in their language might be all handwritten and rare. If other regional villages started schools recently, they might have gathered most of the available books for themselves, leaving few for other schools.

NF6. The teachers are not paid or materially supported for their work and have to make ends meet through some other avenue. The villagers might be leading an existence that is too impoverished to pay teachers. Or, the national government's promises to pay the teachers and other civil workers could go unmet due to political standoffs. Or, national government pay to teachers could be prevented because all deliveries must ford a narrow valley entrance that separatists from another ethnic group might block.

NF7. The children do not attend the school frequently or consistently enough to learn and retain the information. If the villagers live near the subsistence level and cannot afford to go without help from the children at home, it could require the children to skip school for days or weeks in a row.

NF8. The children do not take the schooling seriously. The national economy holds no appealing prospects for people who are well educated, which lowers the stature of obtaining an education. Or, the national sports idols all tend to be poorly educated, which affects the children's perceptions.

NF9. The children often arrive at school with extreme distractions such as hunger, fatigue, or fear. Three years of drought and one year of crop disease in a six-year period would make hunger the norm in the village. Or, if village men between the ages of 16 and 35 are conscripted into the military to join in a long and bloody war involving four nations, it would leave the village without its strongest men and family members. This would force children to carry larger burdens to ensure day-to-day survival.

5.12 The Village School Example, continued

That takes care of feedback loops and independent forces. Our lists for functionality contained eighteen total positive and negative factors. But lists for sustainability will be longer because that involves many more factors at work over an extended period of time.

We could have included fewer hypothetical scenarios in our sustainability analysis, which would reduce the number of factors we eventually list. However, those hypothetical scenarios are important in order to help establish what type of situations will impact the school's sustainability profile, and thus what designs or actions might be considered in conjunction with the project.

Beyond the nine criteria for functionality (upon which sustainability is based), the twenty-three situations we project to support sustainability are:

- PS1. Increasing teacher knowledge levels as time passes

- PS2. Increasing ability of the teachers to coordinate as time passes

- PS3. Increasing ability of the teachers to translate information to children as time passes

- PS4. An increase in books, equipment, and tools *if funding allows*

- PS5. An increase in teacher pay *if the villagers are very pleased with the school's results*

- PS6. Improved attendance *if the parents and villagers are pleased with the school's results*

- PS7. An improvement in the children's opinion of the importance of school *if the parents and spiritual leaders are pleased with the school's results*
- PS8. An improvement in stress levels of the pupils *if the school helps the children become financially successful*
- PS9. The ability to use high-quality, surplus facilities
- PS10. The ability to obtain building materials and obtain them at low cost
- PS11. The presence of post-secondary educational institutions
- PS12. The availability of grants or other public benefit money pools to fund school-related projects
- PS13. The presence of a post-secondary training institution for teachers
- PS14. The presence of wealthy patrons or philanthropists
- PS15. Open trade relations with neighboring countries
- PS16. Nearby sources of extractable wealth, such as precious metals
- PS17. Laws that directly support and benefit teachers, schools, or their employees
- PS18. Laws that require schooling
- PS19. Inventions that reduce home and village labor requirements
- PS20. National or military demand for children to be educated in specific fields
- PS21. The introduction of technologies or practices that reduce disease
- PS22. The introduction of technologies or practices that reduce hunger
- PS23. Good national relations with neighboring countries

However, sustainability is likely to be reduced or hampered by the nine negative conditions for functionality, and the following twenty-five scenarios:

- NS1. A deteriorating school building *unless it is maintained*
- NS2. A decrease in coordination between the teachers as time passes *if they feel undervalued, undercompensated, or do not have the same belief systems*
- NS3. A decrease in books and other teaching aids *if there is no ongoing funding for these items*
- NS4. A decrease in teacher pay *if the school is draining money or energy from a people of limited means*

- NS5. A decrease in attendance *if the school is not enjoyable to the children, not contributing to economic prosperity, or not valued by parents*

- NS6. A lack of respect for the schooling *if the school is detrimental to economic prosperity, or if it is not in line with village values*

- NS7. An increase in stress levels among the children *if the villagers already live near the margin of survival, or if the village exists in a conflict area where there are ongoing battles for resources between different groups*

- NS8. Usage of inappropriately designed or uncomfortable building space, even if it is available for free

- NS9. Free usage of a substandard or imperfect building *if there is strict government oversight and enforcement*

- NS10. Environmental degradation occurring to the property used or needed by the school

- NS11. National-level budget cuts in teacher training programs

- NS12. School administrators with management strategies that significantly interfere with teacher interaction or methodology

- NS13. Teachers from families or clans that are in conflict

- NS14. Inability to teach using the native written or spoken language

- NS15. State intolerance of the local culture

- NS16. A relatively high national tax burden upon the local community

- NS17. Competition with neighboring communities for limited resources

- NS18. Impoverishment in the village

- NS19. A national-level political stalemate or power struggle

- NS20. Local or regional conflict between ethnic groups and the state government

- NS21. A high village reliance upon the children's chores

- NS22. A national economy without job opportunities for educated citizens

- NS23. Cultural adulation for poorly educated heroes

- NS24. A disruptive or unpredictable climate

- NS25. Major national military conflict

That makes forty-eight feedback loops and independent forces that could hypothetically impact our eighteen positive and negative functionality factors, at least by this analysis. So if we are trying to assist the school proponents to get the school going, or keep it going sustainably, this is our chance to propose certain steps on its behalf.

Determining these steps can be accomplished through a sort of brainstorming exercise, in which we imagine how to encourage the positive functionality and sustainability factors and reduce the negative functionality and sustainability factors. In addition, determining these potential actions ahead of time will then help us to evaluate compatibility. With our hypothetical school, we envisioned many potential scenarios which, of course, won't all occur, and some are mutually exclusive (cancelling the possibility of others). But we can still consider a potential response to each scenario.

Focusing more upon things we might be able to exert some control or influence over, we will start brainstorming with what we thought would imbue functionality in the first place.

Each of the following school design aspects would be an attempt to either support a positive functionality or sustainability trait, or reduce a negative functionality or sustainability trait.

1. In general relation to a school building and location:
    a. **Provide a secure and comfortable building in which the school can operate.**
    b. **Seek out a surplus building** prior to seeking funding for a new one.
    c. **Site the school building away from areas undergoing environmental change.**
    d. **Hold school outside when weather allows,** if no suitable buildings are available.
    e. Attempt to **bargain or barter for building materials.**
    f. **Reuse building materials** from buildings that are being torn down in order to save money.
    g. **Maintain the building on a frequent basis.**
    h. **Bribe inspectors** in order to pass public building inspections.
2. In general relation to level of teacher expertise:
    a. **Hire teachers that are trained or experienced** in the areas of math, science, history, social studies, language, and other relevant topics.
    b. **Retain teachers as long as possible** since that allows them to become more skilled in their subject matter and in conveying information to the children.
    c. **Advocate for the establishment or funding of national or regional post-secondary institutions.**
    d. **Advocate for permanent nondiscretionary funding of teacher training programs.**
3. In general relation to teacher coordination of curriculum
    a. **Require that the teachers coordinate their curriculum** with each other and between different classes and different grades.

b. **Hire teachers in groups** so they learn to work together as a team over months and years.
   c. **Hire teachers with similar political or professional values** to help set the stage for good coordination.
   d. **Hire teachers from the same clan or group.**
   e. **Hire a school manager or principal with a decentralized management style,** or one who is likely to align well with teacher values.
   f. **Praise the teachers for school successes** whenever possible.
   g. **Seek out grant monies or funding channels** for the school.
4. In general relation to teachers translating their knowledge
   a. **Hire teachers that know how to translate their adult-level knowledge** into child-level understanding.
   b. **Train new teachers on effective teaching methods** if no trained individuals are available for hire.
   c. **Advocate for the establishment or funding of teacher-training institutions.**
   d. **Advocate for a cultural tolerance law** at the national level.
   e. **Hold school in small secret groups to allow teaching in the native language** if its use is being officially barred.
   f. If there is a government assimilation program, **encourage the villagers to abandon their native language** for the national language.
5. In general relation to books and other teaching aids
   a. **Provide the teachers with books and exercises** that are appropriate for children of various age levels.
   b. **Provide an annual fund for new books or other learning tools.**
   c. **Contact wealthy former villagers or well-known philanthropists** to seek financial support.
   d. **Attempt to establish a voluntary village fund for annual purchase of books and school items.**
   e. **Seek charitable organization assistance with managing limited village resources.**
   f. **Help establish a council in the village to promote tourism and economic growth.**
   g. **Help the villagers fight high tax rates or to find tax loopholes.**
6. In general relation to teacher pay and focus
   a. **Pay the teachers well** to prevent them from being distracted by another job.
   b. Make sure that the **material taught is strongly approved of by the villagers.**
   c. **Advocate for laws that provide more benefits for public employees or schools.**
   d. **Advocate for the extraction and sale of resources** if any are on lands controlled by the village.
   e. If they are being paid, **bar the teachers from holding a second job.**

f. If they are not being paid enough, **cut back on school hours** so that the teachers have adequate time to hold a second job.
   g. **Establish the school to be as independent and non-reliant as possible** upon the national government.
   h. **Pay separatists to allow important deliveries to reach the village.**
7. In general relation to attendance levels:
   a. **Schedule the school to run at least four days a week and over half of the year** so children attend the school frequently.
   b. **Advocate for a law requiring children within a certain age range to attend school.**
   c. **Make sure the children have plenty of time for play and games while in school.**
   d. **Encourage the villagers to have more children** in order to lower the per-child chore requirements.
8. In general relation to children's perception of school
   a. In hopes of convincing them, **tell the villagers that their children are better off being schooled.**
   b. **Align the school's curriculum with teachings of local spiritual leaders.**
   c. **Bring prestigious speakers to the school who can demonstrate the value of an education.**
   d. **Find famous athletes to encourage the children** to complete their education.
   e. **Encourage the village children to leave the country to find work** once they are educated, if the national economy holds no prospects for them.
9. In general relation to children's levels of distraction
   a. **Allow highly stressed children to have extra time for play, games, and exercise.**
   b. **Feed any hungry children** that arrive at the school.
   c. **Give counseling to any children who are disturbed or traumatized** by events.
   d. **Tailor the education so that it gives the children an opportunity to become financially successful.**
   e. **Encourage villagers, NGOs, or the government to install infrastructure that reduces risk of disease.**
   f. **Encourage villagers to adopt practices that make their food supply more reliable.**
   g. **Encourage the villagers to enter negotiations with their enemies.**
   h. Help to **hide military age men and boys** when military officers are rounding up soldiers.

Next we should do an analysis of compatibility—remember that this entails a look at effects that the school might have on systems, structures, and institutions, or how it might be affected by them.

As with the watermill analysis, we currently have little information on nearby cultural, political, and economic systems. So for now let's just analyze compatibility issues that the school might create for these other systems.

For the compatibility analysis we need to know "inputs" to the school and "outputs" from the school. We already know many of the possible school inputs (the 57 action items in the list above). For convenience in the compatibility analysis, we will continue numbering where we left off for additional items.

1. Inputs: List all materials, energy, and activities that have to be (or could be) redirected, created, or initiated in order to create and sustain the system/structure/institution. This can include anything that is directly required, or likely to be indirectly required.

   We went through a functionality and sustainability analysis for the school already, so we have a sense of what it might require to initiate and operate the school. In addition to the 57 items listed above, here are possible items:

   a. We will need **village agreement or consultation** to initiate the school. Even if the school is essentially imposed from outside, the villagers will have to participate to some degree.
   b. The new school building, if there is one, will require us to **obtain construction materials such as lumber, bricks, stone, cement blocks, etc.**
   c. We will need to **obtain books, desks, furniture, chalkboards, and other appropriate classroom items.**
   d. Depending upon the size of the building, we might need to **hire a janitor or groundskeeper.**

2. Outputs: List all material, energy, or conditions that result as an intentional or unintentional consequence of creating or sustaining the system/structure/institution. (This can include anything that is a direct product of the school, the purpose for creating the school, or the actions we conduct to make or operate it.)

   a. **Children will be educated** in areas of reading, history, math, science, etc.
   b. **Children will spend much of their day away from their nuclear family.**
   c. Due to changes in how their time is spent, it is probable that **children will have fewer teachings and values instilled by their nuclear family.**
   d. **Children** will probably **view their native culture in reference to a national culture or a planetary population.**
   e. **Children will spend much of their day away from their agricultural or subsistence lifestyle.**
   f. Some part of the village, such as a **pasture or crop field will have to be replaced by the school grounds.**

3. Compatibility: With fifty-seven possible action items, plus a total of ten inputs and outputs, we then list any system (excluding our own), or conditions that (1)—would tend to be directly strengthened or gain support because of our system; or (2)—would tend to be directly weakened or lose support because of our system.

Here are the compatibility results in tabular format:

Inputs and Outputs	<u>Supported</u> Systems, Structures, Institutions, or Conditions	<u>Unsupported</u> Systems, Structures, Institutions, or Conditions
1a. Provide a secure and comfortable building in which the school can operate	Construction crew, if one will be hired to build the building	Village or taxpayer/ national funds, if they are used to construct or purchase the building
1b. Seek out a surplus building	The owner of a surplus building, if it is sold to the school	None in particular
1c. Site the school building away from areas undergoing environmental change	None in particular	None in particular
1d. Hold school outside when weather allows	Social connections within the village	Children's health, if weather or insects are challenging
1e. Bargain or barter for building materials	None in particular	Village resources such as livestock, if used for barter
1f. Reuse building materials	The environment	None in particular
1g. Maintain the building on a frequent basis	A handyman	Village or taxpayer/ national funds
1h. Bribe inspectors	Government corruption	The children's well-being if the building is unsafe
2a. Hire teachers that are trained or experienced	Professional teachers	None in particular
2b. Retain teachers as long as possible	Job stability for employed professional teachers; teachers working until retirement	None in particular
2c. Advocate for the establishment or funding of national or regional post-secondary institutions	Social advocacy; post-secondary institutions	Taxpayer/national funds

Inputs and Outputs	Supported Systems, Structures, Institutions, or Conditions	Unsupported Systems, Structures, Institutions, or Conditions
2d. Advocate for permanent non-discretionary funding of teacher training programs	Social advocacy; teacher training institutions	Taxpayer/national funds
3a. Require that the teachers coordinate their curriculum	Interaction between the teachers	None in particular
3b. Hire teachers in groups	None in particular	None in particular
3c. Hire teachers with similar political or professional values	A particular set of political or social values	Relative neutrality, or different political or social values
3d. Hire teachers from the same clan or group	One clan or group and its values	Values of different clans or groups
3e. Hire a school manager or principal with a decentralized management style, or one who is likely to align well with teacher values	A particular set of political or social values	Relative neutrality, or other political or social values
3f. Praise the teachers for school successes	None in particular	None in particular
3g. Seek out grant monies or funding channels	The influence of grantors upon the curriculum	Public monies
4a. Hire teachers that know how to translate their adult level knowledge	Trained teachers, professional teachers	None in particular
4b. Train new teachers on effective teaching methods	None in particular	Village or taxpayer/ national funds used for training
4c. Advocate for the establishment or funding of teacher-training institutions	A university and a teacher training school	Taxpayer/national funds
4d. Advocate for a cultural tolerance law	Social advocacy; ethnic and cultural diversity	National cultural unity and assimilation
4e. Hold school in small secret groups to allow teaching in the native language	The native language and culture	National cultural unity and assimilation; the national laws; the safety of teachers

Inputs and Outputs	Supported Systems, Structures, Institutions, or Conditions	Unsupported Systems, Structures, Institutions, or Conditions
4f. Encourage the villagers to abandon their native language	National cultural unity and assimilation	The native language and culture; cultural diversity
5a. Provide the teachers with books and exercises	Book publishers, educational product developers, and distributors	Taxpayer/national funds
5b. Provide an annual fund for new books or other learning tools	Book publishers, educational product developers, and distributors	Village or taxpayer funds
5c. Contact wealthy former villagers or well-known philanthropists	Influence of a single patron upon what the school might teach	Personal interest or subject areas that the teachers might have
5d. Attempt to establish a voluntary village fund for annual purchase of books and school items	Influence of the villagers upon the curriculum; book publishers	Village funds
5e. Seek charitable organization assistance with managing limited village resources	Outside cultural contact or influence upon the village	Charitable funds
5f. Help establish a council in the village to promote tourism and economic growth	Outside cultural contact or influence upon the village	The native language and unmodified culture
5g. Help the villagers fight high tax rates or to find tax loopholes	Village wealth	National taxpayer funds
6a. Pay the teachers well	Professional teachers; a middle class or professional social class	Village or taxpayer/national funds; income parity if villagers are poor compared to the teachers
6b. Material taught is strongly approved of by the villagers	The culture, traditions, and cohesion of the village	National or "standard" versions of science, history, art, language, or other topics
6c. Advocate for laws that provide more benefits for public employees or schools	Professional teachers; a middle class or professional social class	Village or taxpayer/ national funds; income parity if villagers are poor compared to the teachers

Inputs and Outputs	Supported Systems, Structures, Institutions, or Conditions	Unsupported Systems, Structures, Institutions, or Conditions
6d. Advocate for the extraction and sale of resources	Village wealth	The natural environment
6e. Bar the teachers from holding a second job	General ease of obtaining a job	Professional freedom for teachers
6f. Cut back on school hours	Children's time with family and ability to do chores and learn village customs	Likelihood of children later having more advanced careers
6g. Establish the school to be as independent and non-reliant as possible	The self-determination, culture, traditions, and cohesion of the village	Influence of the national government
6h. Pay separatists to allow important deliveries to reach the village	The separatist movement	The national government
7a. Schedule the school to run at least four days a week and over half of the year	Parental ability to work, undistracted by younger children that are old enough for school	Children's immersion and learning about nature, farmstead, and/or household activities; time with extended family; nomadic or migrant patterns of families; children's activity and fitness levels
7b. Advocate for a law requiring children within a certain age range to attend school	A national culture; an educated populace	Cultural diversity
7c. Make sure the children have plenty of time for play and games while in school	Children's social skills, happiness, and fitness levels	The children's knowledge levels and ability to later obtain higher level careers
7d. Encourage the villagers to have more children	The village population level; professional teachers	Per capita village wealth; the natural environment
8a. Tell the villagers that their children are better off being schooled	None in particular	None in particular
8b. Align the school's curriculum with teachings of local spiritual leaders	The religion, culture, traditions, and cohesion of the village	The national or "standard" versions of science, history, art, language, or other topics

Inputs and Outputs	Supported Systems, Structures, Institutions, or Conditions	Unsupported Systems, Structures, Institutions, or Conditions
8c. Bring prestigious speakers to the school who can demonstrate the value of an education	The fame level of particular prestigious speakers	Village funds that are used to pay speakers' costs
8d. Find famous athletes to encourage the children	The adulation of athletes	None in particular
8e. Encourage the village children to leave the country to find work	Emigration from the village and the nation	The culture, traditions, and cohesion of the village; national identities
9a. Allow highly stressed children to have extra time for play, games, and exercise	Children's social skills, happiness, and fitness levels	None in particular
9b. Feed any hungry children	Health of otherwise malnourished children; prepared food providers	The religious or social networks that would otherwise feed hungry children
9c. Give counseling to any children who are disturbed or traumatized	Children's mental health; possibly a trained or professional counselor	None in particular
9d. Tailor the education so that it gives the children an opportunity to become financially successful	Future income levels of the children and their families	Children's training in less lucrative subjects—perhaps gym, art, history, writing, traditional teachings, spiritual values, and social education
9e. Encourage villagers, NGOs, or the government to install infrastructure that reduces risk of disease	Outside cultural contact or influence upon the village; village population level	The natural environment
9f. Encourage villagers to adopt practices that make their food supply more reliable	None in particular	None in particular
9g. Encourage the villagers to enter negotiations with their enemies	Peace	Sellers of warfare items
9h. Hide military age men and boys	The culture, traditions, and cohesion of the village	The national military; national security; the welfare of apprehended subverters

Inputs and Outputs	Supported Systems, Structures, Institutions, or Conditions	Unsupported Systems, Structures, Institutions, or Conditions
10a. Village agreement or consultation	Village self-determination; the village process of decision-making	People in the village that object to the school decision
10b. Obtain construction materials such as lumber, bricks, stone, cement blocks, etc.	Sellers of construction materials	None in particular
10c. Obtain books, desks, furniture, chalkboards, and other appropriate classroom items	Furniture and implement makers; book publishers, educational product developers and distributors	None in particular
10d. Hire a janitor or groundskeeper	A janitor or groundskeeper	Village or taxpayer/ national funds
11a. Children will be educated	An educated local populace	The children's education in traditional village crafts and guilds
11b. Children will spend much of their day away from their nuclear family	None in particular	Traditional family learning and methods of passing knowledge to succeeding generations
11c. Children will have fewer teachings and values instilled by their nuclear family	None in particular	Village and family knowledge and traditions
11d. Children view their native culture in reference to a national culture or a planetary population	A national culture; the global economy	The religion, culture, and traditions of the village
11e. Children will spend much of their day away from their agricultural or subsistence lifestyle	None in particular	The children's activity levels and physical health; the children's knowledge of agriculture and subsistence activities
11f. Pasture or crop field will have to be replaced by the school grounds	None in particular	An area of natural environment converted in response to this loss

Table 2. Village school compatibility

5.13 Discussion: Designing a Village School and Complex System Analysis

The results of the compatibility analysis are only predictions. Also, given our lack of more contextual information, the compatibility analysis was one-sided (we only looked at the school's effects on other systems, and didn't consider those system's impacts upon the school). Systems such as families and economies would probably have many compatibility implications for the school—implications which we have not even begun to investigate.

Despite limits to our inquiry, our analysis still provides numerous informative results. The effects of the school upon other systems, structures, and institutions seem to vary quite widely. Many effects would depend upon the strategies we take in establishing the school relative to the particular cultural, political, and economic circumstances present in the nation and in the village.

One theme seems to be the economy—both local and national economies seem very intertwined with the school's impacts and eventual success. In some potential situations the village might be economically weakened by the school and its formative decisions, but under other scenarios it could be strengthened.

Another theme that emerges is culture. The local culture and values seem to have many opportunities to be weakened if the school opens, although there are some situations in which they could be strengthened. Some of the situations we construe indicate the school might be just a partial cause or expediting factor in changes that would have occurred anyway (such as with the implementation of a national assimilation policy).

Along with culture, it is interesting how politics could get mixed in with the school. For example, the teachers may be a like-minded bunch because it is convenient for school operation, but an unintended consequence of this could be that they present one-sided political views to the children. Another pattern seems to be how important it is that the parents and villagers accept and value the schooling, though they may be understandably cautious to accept it, given the possible impacts upon their families, economy, and social continuity.

Another issue that surfaces is the fate of the village, which seems to be uncertain if the school opens. The children might leave Landesby once they are educated, and thus the village could decline in numbers or energy, especially if its most creative and ambitious tend to leave.

Finally, in some ways school and *local* interests seem aligned (as with reducing disease levels), and in other ways school and *national* interests seem aligned (such as when the national government has ample revenue). But when the local and national interests seem divergent, the school appears to be almost caught in the middle or forced to take sides. For example, when it comes to national military service, tax breaks, and use of the native language, positions taken by the school would seem to be taking sides with local interests and against national interests, or visa versa. If the school takes sides against the village or the nation, will that make the school more sustainable, or less sustainable?

Our detailed sustainability review of the school was conducted without complete information, but as assigned advocates, it has still prepared us to act through advocacy, policy, or economy to promote the needs of the school. Some actions seem risky, such as bribing inspectors or hiding males from military conscription. Other strategies seem weak, such as essentially telling the villagers, "School is important," and hoping they will be convinced. Some strategies seem to boost the school in one respect while running counter to other school interests. For example, cutting back on school hours so that teachers can hold a second job might support teachers by detracting from learning. All told, this analysis process for functionality, sustainability, and compatibility has been extremely informative, and has yielded many topics for our consideration that we, as amateurs in this situation, undoubtedly would have otherwise overlooked.

It's time to step back now though, and look at why we have engaged in this analytical process, and what it has accomplished. What brought us to this point anyway? It was the idea that so many things within our own nation don't seem to be working well. We began looking for a way to methodically deconstruct and reconstruct any system so that it works properly. We learned that to work means to be functional, and that to continue working means to be sustainable. And with this, we set out to develop a way of creating that functionality and sustainability for any system, structure, or institution.

Though the concepts might seem obtuse at first, by staying focused, we were apparently successful in developing a workable superball situation. We ended up with a handful of clear and simple directives, such as having a weight limit on the ball, and a hard flat surface that won't crack. We were similarly successful at determining functionality of a glacier—it required simple things such as cold winters and ample snowfall.

Then we tried the same methodology on a waterwheel. That seemed pretty successful too. Through the analysis we foresaw numerous important design elements, such as whether the paddles were numerous enough, whether the shaft spun with low friction, and whether the flowing river would change its behavior. Even though largely successful in guiding a waterwheel design, the process still left us with unknown factors related to local economy, politics, and climate. Furthermore, in our compatibility analysis, it seemed that the waterwheel had strands of impact reaching back into the local economy, environment, and political control. These unknown factors and potential impacts wouldn't prevent a well-designed waterwheel, but they seemed to cast a bit of uncertainty over its sustainability.

Then, in trying to develop a functional and sustainable village school, the amount of information our process generates seems to have grown immensely. The analysis helped to hone our thinking and allowed us to envision what initial elements create a functional school (with eighteen "yes's" and "no's"), but the ensuing sustainability analysis resulted in forty-eight issues of support or caution. With that we produced a list of *fifty-seven* things to do, or to consider doing. Aside from the fact that a to-do list of fifty-seven large tasks on behalf of a village school is probably impossible, many of the to-dos were contradictory, meaning we have to favor some options over others. And how do we know which options to favor? Shorten the school day or lengthen it? Teach in the native language or embrace assimilation?

We could try to base our design choices for the school upon compatibility with other systems, such as the village. But our compatibility analysis results in over one-hundred and thirty possible impacts upon other systems. And that analysis was only half complete—to be thorough we would still learn about other systems and conduct an analysis of their compatibility for the school. It's almost as though with the school, the harder we try to analyze the situation the more complex it becomes, and the less certainty we have about which design strategies to embrace and which to forego. Although a "deconstruction" and "reconstruction" of the school has been very informative, it's as though a level of analytical chaos has begun to show up.

Analyzing the superball was like hopping onto a train and getting let out on the doorstep of our destination. Analyzing the watermill was like getting dropped off close to our goal, and having to climb a couple flights of stairs to reach our final destination. Analyzing the school is like getting into a car for a 600-mile drive to a resort, but getting let out at a random intersection with 300 miles left to go. That car ride is better than nothing, but where are we supposed to go now? And if this is how complicated the analysis of a village school becomes, how complicated would it be to determine sustainability for a large corporation, a regional economy, or a whole culture?

Why should it be this complicated? This is the nature of complex systems. Complex systems don't have clear boundaries or centralized control points, and they don't have tightly-defined futures. In contrast to our analysis of the superball, by the time we're done thinking about the school, we seem to have lost a huge amount of the control we had, and instead we've encountered valuations that we aren't well positioned to make (school versus economy) and faced with many ancillary cause-and-effect situations that we probably can't afford to ignore but can't afford to address.

The further we go with the analysis, and the more complex the system, the weaker our power of prediction becomes, consequently weakening our ability to effect directed control. An analysis of cause-and-effect relationships can allow excellent control of simple systems, but the control obtained through this process becomes weaker and weaker as our systems become more and more complex.

Careful system analysis accomplished by thorough deconstruction, review of every system component, and reconstruction is highly informative, but it cannot tell us how to make our complex systems, structures, and institutions work just as we'd like. The more complex our system becomes, the more limited our predictive abilities and influence become. With an increasing complexity in human systems, we are witnessing increasing difficulty in predicting our problems, our solutions, and the effects of our solutions, and increasing difficulty in managing or establishing control over important situations.

Can't we just "figure the darn thing out," and then do what is best? People are smart, and they can solve problems, right? Yes, they are smart, but no, they cannot *figure* out how to solve every complicated problem. Sometimes there are too many variables. There are too many connections between those variables. And those connections cause too many impacts.

So how should we respond to this on behalf of the proposed school? A typical response would be to forge ahead with our own agenda, despite many uncertainties, trying to "right the world" in line with our personal cause. And with the help of our careful analysis, we will probably experience *some* success. With sweat and determination, we can shape a corner of the world to fit our values and needs. And as we press ahead, while our true goal is a successful school, we have become accidental stewards of the economy, the culture, and arbiters of power between local and national interests.

Being school advocates, maybe we don't care that much about how the local economy or culture is altered, as long as the children are educated—"Just stick to our own job," we tell ourselves. And maybe just the same, the cultural protectors, economic interests, and the military powerbrokers are in a similar situation. Are these other people going to conduct a compatibility analysis before they make decisions? Are they going to make sure the welfare of a school is insured under their plans and schemes? Probably not. But even if they did, they would find, like us, that they have a limited ability to predict their own problems and solutions, limited leverage to affect solutions, and limited ability to anticipate the wider effects of their actions. So, like us, they will take default actions that serve *their* bottom line. As long as the culture is protected (or their version of it), as long as the economy works for them, and as long as power is maintained for their purposes—they have accomplished *their* job, or so they will tell themselves. Where will that leave our school? Where will our school leave them?

As groups and institutions grow and interact more frequently, their new complex formations and interactions become more difficult to predict, to make preparations for, to manage, and to operate without mutual interference. Doing more and more analysis might help to a degree, but it won't deliver a nation that works the way we'd like. As for our *original* question, "Are things working as planned?" they are working, but not working especially well, and the trend lines, which many people might prefer to ignore, are becoming discomforting.

Given the way complex systems behave and interact, and our lack of an overall plan or history of dealing with them as such, is it any wonder that our systems seem to be bearing off into some unplanned and unwanted future? As this situation progresses, we now find ourselves in the beginning stages of an unintentional, collective sabotage.

We have a nation filled with intelligent, well-meaning, and skilled people. Yet something or some things seem out of synch and getting worse. Can't we debate our way out of this, or argue ourselves into the clear? If it were just a matter of arguing to the point of victory, we—or someone—would have won the battles long ago. A state of permanent analysis won't work. Identifying the right and the wrong people or arguments won't work (there aren't enough of them). Perhaps it is time for us to find a different way of doing things. But what could that be?

We can't know much of the future. We can't control much of the future. But we might be able to create systems, structures, and institutions that work the way we need them to, however the future unfolds. What would they look like? How would they behave? The remainder of this text is dedicated to addressing these questions.

PART III

NATURAL SYSTEM SUSTAINABILITY

Chapter 6 Natural Systems

If only we had an owner's manual for complex systems, structures, and institutions, it would guide us in mending the things that aren't working. Then again, what if we could create the manual? Can this be done? How? Maybe we already have what we need to make the manual. Maybe we've had it all along, but haven't recognized it.

According to the fossil records, there has been life on Earth for over three billion years. The earliest forms of life for which we have evidence were relatively simple one-celled organisms. In time, according to evolutionary theory, some of these single-celled organisms transformed or evolved into larger plant and animal life forms.

Eventually these larger plants and animals transformed themselves further into more new organism types. As this formation and evolution advanced, interactions between these different organisms became more complex, with new life strategies hinged not just upon the natural environment, but upon the other life forms. For many millions of years on Earth, these life forms have developed, adapted, interacted, and reshaped themselves and each other. While over very long time periods the specific types of plants, animals, and microscopic life forms have ebbed and waned, they continued to adapt, interact, and develop into complex and interdependent groupings.

Today, these plants, animals, fungi, and microorganisms inhabit essentially every nook on and near the Earth's surface. The number and variety of plant and animal types, or *species*, that now occur is *astounding*—including land and sea, several million are believed to exist. Even now, these millions of species continue to live and interact in complex, interdependent groups that are often referred to as *ecosystems* or *natural communities*. Like the ability of each species to adapt to changes, these ecosystems and natural communities also change through time, adjusting and readjusting to each other and to the environment. The Earth's ecosystems and natural communities have probably always been a source of survival interest and curiosity for humans, as they are now a continual source of scientific intrigue.

When it comes to complexity and survival through changing conditions, there is probably no equal to these ecosystems, natural communities, and the organisms that inhabit them. They display perhaps the most widespread and wondrous abilities to maintain or sustain themselves in the midst of uncountable situations, interactions, and changes. Fundamentally and over long periods, these systems and organisms are tightly held to the laws of nature. That which works best, spreads. That which works least, disappears. Systems that can sustain themselves remain. Systems that cannot sustain themselves transform or disintegrate.

This natural push toward functionality and sustainability has led to many amazing system designs that have an ability to persist and reproduce themselves, despite what might look like challenges. To view nature is to behold a university that has stood—not for an admirable three hundred years—but more like a humbling *thirty thousand* or *three-hundred thousand* years.

Just what do these natural systems do, or how is it that they have "learned" to be so sustainable? Herein, there will be an attempt to understand the wisdom of these complex systems. With that wisdom, perhaps we might begin to develop a rudimentary picture of complex system sustainability.

Because complex natural systems have so much variety, discussing their properties will require some consistency of terminology and definitions. Before giving those, an ecological system overview might be helpful:

From a modern scientific perspective, all living things belong to some type of species of animal, plant, fungus, bacteria, etc. Each individual being of each species usually attempts to produce offspring while surviving in the context of a natural community of other living species. Areas where several natural communities are adjacent, interacting, and dependent upon a similar climate or environment can be thought of ecosystems. The ecosystems bound other ecosystems across the Earth to form the Earth's living web, or biosphere. Here are some definitions presented as a natural system hierarchy, given in the order of upper or larger levels last:

1. Individual: A single plant, animal, or other creature. An example would be an individual ladybug.
2. Offspring: The seeds, eggs, babies, or young of plants, animals, or other creatures. An example would be a litter of fox pups.
3. Species: Plants, animals, or other creatures that primarily have the same biological requirements, survival strategies, and overall appearance, and can reproduce together. An example would be the saguaro cactus of the American Southwest.
4. Species Classes or Suites: Groups of species that live with or near each other and have somewhat similar forms and life strategies. An example would be the trees in an oak-pine forest on an Ozark Mountain ridge.
5. Natural Community: The local aggregations of species that occur when particular environmental conditions occur in a particular region or location. An example is a marsh located in the central Great Plains.
6. Ecosystem/Ecoregion: The aggregations of natural communities formed by species in a particular portion of the biosphere, with its associated broad climate and environment. An example is the central Great Plains.
7. Biosphere: The planetary network of living things and the natural environment that they affect and interact with. This includes the living things on planet

Earth, including in the ocean, on the land, in the air, in the soil, and underground.

We will be looking at and discussing the properties or patterns displayed by these sustainable natural systems—individuals, offspring, species, species classes, natural communities, ecosystems, and the biosphere. Before doing that however, it is important to understand some inherent limitations in attempting to do this.

There are so many ways in which the Earth's natural systems operate. Nature has countless forms and it has found innumerable solutions to innumerable problems. Due to this variety, there are probably exceptions to every pattern in nature. The emphasis herein will be upon the predominant patterns, or what is true in most situations most of the time, recognizing that there are probably exceptions or atypical situations for which the predominant pattern does not hold. It should be understood that the natural system patterns which will eventually be presented herein are *probably not true all of the time*.

There is not necessarily any limit to the number of natural system patterns that might exist, and that one could choose to highlight. The goal here will not be, and cannot be to find all patterns by which natural systems operate. It should be understood that the patterns presented herein are probably a subset of, or possibly a minority of the number of natural system patterns that actually exist.

Natural systems are complex, and like many complex human systems, they generally lack sharp boundaries. For example, a "forest" could blend very gradually into a "grassland." If many of the grassy areas contain scattered trees, and many of the forested areas are grassy, then who can say where the exact line is between the forest and the grassland? Won't ten people have ten different opinions? There are often different beliefs or interpretations of where natural systems begin and where they end. The resulting lack of a uniform definition makes it more difficult to study them, describe them, and to communicate information about their characteristics.

Even further, while it may be helpful for us to view natural systems as nested hierarchical organizations (e.g., discussing an *individual* of a certain *species* living in a certain *ecosystem*), there are quite strong philosophical arguments that can be made that no such groups or hierarchs exist. For example, the concept of individual species can be challenged on many levels, including on genetic, population contact, and behavioral grounds. Many, many of the plant forms known to science, for instance, do not quite seem to match a "species" concept, because they do not completely separate or fully combine with similar plant forms. The concepts of natural communities and ecosystems can be challenged based upon a lack of precisely repeating patterns, and based upon the lack of correlation with other factors, such as species' ranges.

Philosophical debates regarding the "true" existence of individuals, species, natural communities, etc., are scientifically important because they help to keep scientists from being trapped in overly simplified conceptual models of how the natural world is organized. While natural system groupings such as natural communities may not necessarily exist from some philosophical perspectives, accepting the existence of these groups allows science to interpret and understand much of the non-uniformities (i.e., patterns) that exist in nature.

The same points can be raised for *all* complex systems. Exactly where is the boundary of a particular culture, a certain local economy, or a social network? There are rarely strong lines that separate these entities, and clear-cut definitions of their boundaries tend to fall apart upon detailed review. So, if their boundaries aren't clearly defined, can we even really say the culture, the local economy, or the social network objectively exist? Actually, probably not! But being able to discuss these groups anyway allows us to communicate and understand far more than we otherwise could. Herein terms for natural and human grouping are used to develop a better understanding of factors that are associated with sustainability—not in order to argue for their objective or intrinsic existence.

In summary, it should be recognized that there will be exceptions to the natural system patterns presented, and that there will be additional patterns that are not presented here. Also, terms for natural and human groupings and hierarchies are being used for communication and modeling purposes, not in order to argue that they positively exist in the objective sense.

That said, let's take a look at a fairly straightforward pattern that occurs in nature that applies to all seven levels of the natural system hierarchy. It is a good one to start with because it gets to the big picture of living sustainability, and it relates to the way in which stability is maintained within complex systems.

To help exemplify this first pattern, visualize a forest. There are many forests on Earth that have stood for decades and decades. These forests would appear stable to the casual observer. However, as stable as they may seem on the outside, these forests are full of activity. Imagine a floodplain forest along a river. The casual observer might pass by this forest, and believe that the forest is essentially "sitting there," without doing anything.

In contrast, a naturalist tracking the status of the forest would note an unending level of activity. For example, the floodplain forest will undergo seasonal changes in wildflower species as spring progresses to fall; changing tree density and height as trees are killed by floods, beavers, and wind; changing numbers of seedling trees as new trees sprout and are eaten by deer; changing total numbers of species as birds, mammals, reptiles, fish, and insects periodically inhabit, and then migrate out of the forest.

Despite these evident changes, the forest community as a whole is remarkably dependable, persisting month after month, year after year, and century after century through floods, droughts, high temperatures, ice flows, windstorms, and hungry animals. Each day, in many small ways, the forest is different from all previous days and all future days, but through the years it may not change much at all. In the long-term, even major events such as fires or 100-year floods may be little more than bumps in the road for the forest. Sustainability for these natural complex systems means continuity at the whole of the level being considered, but with a simultaneous and somewhat paradoxical constant internal flux.

The next step is to express this concept as a general pattern. The pattern needs to convey what form sustainability takes in nature. The forest is a natural community, but a similar internal dynamism occurs within offspring, individuals, species, species classes, ecosystems, and biosphere levels, so the words of the pattern need to account for all levels:

Pattern 1: In sustainable form, the biosphere, ecosystems, natural communities, species classes, species, individuals, and offspring have ongoing, changing internal processes and states. These changing processes and states may include changing behaviors, chemistry, locations, interactions, and relationships. While important, these changing processes and states do not typically represent, or do not necessarily represent, net system shifts or long-term permanent changes for the biosphere, ecosystems, natural communities, species classes, species, individuals, or offspring.

This natural system pattern may seem interesting, but we need to give it more value for a human perspective. By writing the pattern in terms that aren't so specific to natural systems, we can begin to understand what the pattern means in other realms. Here is an attempt to rewrite the pattern, but without direct references to nature:

Essential Translation of Pattern 1: Sustainable complex systems have ongoing, changing internal processes and states that are important for system operation but often do not represent, or do not necessarily represent, net system shifts or long-term permanent changes.

So that's it. But you might be wondering what this means in practical terms. What would it apply to? Would it apply, for example, to human bodies? Certainly it does. Throughout each day, you constantly change your position and location. Your thoughts and feelings change minute to minute. Meanwhile, your body is continually taking in new material (air, water, food) and getting rid of old and used material. Your body parts are constantly being used up or healed, or destroyed and recreated (over a million red blood cells in your body are replaced per second!).

Despite these changes, your body continues day after day, week after week, and year after year. In other words, through an endless series of changes, your body is sustainable for many years. We can even measure things such as your temperature, heart rate, and hunger level. Although these measurements will "bounce" around through each day, and from day to day, your long-term averages will stay remarkably constant.

Would the pattern also apply to a big train station? Yes. Each day there are dozens of incoming and outgoing trains, and thousands of incoming and outgoing people. This goes on and on, cycling through day and night, weekday to weekend, and workweeks to holidays. Yet through it all, the train station (physically, and in the transport role it plays) stays pretty much the same.

There are many other examples for which we could imagine the above pattern to be true. But just trying to figure out situations that either fit or don't fit the pattern is still too open-ended. To be even more useful, there needs to be something more tangible, something that clarifies when the pattern is expected to hold true. The next section addresses this.

6.1 Translation to Complex Human Systems

In order to translate complex natural system properties into a more meaningful human context, we need to have a more defined sense of how people interact with the world. We will look at why people establish complex systems, and from there we will move into the types of systems that they establish.

A lot of what people do is an attempt to achieve material needs. The material or physical needs of each person include adequate shelter, water, food, air, and fire or other energy sources. But material requirements don't explain all of our behaviors and apparent interests. Even when material needs are met, people have additional personal goals and aspirations. There have been different attempts to explain or model just what these are. Herein we consider a relatively simple model.

To borrow a concept from naturalist Tom Brown Jr., our goals and aspirations all come down to just a few universal things—a pursuit of *peace*, *love*, *joy*, and *purpose*. If the meanings of these words are considered in the fairly broad sense, it does seem that outside of our immediate material needs, most of our interests fall into these categories.

In striving for peace, people try to prevent disasters, secure their surroundings, conduct diplomacy, fight wars they can't prevent, and avoid personal conflicts. In striving for love, people marry, maintain close friends, have pets, and cuddle babies. In striving for joy, they play sports, dance, see plays, drive fast on the open road, try to stay healthy, and listen to music. In striving for purpose, they pick careers, go to church, take care of families, and volunteer for charities. Of course, there would seem to be lots of overlap between these words and their interpretations, but they seem able to broadly capture what underlies so many human endeavors.

Then for our purposes, we need to develop a defined sense of *how* people attempt to survive and have *peace*, feel *love*, experience *joy*, and find *purpose*. What do they actually do in a typical day? A typical person might buy groceries, call a friend on the phone, show up to work, follow traffic lights on their drive home, mow the lawn, say grace before dinner, and go to bed at night. These familiar and simple tasks represent *seven* ways in which humans try to survive, have peace, feel love, experience joy, and find purpose.

When someone buys groceries, they are transferring and acquiring material goods and energy. When they call a friend on the phone, they are developing relationships. When they show up to work, they are performing specialized skills and work patterns. When they stop at a red light, their behavior is being influenced. When they mow the lawn, they are modifying the environment. When they say grace before dinner, they are engaging in traditions and disseminating information regarding life's meanings. And when they go to bed at night, they are responding to the wants and needs of their bodies.

These seven ways of achieving our needs, aspirations, and goals can serve as the basis for classification of complex human systems. The categories can be seen as:

1. Economic systems: affecting the distribution and quality of materials and resources.
2. Community systems: affecting the distribution and quality of human relationships.
3. Occupational systems: affecting the division and distribution of skills and labor.
4. Governmental systems: affecting the quality of behavioral management.
5. Environmental systems: affecting land usage and patterns.
6. Cultural systems: affecting behavior patterns and interpretation of information.
7. Physiological systems: affecting human physiology and bodily condition.

These seven categories, or complex system "modalities," can be viewed as the means all humans have and use to affect world order and (hopefully) provide certain outcomes for themselves. Here is an example of how our activities can be viewed through these seven complex system categories:

Maybe a group of people want to build a bridge across a large river in order to increase commerce; their interest lies in *economy*. They will need to gather local support; this support will be gathered through connections within the *community*. Agreement to build the bridge will have to be developed, including who is responsible, and who is liable; that requires *government*. Skilled engineers will design the bridge; this means people with a special *occupation* are needed. The bridge will affect the flow of the river and its local usability for boating, fishing, and freight; thus the new bridge will shape the *environment*. Additional traffic across the bridge will bring people into contact with other towns or groups with which contact had once been sporadic; this might change the *culture*. The nighttime lights and rumble of trucks over the bridge could affect sleeping patterns of people living nearby; thus the bridge could have some effect on human *physiology*.

The bridge requires skill to design and construct, but it alone is not a complex system. It is the planning, construction, and operation of the bridge that will involve and indelibly affect the seven complex system modalities of economy, community, occupation, government, environment, culture, and physiology. Will it have positive, negative, or mixed impacts upon these systems? We don't know without developing a better understanding of how sustainable complex systems work, which is what brought us here in the first place.

With these seven complex system groups or modalities, we are much closer to translating properties of complex natural systems for human purposes. The key to doing this is to understand the essence of the property being considered. Let's use a fictitious example just to show how this can work:

For example, say we learn that "animals in herds alter their social organization (i.e., the way the animals interact with each other) when their populations rise." On learning this, our job would be to consider what human activity or system this herding behavior could be analogous to. This is a very simple example, and we are limited to knowing that the interactions between the animals are reorganized when their numbers increase.

Based just on this, can we say that the animals have altered their distribution and quality of materials and resources (economy)? No, we have no information to indicate this. Can we say that the animals have altered the distribution and quality of their relationships (community)? Yes, something has changed regarding which animals interact or how they interact. Have the animals altered their distribution of skills and "labor" (occupation)? No, we haven't been given any information that they specialize in behavior any differently. Have the animals altered their "quality of behavioral management" (government)? This one is tricky to understand, and we'll address government more later, but *not necessarily*; it isn't clear that their behavioral guides have changed. Have the animals altered their land usage and patterns (environment)? No, we lack definite indications that they are changing relative to their environment when their interactions with each other change. Have the animals altered their behavior and interpretation or response to the world (culture)? Not especially. Although their animal-to-animal interactions somehow have changed, it doesn't mean that they have learned anything, or modified themselves or their activities beyond the animal-to-animal level. What about their physical bodies—are the animals' bodies or physiology changing in response to their higher numbers (physiology)? Again, though this might happen, our information doesn't show this. In all it looks as though the animals' response of altering their social organization is clearly analogous to a change in community, weakly or uncertainly analogous to a change in government and culture, and not particularly analogous to a change in economy, occupation, environment, or physiology.

With community as a clear link in this case, we could translate the pattern into the human realm and conclude that sustainable human systems could include humans altering their social organization (i.e., the quality and distribution of their relationships) when their populations rise. Though the animal behavior might also be saying something about governmental or cultural response to population rise, this link is more confusing or uncertain, so it might be best not to venture forward in this (since indeed, we don't really know what the herd's organizational change means for animal governments and culture anyway, so translating with these modalities would probably have a higher risk of providing an untrue result).

But, there's still one more issue we need to address to best translate patterns from natural systems into human systems. This is a *scale* or *level* issue. Imagine there is a pattern within natural communities, and we are trying to translate the pattern into a human context. Would the natural community pattern translate to the level of a human individual, a family, a neighborhood, or the nation? In other words, to what size or level of human group should we try to apply the pattern to? A solution to this can be to translate the natural system pattern to similar *human scales* or to similar *human relational levels*.

As an example of translating based upon scale, a natural community might occur in many patches that are a few square miles each. These patches might be scattered across an ecosystem that covers several thousand square miles. If we tried to imagine a human group that occurs at a very similar scale, it could be in the form of towns that are scattered across an area of several thousand square miles; this might be termed a "regional community."

Or, if we want to conduct the translation according to relational levels, we'd first consider the relational levels represented by a natural community. A natural community is a local grouping of many individuals of several species—all seeking to survive in each other's context. Each species is composed of individuals with similar needs, and all the individuals within a natural community share an interest in the natural community's survival. Human families are groups of people with similar needs; and groups of families that band together with a common interest, such as tribes, local communities, or regional communities could then be considered relationally analogous to natural communities.

Considering these scale and relational-level similarities, translating a natural community pattern into a human context might best reference an approximate tribal, local, or regional community.

Similarly, the biosphere can translate roughly to the human global or national level; ecosystems to the approximate state or regional level; species classes to the approximate neighborhood or local level; species to the approximate family level; individuals to the approximate individual human level; and offspring to the human offspring level. Of course, the terminology of these grouping needs to account for the modality under consideration (e.g., a small human community group is a *family*, but a small occupational group is a *profession*, and a small area of land use is a *parcel* or *site*.)

Table 3 (over) is an attempt to provide the approximate complex natural system levels that correspond to levels of the seven complex human system modalities.

Earlier, the lack of an objective existence of natural system groupings was noted. However, being able to discuss natural systems as though they exist does allow us to communicate and develop better understandings. Similarly, while there is no preordained correlation between natural and human system levels, drawing an approximate correspondence between them allows us to more clearly envision the way that human systems might work better.

Natural Systems	Economy	Community	Occupation
Biosphere	Global Economy	National or Global Population	Global Industry Sectors
Ecosystem/ Ecoregion	National Economy	State or Regional Population	National Industry Sectors
Natural Community	Local Economy	Neighborhood	Regional Business Sectors
Species Classes or Suites	Neighborhood Economy	Extended Family	Local Business Sectors
Species	Family Economy	Nuclear Family	A Single Business or Profession
Individual	Individual Economy	Adult Person	A Single Job
Offspring	Dependent Child Economy	Child	A Job Apprentice

Table 3. Translation between natural systems and human systems

However useful translations from natural to human systems are, it is important to refrain from translating blindly, without discretion. Translating without discretion can lead to nonsensical results. For example, if we know that "animal species with a higher male-to-female ratio exhibit higher breeding competition," then an unconsidered translation could suggest, "human families with a higher male-to-female ratio exhibit higher breeding competition." This is a confusing statement that doesn't really correspond with our contemporary understanding of what a human "family" is—indicating that we have patched together a quick translation without understanding what male-to-female ratio means in animal species versus people.

Government	Environment	Culture	Physiology
Federal or International Law or Treaty	Planet or National Homeland	Human Culture or National Culture	Human Physiology and Genetics
State Law	National or Provincial Homeland	National or Provincial Culture, or Cultural Groups	Racial Physiology and Genetics
County Law	Regional Landscape and Land Use	Regional Culture or Tribe	Tribal/Ethnic Physiology and Genetics
Town or Precinct Ordinance	Local Landscape and Infrastructure	Clan	Extended Family Physiology and Genetics
Institutional or Business Code of Conduct	Parcel	Household Practices and Beliefs	Family Physiology and Genetics
Personal Behavioral Boundaries	A Location and its Use	Personal Practices and Beliefs	Single Human Physiology and Genetics
Child Behavioral Boundaries	Potential or Future Land Use	Teachings	Baby/Child Physiology and Genetics

In such cases where an apparently nonsensical result occurs, it can help to assess whether the property translates correctly across levels as shown in Table 3, or if the translation needs to slide upscale or downscale to make more sense. In this example, we could try to retranslate upward, yielding "human tribes and ethnic groups with higher male-to-female ratios exhibit higher competition during marital partner selection." This is a hypothetical example that might not be true, but at least it has more clear meaning, and it could make sense that when single men outnumber single women, the men could be forced into competition with each other. For this reason, the translation from natural systems and human systems should always be done with discretion so as to apply the nuance of the natural system pattern to the most meaningful human system category. With that, let's get back to translating our first pattern.

Chapter 7 Dynamics and Hierarchy

Recall our first pattern:

> Pattern 1: In sustainable form, the biosphere, ecosystems, natural communities, species classes, species, individuals, and offspring have ongoing, changing internal processes and states. These changing processes and states may include changing behaviors, chemistry, locations, interactions, and relationships. While important, these changing processes and states do not typically represent, or do not necessarily represent, net system shifts or long-term permanent changes for the biosphere, ecosystems, natural communities, species classes, species, individuals, or offspring.

You might recall that after clarifying this pattern with examples of a floodplain forest, the human body, and train stations, we could express the pattern in a general form that could perhaps apply to any complex system. In this case, the translation of the pattern to its *essence* was:

Sustainable complex systems have ongoing, changing internal processes and states that are important for system operation but often do not represent, or do not necessarily represent, net system shifts or long-term permanent changes.

This essential translation is our hypothesis. If the hypothesis is true, it can be translated into the seven modalities so that we develop insight into sustainable human systems. Pattern 1 is very broad, encompassing scales from smallest to largest, and is not limited to only resource use, or relationships, government, etc. It appears that this pattern applies more or less to all levels of all seven modal categories.

What follows is an attempt to translate the pattern into words for each modality. Brief fictional and hypothetical examples then follow each modal translation. The simple examples are merely intended to help provide an image of the pattern in action. However, many other examples could be supplied, so they should be seen as just one of many possible pattern scenarios:

Economic Translation: At all levels, a sustainable economy may remain relatively stable in overall magnitudes and measures, but this stability is an overall feature that occurs or is comprised of dynamic underlying processes.

For example, a national economy has a relatively stable GDP year on year, but that GDP is produced by millions and millions of ongoing economic exchanges, product outputs, and consumption events. In another example, one family's economy has a fairly steady overall rate of income and consumption. But these steady values are the result of hundreds of ongoing or regular acts of consumption and production.

Community Translation: At all levels, sustainable populations, communities, and families maintain relatively stable forms and relational structures, but this stable form and structure at each level overarches dynamic interactions and events.

For example, a regional community has a stable demography created through births, maturation, and death. In another example, one family's stable relationships are upheld by frequent interactions.

Occupational Translation: Sustainable industry and business sectors may be stable across regions, or stable in overall numbers in a region, but within a given region, the activity rate of business and industry sectors will ebb and wane as opportunities expand and contract. Also, sustainable industry and business sectors are continually working to meet economic demand and exploit new opportunities. Sustainable individual professions can also expect demand for their skillset to ebb and wane to some degree.

For example, a rare metal production facility experiences cycles of demand and supply requests as a result of economic cycles. Its sales are steady decade on decade, but they shift year to year as a continual review for new opportunities occurs.

Government Translation: Sustainable laws, rules, and behavioral expectations at various levels (federal to neighborhood) may be quite stable year on year, but they may overarch numerous activities that were modified in order to adhere to these laws, rules, and expectations. As a result, continually dynamic systems of economy, community, occupational groups, environment, culture, and physiology would have behaviors and properties as modified by these laws, rules, and expectations.

For example, a nation's laws against selling dangerously defective goods result in few such goods being sold, but this overarches lots of internal manufacturer and seller activities that ensure their goods aren't defective.

Environmental Translation: National to local level sustainable land uses may normally be stable across the landscape through decades. However, many activities and environmental forces at work within each area maintain conditions on those lands or for those uses.

For example, an agricultural landscape is subject to planting, tending, and harvesting crops; and to raising livestock. Through each year, the growth and management on each plot of land moves through predictable cycles. Crop and grazing rotation are implemented so that patterns vary through any three consecutive years. Nevertheless, over ten or fifteen years, the patterns repeat, and over the span of decades little changes.

Cultural Translation: Sustainable culture at the national, regional, and family levels could appear to be stable, not shifting greatly from one generation to the next. However, that stability would be the result of continual activities and teachings that regenerate the culture(s) and transfer it to the next generation.

For example, one extended family has a unique sense of identity and beliefs that distinguish it from other families. Its identity and beliefs are passed to each new generation through continual engagement with stories, traditions, and rituals that involve all generations.

Physiological Translation: Sustainable human physical health—from the global level to the personal level—may be typically stable over extended timeframes. However, many events and activities underlie this overall stability. The events and activities center upon the underlying design of the human body, which can vary somewhat between any two people and any two groups of people. Also, instability or change is addressed through stabilizing or healing mechanisms that include internal bodily abilities or processes.

For example, a particular family member's strength levels are known to be stable over the medium and long terms. This overall stability occurs through cyclical internal and external changes, including temperature, eating, sleeping, and exercise changes.

Notice that the modal translations are each a bit different. In other words, they aren't just word substitutions into a standard sentence. This is partly due to the fact that the modalities have different forms. For example, most human law is essentially on paper and doesn't really exist in physical form, but land use is predominantly a physical attribute or activity, so the wording can take this into account. In addition, the wording for each modal translation is an attempt to represent an attribute *as it occurs in nature*. So, you may notice that the occupational translation connotes more of a shifting element, with the demand for each occupation possibly being somewhat fluid or cyclical in a given area. This is an attempt to capture the nuance of "occupation" in nature, wherein it is often common for the habitat conditions pertaining to any species or group to shift in time and space along with weather, season, and other events. While it is beyond the scope of this text to discuss ecological processes in high detail, translations herein attempt to recognize the nuance of each pattern across modalities in nature, and to translate that pattern accurately into human system modalities.

That gives us one broad natural system pattern, including translations into human system contexts. Now let's look at a second broad pattern that addresses the overall form that sustainability appears to take in natural systems.

Our first pattern described how the form of complex systems in nature is upheld through continual internal activity. What was not discussed, though, was that the overall form itself (whether a forest, a frog species, or a single wildflower) is never completely fixed either. Complex systems in nature often change their form, even if this change is gradual. These changes can be prompted by internal or external events.

In an example of an internal shift and refinement, perhaps an insect species near the northern edge of its range makes adaptive hormone adjustments that allow it to survive better in forests on northern mountain slopes. This insect has one host tree species (it feeds on just one tree species). As a result, the host tree species declines in density, which allows three other tree species to increase on the mountain slopes where the insect's host has declined. The mountain slope forests have shifted their overall form in response to an internal event.

An externally prompted example of this pattern could be a mammal species that develops thicker fur over several thousand years as the climate in a particular location slowly cools. This mammal has shifted its form in response to an external event. The next pattern reflects this ability of natural systems to undergo shifts in overall form.

> Pattern 2: Sustainability for the biosphere, ecosystems, natural communities, species classes, species, and individuals is generally *not* represented by a fixed endpoint. Sustainability is more often an adaptive state of refinement or shifts (even if small) caused by internal adjustments or responses to internal or external changes.

A translation of the essence of this pattern would be: Sustainable complex systems may be largely stable in overall form, but they will tend to shift gradually as they refine their internal and external relationships, or adjust to shifting internal or external conditions. While there may be additional wandering shifts that have a more random or oscillating component, the shifts in this case are directional and in effect, may appear purposeful.

This pattern can now be translated into human system modalities, with the translation capturing the nuance of each modality and the form of the pattern in nature to the degree possible. Recall that the examples which follow are merely intended to help provide an image of the pattern in action. However, each example should be considered just one of many possible pattern scenarios:

Economic Translation: Sustainable economies, from the global to neighborhood to individual level, may be largely stable, but will tend to make periodic adjustments or gradual shifts. These adjustments and shifts will tend to be responses to internal conditions or changes (within the economy) or to external conditions or changes (responses to outside economies or factors). These shifts and adjustments will generally help to maintain overall economic system stability and efficiency.

For example, over time, a family (family economy) slowly lowers its purchases of food as the family members develop a more finely tuned understanding of what will be eaten versus thrown away due to disinterest or spoilage. The family (economy) shifts money saved on food toward vacation spending.

Community Translation: Sustainable human communities at all levels, from the global to local to family level may be largely stable, but will tend to make periodic adjustments or gradual shifts in response to internal or external changes or conditions. These shifts and adjustments will generally help to maintain overall community stability, relationships, and value of communities for people.

For example, the structure of nuclear and extended families adjusts to slowly increasing lifespans by creating stronger relations between the very old and very young, but otherwise these communities are rather stable through time.

Occupational Translation: Sustainable local to global industry sectors, professions, and single jobs may exist as entities over long periods. However, the ways in which they operate—or the intensity with which they operate in different locations—may shift to become more efficient (take better advantage of conditions), or adjust in response to economic changes, environmental changes, etc.

For example, over several decades, gradual improvements in gas mileage cause a reduction in support for the petroleum industry while raising the affordability of driving and thereby raising demand for the auto industry.

Government Translation: Sustainable laws, rules, agreements, and behavioral expectations—from the international to local to individual level—would be established over relatively long periods, but subject to occasional change based upon internal shortcomings (inadequacies or undesirable consequences) or changing external conditions. The amount of change in laws, rules, agreements, and behavioral expectations would be small compared to the amount that was left unchanged. What's more, the changes that do occur would tend to make the laws, rules, agreements, and behavioral expectations more focused upon achieving a particular outcome.

For example, advances in weaponry eventually make it relatively easy for any nation or group to attack and destroy another nation. International laws shift to align with this fact and the focus on outlining ethical warfare shifts toward a punitive stance toward war activities. At the same time, wider economic and cultural protections are established to reduce the number of grievances between nations and groups.

Environmental Translation: Sustainable landscapes and land uses would be rather stable at the planetary, national, and more local levels, but shifts would occur in order to refine uses according to changing conditions or improvements in stewardship or management techniques.

For example, the amount of land under agricultural production is fairly stable in one region, but refinements in farming strategies and management techniques, perfected over centuries, enable small reductions in the overall amount of land being farmed in favor of other land uses.

Cultural Translation: Sustainable cultures would be well established across nations and the planet, each in relative equilibrium with its particular human history, landscape, economy, etc. Though each culture would be stable for the most part, each would shift in order to maintain or improve its state of balance between internal and external conditions and forces.

For example, in a town without an especially distinct heritage or sense of identity, what begins as a local gathering by a few people to share blueberry pies evolves into a three-day festival every summer, complete with food, games, and activities relating to the local blueberry crop. Through this festival, the town develops a stronger self-identity, self-esteem, and sense of place for its unique activities and landscape context.

Physiological Translation: By any measure, sustainable human health and physical condition across the planet, and within any nation or tribe, race, or group would undergo only small changes through time. However, some changes in human health might occur gradually in response to changes in occupations, communities, the environment, etc.

For example, gradual but substantive gains in understanding regarding the causes of several diseases lead to behavioral changes that improve health, thereby slowly raising the average life expectancy.

That makes two fairly broad sustainable complex system patterns. They apply to all seven levels of all seven complex human system modalities. They address what changes and what does not. In a nutshell, these patterns are saying that sustainable complex systems change little through time in an overall sense. But this broad stability is maintained by, or is overarching, very dynamic internal conditions. When a system does undergo overall change, it is usually a directional shift that takes place in order to refine or align internal conditions in response to internal inefficiencies (misalignments), or to internal or external changes. Another way to think of this is that when the systems do change, it is for purposes of improvement (here "improvement" is not intended to imply a positive or negative value judgment).

The next pattern is also broad in that it applies to all seven complex human systems. It has to do with the level of influence that lower and higher hierarchical levels have upon each other.

In nature, individuals come and go—live and die—and their condition and fate are virtually unnoticed by the overall species to which they belong, to the natural community to which they belong, to the ecosystem in which they occur, etc. But quite differently, a change in the status of the species to which individuals belong, to the natural community to which individuals belong, to the ecosystem in which individuals occur, etc. generally has a high chance of affecting each individual.

A hypothetical example of this pattern in nature could occur in the longleaf pine woodlands of the Southeast. Gopher tortoises are one of the unique inhabitants of these open pine woodlands. A few dozen burrowing gopher tortoises in a 20,000-acre patch of woodland might kill a few grass and wildflower plants—by tearing some plants out and covering others with dirt during burrow excavation. The results of this plant killing will be extremely localized, and this localized disturbance won't lead to strong effects on other grasses and wildflowers or upon the whole longleaf pine forest. Certainly the tortoises would have more effect by repeating their digging over several years, but their impacts would still be patchy in scale.

Now contrast those local impacts with a system-wide alteration, like settlement of the landscape with roads, periodic homes, and farmsteads. Roads and farmlands might prevent fires from spreading naturally across the forest floor. Farm animals or feral domestic animals might root around, consume and destroy swaths of vegetation, and spread diseases. The net result of the installation of roads and farmsteads in what was once an entirely forested land will be the loss of many, or perhaps eventually most, plants and animals of the forest. This would hold true even for plants and animals that were not removed directly to make way for roads and farms.

So here, a tortoise (a local effect), has not greatly affected the system by disturbing a small area. But the broad-scale, system-wide effect of human development has degrading consequences for plant and animal individuals and species across the forest.

An analogy to this pattern could occur with a tin coffee can that has a bunch of marbles inside. If you shake the can, you can hear and feel the marbles banging around. If you take out one marble and shake it again…you won't notice a difference; it's still a coffee can with a bunch of marbles. If you were to add three or four extra marbles, you still wouldn't notice a difference; it's still just a coffee can with a bunch of marbles.

The addition or subtraction of a few marbles from the can doesn't affect the overall situation to any discernable degree. Likewise, the other marbles in the can wouldn't "feel" the difference of plus a few or minus a few. On the other hand, if you were to drop the can from a third-story window, not only would the can experience a bang and dent, but the marbles would collide with each other and scatter in all directions if the lid popped open.

In these examples, it seems that the action of one marble doesn't matter much to the other marbles, but the action of the *can of marbles* matters a lot to each marble. Is this merely pointing out what is mathematically obvious? Sure, the loss of one marble from a group of 100 is only a 1% change, while dropping 100 causes a 100% marble impact. But in nature, this pattern is more than just a restatement of simple math. Natural systems have many parts with subtle interconnections. Natural systems are probably adapted to gain and lose individuals without undergoing great change.

In contrast, system-wide changes have immediate effects on many individuals, *and* effects on dynamics and interactions, which sends aftershocks through the system, affecting individuals further and further or over and over. In the above example, the plants killed by burrowing tortoises are like marbles in the can—whether you lose or gain a few, nothing really changes for most individuals. But the incursion of human development, even if spread out, is like dropping the can from the third-story window—everything is eventually affected.

> Pattern 3: The status or behavior of one (or a few) individuals, and the events that affect one (or a few) individuals within a sustainable species, species class, natural community, ecosystem, or the biosphere tend to have no significant or appreciable effect upon the species, species class, natural community, ecosystem, or the biosphere. In contrast, the status or condition of a sustainable species, species class, natural community, ecosystem, or the biosphere, or the events that broadly affect them, tend to have widespread, appreciable effect upon many individuals. This pattern would hold true between higher and lower levels in general, not just between higher levels and individuals.

Essential translation: Sustainable complex systems display a general pattern in which hierarchically lower levels have a relatively weak influence upon hierarchically similar or higher levels; and in contrast, hierarchically higher levels have a relatively strong influence upon hierarchically lower levels. This means that localized impacts upon any system might have little impact upon its relatively higher hierarchical levels, and significant impact upon its relatively lower hierarchical levels.

Economic Translation: In a sustainable context, economic changes tend to have effects that propagate laterally and downward toward smaller economies, but the effects upon even larger economies tend to be weak.

For example, the loss of a single blue-collar job at a local factory had no significant impact upon the local economy. In contrast, when the factory laid off half of its workforce, there was a significant impact upon many individuals in that locality, including many who were not even employed at the factory. Still, those layoffs did not significantly impact the national economy.

Community Translation: In a sustainable context, changes to a community tend to generate impacts that propagate laterally and downward toward smaller communities. But changes in smaller communities tend to only generate weak effects upon larger communities.

For example, most people in a community were essentially unaffected by the arrival of a few immigrants from a nearby nation. But when emigration of one-quarter of the community's population occurred due to economic forces, individual people in the community were greatly affected by the loss.

Occupational Translation: In a sustainable context, changes to an occupational level such as an industry sector or profession tend to generate effects that propagate laterally and downward toward smaller occupational levels, but effects of these changes upon even larger occupational sectors tends to be weak.

For example, one firm's decision to give extra training to its land surveyors had little impact on the surveying profession, or on people outside of that company. But a later ruling that allowed land surveyors to unionize significantly impacted thousands of individual surveyors/positions and their work hours, pay, and job stability. It also affected the economics of surveying as a profession.

Government Translation: In a sustainable context, changes to higher level (state or federal) laws or behavioral expectations tend to generate effects that propagate laterally and downward toward local laws and situations. But the effects of changes to local laws or behavioral expectations upon higher levels such as international law, tend to be weak.

For example, the change in one family's agreements over who gets to use the car affected the members of that family, but it did not have further significant repercussions. In contrast, a change in federal law reducing the minimum driving age affected families across the nation, including many people that weren't directly targeted by the law change.

Environmental Translation: In a sustainable context, the effects of widespread environmental changes or land use changes tend to propagate laterally and downward toward the lands they contain, but the effects of these changes upon the even larger landscape is weak.

For example, the configuration of local roads in one area affects local drivers but doesn't have much impact upon national highway routing. In contrast, the highway layout influences their local roadways—by overpass, underpass, or dead-end status, and by the disproportionate use and maintenance of roads that access the highway. Nearby towns themselves are also impacted by their highway connections, or lack thereof.

Cultural Translation: In a sustainable context, the effects of local cultural changes tend to propagate laterally and downward toward smaller groups or cliques, but the effects of these same changes upon the larger culture tend to be weak.

For example, the way one group of friends communicates affects those friends, but has little widespread effect. In contrast, when a more widespread reliance upon text messages rather than verbal communication occurs, the way many people experience their lives is greatly affected.

Physiological Translation: Sustainable human individual or group biology and physiological trajectory is little affected by the outfall or outcomes associated with one (or a few) individual(s). However, group-scale disruptions can profoundly affect many individuals.

For example, the loss of one individual in a clan setting was not extremely difficult for the other members of the community to withstand. But when the loss of many individuals occurred due to disease, it put a large physical strain upon remaining members—who had to face changed conditions and loss of assistance in accomplishing tasks. This strain translated to disease and weakening of the surviving members.

The previous pattern suggests that the system trumps the individual, and while each individual really needs to be in touch with the system to thrive, the system can do with or without any given individual. While that may be so, this next pattern is sort of a corollary to the prior one. It indicates how individuals actually can become important at the system level.

In natural systems, the actions of any one individual rarely have a large and system-wide effect. But, when many individuals act in concert, or act similarly over extended periods, then they often do have large effects. Similarly, the condition of one small area usually has little influence over the greater environment, but many small areas sharing a similar condition can become an influential weight that helps steer large-scale processes.

An example of this pattern in nature could be the spawning of salmon. A single fish swimming up a river will have virtually no effect on the river, or on the river's surroundings. Certainly, that fish will eat a few insects or minnows, and might make a fine meal one day for a river otter, but thereafter, the riverine ecosystem will continue on the same trajectory it was already on. On the other hand, when the number of spawning salmon are multiplied out into the thousands, these individuals (acting together) will impact the annual survival and habitat conditions for dozens of predators (bears, eagles, seagulls), egg-eating fish, and plants within miles of the river (yes, even plants as they utilize nutrients from the consumed fish being dispersed through the landscape). By acting together in large numbers, the salmon can actually alter the entire natural community.

The same pattern would also be apparent back in longleaf pine woodland. A few dozen tortoises and a few dozen burrows are incapable of making a forest-level change. But if those turtles operate decade after decade in large numbers, then they certainly will impact the whole woodland.

Similar relational patterns occur between many different natural system levels, such as the impact of one species versus many species on an ecosystem, the impact of one offspring on a species, etc. This concept also aligns with our marbles-in-a-can concept. One or two marbles don't change the state of the marbles-in-a-can, but 20, 30, or 40 of them do. Get 30 of the marbles moving simultaneously in a single direction, and they can cause the whole can to move.

> Pattern 4: Individuals, even if physically large, do not have significant formative influence upon the natural communities and ecosystems that they inhabit, unless they are acting in concert with a larger group of individuals. This generalization holds true as well when looking from any smaller scale to any larger scale. Also, the smaller the individual, the larger their numbers generally need to be before they exert formative influence upon the system.

Essential Translation: Entities or subunits that constitute part of a significantly larger sustainable system usually have to be behaving similarly and repeatedly, or simultaneously, in order to have a significant impact upon the larger system.

This pattern can now be translated into human system context:

Economic Translation: In sustainable situations, individuals or subunits of a given economy tend not to have significant impact upon the larger economy unless they act similarly in large numbers.

For example, a scattering of people who like a certain tropical fruit in a northern nation have to pay a high cost to obtain it as a health-food store novelty. But as interest for the fruit grows to a certain threshold, competition to meet market demand develops, causing additional cultivation and bulk transport among growers and importers, and lowering the price of the fruit for everyone in that country.

Community Translation: Individual people or families within a sustainable state or regional community do not greatly shape those communities unless acting similarly along with other individuals or families.

For example, a medium-sized town essentially serves as a bedroom community for professionals who work for dozens of different employers. Within that community there is relatively little interaction between most residents, resulting in a weak sense of identity, little overall willingness to make long-term infrastructure and cultural investments, and few children growing up who are interested in staying within the community.

Another town with two primary long-time employers has frequent social interaction between most residents, resulting in a strong sense of identity, an overall willingness to make investments in the future of the community, and many children who hope to stay within the community. The difference between the two towns is not goodwill among residents, but the frequency and persistence with which friends and families socialize and become intertwined in each other's lives.

Occupational Translation: No single worker or small group representing a profession has a large impact upon other sustainable professions, business sectors, or regional economies; but a large group of such individuals representing an occupation can have this large impact.

For example, in one nation there are very few shoe repairmen and the culture is geared toward disposable shoes. Well-made shoes are hard to find, and few youths are interested learning the shoe repair trade. An adjoining nation has many shoe repairmen, and in seeming reflection of that, there is a strong market for durable, repairable shoes, and a continual stream of youths that enter the shoe repair trade.

Government Translation: In a sustainable context, rules and behavioral expectations do not significantly affect the trajectory of societies or behaviors unless they are applied and/or followed and/or enforced on a consistent or widespread basis.

For example, a state with limited revenue sources declined enforcement of income tax payments, and a few individuals chose not to pay their taxes. Knowing this option was available, many additional people and businesses were attracted to that state on the basis of avoiding taxes. Eventually, people who were otherwise inclined to pay their taxes began to feel penalized, and stopped paying. Without much income, the state bureaucracy shrank, and certain government functions were taken on by charities and corporations. The state's income tax law, being largely ignored, eventually did not significantly impact conditions within the state.

Environmental Translation: Localized or minority land uses or land cover types do not greatly modify or determine the condition of the larger sustainable landscape unless they occur in large groups or occur repeatedly across the landscape.

For example, in one nation, an originally vast prairie covering over 100,000,000 acres was essentially eliminated from the landscape for agriculture by farmers and laborers on individual homesteads covering 200 acres each. With the assistance of oxen and horses, this was accomplished, because many thousands of people worked toward the same goal year after year, and square foot by square foot.

Cultural Translation: Isolated behaviors of individuals or small groups of people do not significantly impact views or conditions for a larger sustainable society unless there are many more individuals or groups acting in concert, at which time they can greatly shift the larger group.

For example, for many decades, interest in longball was held by few people. It was enough to keep the game alive as a niche interest, but the sport was not popularly known or played, and it was even hard to find longball equipment. But gradually, the number playing the sport grew, and it gained attention. Eventually longball began to receive media coverage, take up shelf space in the sporting goods store, support professional players, displace other sports in high school, and become entwined within the culture (how people spent time and what they cared about).

Physiological Translation: In sustainable situations, specific internal or external environmental conditions may have temporary health impacts but don't especially determine overall physiological health or wellbeing unless repeated regularly or in large amounts.

For example, a team of researchers found that small amounts of refined foods or additives eaten as part of a normal diet were unlikely to have a major effect on the human body, because bodies use and process those products as part of a larger diet. The same team also found that if refined foods or additives were consumed in large quantities, a body had to balance the excess of those products along with a relative insufficiency of regular foods. This consumption imbalance was found to tax certain organs and increase a body's need for certain hormones, enzymes, and vitamins. In turn, this imbalance altered body chemistry and functions, and led to additional health problems that on the surface, did not appear related to diet.

Now we have four rather expansive complex system properties. What they generally say is that sustainable complex systems change little through time in an overall sense. However, this broad stability is maintained by, and is overarching, very dynamic internal conditions. Despite overall stability, these systems can undergo overall change, usually in the form of a directional shift—which takes place in order to refine or align internal conditions in response to internal or external misalignments (inefficiencies), or in response to internal or external changes. Also, hierarchically higher levels have a relatively strong influence upon hierarchically lower levels, but lower levels have a relatively weak influence upon similar or higher levels *unless* large numbers of the lower level entities (perhaps individuals) are behaving similarly and repeatedly, or at the same time, in which case their influence upon higher levels can be very substantial.

Let us quickly recap everything up to this point. First of all, many "things" seem to not be working so well. There are problems arising throughout every sector of people's lives, including within government, economy, society, and environment, and there is rarely consensus on how to correct these problems.

So, hoping to eventually figure out why something works or not, we began to look at how things work in the most basic way. We began developing a logical deconstruction and reconstruction process for anything. To work, or to function, every system or thing needs internal and (or) external support. These supports can be broken down, analytically, by considering feedback loops, independent forces, and compatibility to predict success or failure of a system or thing through time, i.e., to predict its sustainability.

But when we extend an analysis of sustainability into complex systems, specific events and outcomes become almost impossible to reliably predict, meaning that we can't really tell ahead of time what is sustainable and what isn't. This means that we cannot precisely understand and predict the failure of our large systems.

So, we turned away from the direct analytical approach, to take another approach to the problem—trying to figure out if there are patterns or principles that associate with, or govern the sustainability of, complex systems. One place to look for such patterns is in nature where plants, animals, and other organisms have coalesced with each other to form apparently coherent and sustainable organizations in their thousands of respective environments.

With these first four patterns we have begun to describe some of those patterns or principles. Doing this isn't a completely straightforward process because the patterns—as found in natural systems—have to be translated so we can understand how this might look in human terms. If the hypothesis is correct, these first four patterns could help to show us what sustainable human systems, structures, and institutions would look like.

Next we will turn away from overall system qualities and focus more on properties that relate to dependence and individual survival.

Chapter 8 Dependence and Interdependence

The next pattern has to do with the basis of survival for all individuals. In nature, it seems that each plant and animal depends upon its own efforts to survive. But on the other hand, there is no plant or animal that can survive without some degree of support from others or from the environment.

A simple example of this in nature would be a tree in a forest. This tree must compete for resources such as light, water, and nutrients; and it must grow, photosynthesize, and produce chemicals to ward off insect and fungal attack. We may not always think of these plant activities as "work," but like an animal, the tree is undertaking otherwise inconvenient steps for itself in order to survive. Without this work or effort on its own behalf, the tree would die. At the same time, this tree depends on the nearby forest to moderate its climate and provide soil; it depends on freely available air, water, sunlight, and warmth for chemical reactions; and it depends on nearby animals to eat insect pests. These forms of support all come from the tree's outer environment. Without this support, the tree would die. All of this means that the tree, in order to survive, is dependent on its own work as well as the beneficial environment.

This pattern is similar to the notion of what creates functionality (internal and/or external designs, conditions, and forces that support an objective).

However, while it may be possible to imagine a technical case of mechanical functionality in which very strong internal support overrides a lack of external support, or in which very strong external support overrides a lack of internal support, it seems that both internal and external support must be present in order for living things to survive in nature.

> Pattern 5: Each individual survives on the basis of two supporting factors. The first is that each individual works to make/collect/retain its own resources. The second is that each individual survives by the support of its external environment.

Essential Translation: Individuals in sustainable complex systems are only able to survive and thrive with a combination of their own effort and a supportive environment.

This pattern can now be translated into the context of human systems:

Economic Translation: Individual people in sustainable situations maintain a livelihood by their own efforts in the context of a supportive economy.

For example, people in a town that work hard on their own behalf generally have a more prosperous condition than townsfolk who are otherwise similar, but who do not expend much energy on behalf of their own welfare. In another example, people in a county with updated infrastructure, high wages, and low unemployment rates have a more prosperous condition than people who are similar and hard-working, but living in a nearby county that has rundown infrastructure, low wages, and high unemployment rates.

Community Translation: Individual people in sustainable situations depend on the relationships they have with other people and the relationships that those people have.

For example, a group of families living in a large city are tight friends. This includes the adults and children. Although these families don't all live in the same neighborhood, they spend time together doing activities. If any one of them has a problem or needs a favor, like some help getting ready for a wedding, some used sports equipment, or help moving furniture, someone is always willing to lend a hand.

Government Translation: Individual people in sustainable situations depend upon their own self-control and regulation as well as the self-control and regulation of others.

For example, Jeff is an avid biker who likes to spend his Sundays riding on a long trail. Reaching this trail on his bike requires him to ride on a busy suburban road for several miles. Along that road, his safety is tied to his willingness to ride on or near the shoulder and to wear a helmet. At the same time, his safety is also strongly tied to motorists' willingness to watch for bikers, to stay in their lanes, and to drive at moderate speeds.

Environmental Translation: Individual people in sustainable situations depend upon their own impacted environment, and upon nearby environmental conditions or forces. That is to say they are dependent upon the portion of the environment that they control, and they are also dependent upon the portion of the environment that they do not control.

For example, the owner of a 320-acre farm in North Dakota plants wheat in most years. He works hard to keep weeds to a minimum, and to monitor and maintain soil fertility, nutrient levels, and moisture levels. Meanwhile, he is dependent upon larger weather and climate patterns. The difference between a good harvest and a poor harvest can hinge on both his farm management skills and the weather patterns that year.

Pattern 5 suggests that situations in which individuals are expected to prosper without both their own effort and a supportive environment are not sustainable. Or stated in the converse, it suggests that individual prosperity is sustainable if and only if the individual expends significant effort within a supportive environment.

Obviously in nature, the environment is not always completely supportive of each individual. For example, a drought may cause a tree to drop its leaves early just to survive into the next year. In nature, if an environment becomes too unsupportive for a certain individual, the individual will languish and die, despite the individual's own best efforts.

The next pattern speaks to the role that each individual or group plays in maintaining the environment. In nature, it seems that one of the jobs or roles that a given individual has is to maintain the greater environment. Staying with the example of the tree in a forest, the tree's normal activities—standing upright; absorbing sunlight, carbon dioxide, and nutrients, emitting oxygen; and transmitting water—have a significant impact on its local environment. The tree's activities raise humidity, slow the wind, catch rainfall, lower the temperature on summer days, raise temperatures at night, serve as a hiding and perching place for animals, suppress growth under its canopy, provide food for insects, etc. Even when the tree dies, its trunk rots slowly, providing all sorts of support for small animals and new plants.

An implication of this is that the environment is not just an ambient condition being provided by some faraway force; rather, it is modified very locally to, by, and for each individual. Also very importantly, the environment that each individual helps to maintain is the one that supports that individual. Putting this again in the context of our tree, the environment supported by the tree is a forest environment, which is the beneficial home to the tree itself, along with many thousands of other creatures.

This tendency to maintain the beneficial environment also hold true at higher hierarchical levels of species, species classes, natural communities, ecosystems, and the biosphere; the environment that they each maintain is commonly the environment that they require.

> Pattern 6: Each individual's normal activities help to sustain the environment. This holds true from the point of birth to the point of death for each individual, and even includes the process of physical decay. The environmental conditions that each individual sustains are often predominantly local to that individual, and can include the chemical, physical, and competitive environments. This pattern also holds true for species, species classes, natural communities, ecosystems, and the biosphere, but the scale of influence expands with each level.

Essential Translation: By partaking in their normal activities, individuals and groups in sustainable complex systems actively help to create and sustain the environment that they live within.

This pattern can now be translated into the context of human systems:

Economic Translation: By partaking in their normal activities, individuals and groups of people in sustainable situations actively help to create and sustain their economies.

For example, Ron, a home remodeler, improves the functionality and appearance of people's homes for a reasonable cost. While this allows him to earn a fine living, it also helps to maintain the quality of life and the appeal of his own community as a good place to live and work.

Community Translation: By partaking in their normal community-related activities, individuals and groups of people in sustainable situations help to maintain their communities at the local level, and beyond.

For example, Maxine takes her children to the neighborhood barber to get their hair cut. She does this in part because her best friend's brother and cousin work there. In doing this, her children come to know these family friends and acquaintances, which creates a stronger bond between Maxine and her friend and her friend's relatives. These stronger social bonds simultaneously help to form the social network within the community.

Government Translation: By partaking in their normal activities, individuals in sustainable situations help to create and sustain behavioral management patterns at the group level.

For example, on one particular highway there is a curved stretch that is especially dangerous to take at high speeds. Intuitively, the majority of drivers slow down and go a safe speed. When this happens, faster drivers are sometimes blocked and forced to slow down too. Even when they aren't blocked, faster drivers typically slow down when they recognize that they are traveling much faster than the other cars in their field of view. Due to these safer drivers and the driver-to-driver interactions that follow, the accident rate on this stretch is far lower than it would be otherwise.

Environmental Translation: By partaking in their normal activities, individuals and groups of people in sustainable situations actively help to create and sustain land uses.

For example, a farmer plants one field of alfalfa to provide food for his livestock. While this provides the hay he needs, it also enriches the soil to some extent, and provides a degree of habitat for insects and grassland birds. His normal behavior helps to create and sustain his farm soil and habitat for the natural environment.

Cultural Translation: By partaking in their normal activities, individuals and groups of people in sustainable situations help to maintain their own culture.

For example, a group of friends regularly attends square dances, bringing their children along. This maintains a tradition, provides income to musicians, and helps to pass the tradition down to a new generation of people. The normal behavior that these people partake in for enjoyment has the effect of maintaining their cultural heritage.

Physiological Translation: By partaking in their normal activities, individuals in sustainable situations help to maintain their own health.

For example, in a schoolyard, children play variations on "cops and robbers." To win at this game, a given child has to sprint, climb slides and jungle gyms, hide, and wrestle. These physical activities address what their bodies need to grow and develop musculature, control, and coordination. Their normal play activity (plain and simple fun for them) improves their own and each other's physical condition.

The next pattern is a logical outfall of the previous two. Pattern 5 addressed the reliance of each individual upon its outer environment. Pattern 6 noted that the outer environment is altered and maintained by each individual. If both of these are true, then a further conclusion is that each individual is affected by and/or reliant upon each other individual.

For an example of this in nature, we can look back to our tree in a forest. Our tree depends strongly on the trees around it. These other trees raise regional humidity, reduce average wind speeds near the ground, help to capture moisture, reduce temperatures on hot days, raise temperatures during cools periods, house many of the forest's animals, resist takeover by other plant communities, shade out competitors, provide a steady removal and output of nutrients, etc.

We can even consider what will happen to our tree if the other trees around it are removed. Will it experience an improvement in health? Probably not. It will be subject to dry, windy conditions, dried-out soil, possible erosion and nutrient losses, an invasion by new plant species, a loss of the original forest animals, etc. There is a good chance that our tree will not survive these changes. Or if it does survive, it probably won't be quite as healthy.

> Pattern 7: All individual creatures (and groups of creatures) are dependent upon other individuals and groups to create and maintain the outer conditions that they require or that they are subject to.

Essential Translation: Individuals and groups in sustainable situations depend on other individuals and groups to maintain conditions that they operate under.

This pattern can now be translated into the context of human systems:

Economic Translation: A sustainable economy that helps to support each individual person or group is at least partly created and maintained by other individuals and groups.

For example, the strength of the economy in a developing nation enables its people to buy products made in a more developed nation. This helps to maintain economic vitality in one of the factory towns that makes those export products. Profits from those exports in turn help to support better schools, and improves children's likelihood of attending college, which improves their lifetime economic outlook. Though the connection is indirect, the activities and choices of people in the developing country are helping provide a stronger long-term economic footing for children in a faraway nation.

Community Translation: A sustainable community that helps to support each individual person or group is, to a great degree, created and maintained by other individuals and groups.

For example, in a rural region, there are long-standing relationships whereby neighboring families help each other in times of difficulty. During a three-day flood episode, Kyle moves sandbags with his dozer to help save the Hansen's house. Kyle's children spend those three days safe in the hands of the Magnins. The Magnins don't mind watching other people's children. They do it partly out of thanks to still other neighbors that give them butchered meat and plow their driveway all winter. Examples of freely given assistance exist throughout the region. From the perspective of each person, in effect it is the community connections created and maintained by a greater network of other neighbors that allows them to receive assistance from any neighbor at any time.

Government Translation: In sustainable situations, the rules and agreements that help to support each individual person or group are at least partly created and maintained by other individuals and groups.

For example, the local high school emphasizes to its students and staff that they should stay home when they are sick. Nearly everyone follows this request, and because of this, the transmission of disease is lowered. Overall, the school actually experiences higher net attendance as a result of its "stay home" policy. Because the stay-home policy cannot be truly enforced, each individual essentially relies on the compliance of others to receive benefit from this initiative.

Environmental Translation: A sustainable landscape and environment that helps to support each individual person or group is, to a great degree, facilitated by neighboring landscapes that are maintained by other individuals, other groups, or nature.

For example, a city is situated on the leeward side of a mountain range. To its west is forested public land, which provides space for hunting and recreation, clean water, and a very inspiring backdrop. To the south is farmland, providing the majority of the produce eaten in the city. Rangeland sits to the east, providing more food for the city and a good location for a small airport and an off-road vehicle track. To the north of the city sit two precious metal mines, several industrial facilities, and company headquarter buildings. Many of the city's residents are employed at these facilities to the north. Overall, the surrounding landscape is quite supportive of the city's economic, operational, and cultural needs.

Just because the environment of each individual is largely created by other individuals doesn't mean that environment is always highly supportive. Pattern #7, as intended here, allows for qualities that can cause stress on an individual or group. For example, our forest tree depends upon the nearby forest for its survival, but to a degree, it also competes with nearby trees for light, water, and nutrients.

The fact that each individual and group help to maintain their environment doesn't mean they all maintain it in the same way. Clearly, for example, a bird and a bacteria are going to impact their environment differently based upon scale (size) alone. But even creatures of a fairly similar size can have very different environmental impacts. For example, consider the environmental impacts that muskrats and wild turkeys have.

The muskrat's primary natural community role is seemingly to modify marshy wetlands and serve as prey for numerous animals. Muskrats change topography by making burrows, mounds, and swimming lanes; they change plant coverage and composition through consumption of larger emergent plants such as cattails and bulrush; and they provide food as prey for predators such as mink, eagles, and turtles. The turkey's primary natural community role is seemingly to alter plant and insect composition in forests and fields by feeding on seeds, nuts, insects, buds, berries, etc.; to fertilize areas near their roosts with droppings; and to provide food as prey for predators such as coyotes and bobcats.

The muskrat and the wild turkey play very different roles from each other. But what if we compare muskrats to muskrats or turkeys to turkeys? Each muskrat has a slightly different home site, and so its particular impacts will vary from every other muskrat. However, even different-sized muskrats in different marsh types will impact their environment very similarly relative to the different type of impact that a turkey has. Likewise, a turkey living in a forest will have different impacts from a turkey living in hedgerows along fields. Maybe the hedgerow turkey eats more grain and fewer acorns. However, even with some differences, the impacts of two turkeys are much more similar to each other than they are to impacts of any muskrat.

The same thing could be said for other hierarchical levels as well. In a natural community, for example, there is a higher rate of soil build-up and carbon compound storage expected in a wet forest when compared to a dry hilltop forest.

Pattern 8: The impact that individuals have on their external environment varies greatly. The variation is mostly according to the species to which the individual belongs. This pattern also holds true at other hierarchical levels, such as with species classes, natural communities, and ecosystems.

Essential Translation: The effect that individuals and groups in sustainable complex systems have upon the environment that they help to maintain varies mostly by the type of role they play within that system; that environmental effect varies little between individuals and groups that act within the same system role.

This pattern can now be translated into the context of human systems:

Economic Translation: *Within* a given sustainable business, industry sector, or profession, the way in which individuals and groups of people help to create and sustain their economies is relatively similar. But *different* businesses, industry sectors, and professions often have very different impacts upon economies.

For example, a city has a professional baseball team and a large tire plant. On the whole, they both employ about the same number of people, but employment related to the baseball team is very seasonal, whereas the tire plant's employment levels are steady through the year. And while hiring at the baseball stadium is almost the same each year, the tire plant production and annual staffing requirements are affected by business competition and other outside forces. Many of the tire plant jobs pay well, and the plant's corporate tax contributions to the city are larger than the baseball team's contributions, but the baseball team draws in tourists who support many small businesses throughout the city. Through their normal operations, both of these businesses contribute to the city, but the ways in which they contribute are very different.

Community Translation: Individuals and groups of people help to maintain their sustainable communities at the local level and beyond, but the form that this contribution takes tends to vary greatly according to a variety of factors such as aptitude and occupation.

For example, Harvey drives a delivery truck, which doesn't require lots of human interaction. But on evenings and weekends, he coaches basketball. His willingness to do this allows groups of families with children to congregate and get to know each other much more than they would otherwise. Not far away, Marv and Carrie run a corner store. During their work they talk and pass along happenings in the community, such as an engagement or the death of a resident. They help keep a variety of people informed about each other, even when those people go weeks without crossing paths themselves. Due primarily to their different occupations in town, Harvey and the storeowners contribute to their community bonds in significant but very different ways.

Government Translation: Individuals and groups of people actively help to create and maintain their own sustainable government, but the form that this contribution takes tends to vary significantly, often according to occupation.

For example, Marc and Pauline both affect government differently. Marc is a lawyer specializing in the representation of individuals against businesses that have not substantially provided the goods or services that they were paid to provide. Pauline works in quality control at a produce packaging plant. Marc's effect is generally to help enforce consumer protection laws. His assistance makes a large economic difference for a relatively small pool of people. Pauline's actions help to keep the food supply healthy. A mistake on her part could impair the health of many people or cause large financial losses for her company. Though their incomes and work hours are similar, Marc and Pauline have very different ways of contributing to government and behavioral restraint.

Environmental Translation: Individuals and groups of people actively help to create and maintain sustainable land uses and land use patterns, but the form that this contribution takes can vary significantly, often according to occupation.

For example, a teenager mows his family's yard as one of his weekly chores. This is in alignment with his favorite use of the front yard, which is to practice his soccer footwork. In the same town a construction worker helps to lay new road in an area that is being developed. In doing this, he helps to change the route that many people drive, including himself. The very different impacts these two people have upon the landscape are in alignment with the difference between their occupations and community positions. In fact, all regional teens with lawn-mowing duty would impact their environment rather similarly, as would all regional construction workers that build roads.

Cultural Translation: Individuals and groups of people help to maintain their own sustainable cultures, but the contribution that each makes tends to vary greatly according to a variety of factors such as social position, aptitude, and age.

For example, Emily teaches her children how to cook the family foods she had while growing up. Her mother, Lynn, likes to spend time with her grandchildren telling them family stories and teaching them crafts that she learned as a child. She also meets monthly with a quilting group of women mostly her age. Emily's sister, Courtney, teaches classical guitar and mandolin at a small college. All three women are contributing significantly, but very differently to the continuity of their culture.

Physiological Translation: Individuals can sustainably help maintain their own health, but the way they do that varies from person to person, often in relation to age, gender, health, and lifestyle factors.

For example, Rick is slim and in his early twenties. He likes to exercise hard and by himself. During the week, he runs five to ten miles in the early morning before going to his job in data analysis. Denise is retired and in her mid-sixties. She does a water aerobics class three days a week, which lets her exercise with friends and avoid being sore from physical impacts, which would impair her busy days around her house and community.

Now we have another four patterns. Patterns five through eight generally address the relationship of individuals to their environment. More or less, these patterns state that individuals in sustainable complex systems are only able to survive and thrive with a combination of their own effort and a supportive environment. And, as they rely upon the environment, the environment relies upon them, because they create and modify the environment as they engage and work on their own behalf. A logical outfall, then, is that all individuals and groups actually depend on the activities of all other individuals and groups to help maintain their environment, or their conditions. But interestingly, the manner and intensity by which the environment is guided or created by each individual varies greatly according to the role that individual plays (or sometimes chooses) relative to others.

Let's continue seeing where these patterns lead. Next, we look at issues related to material success, wealth, surplus, and waste.

Chapter 9 Income, Wealth, and Waste

"Success" is a subjective concept regarding the amount of accomplishment someone has made. People can view their own success in different ways, such as spiritual (on a spiritual path they believe is appropriate), psychological (feeling satisfied or happy), professional (conducting fulfilling work), material (gaining money, goods, or resources), health related (feeling fit, living long), or reproductive (bearing children).

"Wealth" usually refers to the accumulation of success measures, which means that it is similarly subjective and can also be interpreted from different vantage points. Here we will take a look at success and wealth in natural systems and extrapolate that to human systems. We will focus on particular types of success and wealth—not because they are inherently more important than other types—but because they display patterns in natural systems. The next pattern involves the difference in the level of resource consumption, or *material success* that different individuals enjoy.

In nature, are some individuals more successful than others? If so, by how much? Let's take a look this in sustainable natural systems.

What exactly is "material success" in nature? Material success can be represented by resource consumption. Imagine a common bird such as a robin. Robins fly and run about, eating worms, insects, and berries. A single robin's material success can be thought of as the amount of food it consumes in its lifetime. But how do we know the amount that different individuals consume?

If well fed and prospering in their environment, some animal species or individuals use their success for larger body size, some for more reproduction, and some for a longer life span. Or even more likely, they'll channel their success toward all three (size, reproduction, and longevity), but in varying degrees. Rather than try to figure out which of these most directly constitutes material success, we will lump relative size, reproduction, and longevity together as a composite indication of overall lifetime material success for a given individual.

Any two robins will be born at different times, eat differing amounts of differing diets, reach differing maximum lengths and weights, and probably raise a differing number of chicks. Imagine that we have studied two robins. The first robin was 11 inches long, weighed 3 ounces, and bore 35 fledglings over its 5 years of life.

Our second robin was 10 inches long, weighed 2.8 ounces, and bore 15 fledglings over its 2 years of life. The first robin was larger, bore more young, and lived longer. On that basis, we could say that over its lifetime, the first robin had more material success.

We could then look at other animals such as black bears, gray squirrels, and sandhill cranes, and at plants such as milkweeds, red oaks, and pinion pines. In this case, we don't really want to compare the size and lifespan of a pinion pine to that of a gray squirrel, because these different species are already genetically pre-programmed for different sizes and longevities. But we could compile a list or table of how much bigger, how much older, and how much more reproductive a more-successful individual of each species would be in comparison to a less-successful individual of that species.

This would give us a sense of how much success varies, as a ratio, within a given species. In nature, is a successful individual two times, fifteen times, or ten-thousand times larger/older/more reproductive than a less successful individual?

Before we answer that, we should understand the manner of this analysis. In nature, and to some extent in humans, there is often a big difference between the normal status of adults versus non-adults (offspring). We will look at this topic more later, but for now, suffice it to say that success at the seed, seedling, newborn, fledgling, or immature levels is often of a completely different form than success at the adult level. For example, a common milkweed plant can easily release several thousand viable seeds in its lifetime. Out of the thousands of seeds produced by each milkweed plant, only *one* seed from each adult plant, on average, will survive to maturity. The remaining thousands will be consumed by insects, drowned in water, or dehydrated after failing to reach the soil; or they will successfully germinate but soon die due to weather stress, shade, trampling, predation, etc.

If we try to compare seeds or seedlings to the plants that have reached adulthood, we are really comparing two different classes of organisms. This will not enable us to extrapolate well to human systems, because then we'd be comparing something like the professional success level achieved by adult humans with that of unborn or newborn babies. So as we compare success among plants and animals, we will essentially be considering the relative success between those that reach adulthood, not the success of the generally larger number of offspring that never reach adulthood.

Also, recognize that biological attributes such as how many seeds a tree produces over its lifetime are not commonly measured by scientists, so in considering this topic, we will use broad estimates and typical ranges.

How much does material success differ between adult individuals of the same species? Let's consider North American black bears as an example. Black bear males are normally larger than females, and average size generally varies across the continent. Also these bears gain weight in the fall and lose that weight during hibernation. Given this variation, we'll rely on approximate typical or average values.

Let's say that in a given area, male bears range in weight from about 150 to 450 pounds, and the females range from 100 pounds to 250 pounds. However, a very small adult bear could be just 90 pounds, and the occasional enormous male bear is around 900 pounds. Lengths aren't as variable, ranging from a minimum of about 4 feet to a maximum of about 6.5 feet. But the *typical* variation in length will be even less. We can estimate the typical black bear lifespan to be about 5 to 18 years, but an adult that dies very young could be just 2 years old, and an extremely long-lived individual could be about 30 years old. Female bears usually produce cubs every other year by giving birth to 2 or 3 cubs, though individual litters can range from just 1 cub up to 6 cubs. The females can reach reproductive maturity as early as 2 years old, but this can take as long as 5 years.

Using our typical lifespan of 5 to 18 years, we could calculate that a *less* reproductive female might give birth to just 2 cubs before death at age 6, and a *more* reproductive female might give birth to about 3 cubs every other year from age 3 to 18, producing about 24 cubs. We could envision an *extremely* unproductive female giving birth to just 1 cub before her death, compared to an *extremely* productive female giving birth to four cubs every other year for 28 years (producing 56 cubs).

If we then develop ratios between typical-but-less-successful bears with typical-but-more-successful bears, we get weight ratios of about 1:3 and 1:2.5; length ratios of about 1:1.5; lifespan ratios of about 1:3.5; and reproductive ratios of about 1:12. The average of these ratios is about 1:5.

Of course, most adult bears will fall somewhere between these rather unsuccessful or successful ends of the spectrum (e.g., somewhere between 150 pounds and 450 pounds for male bears), and the bulk of them will probably have size, lifespan, and reproductive ratios in the range of 1:2.

Then looking in the other direction, if we compare the ratios of *extremely* unsuccessful to *extremely* successful bears we get weight ratios of about 1:10; length ratios of about 1:1.7; lifespan ratios of about 1:15; and reproductive ratios of about 1:56. The average of these ratios is about 1:21.

Of course, these extremes could be stretched if we work harder to compare individual bears from across the continent (e.g., between Florida and Quebec). But all this does is compare across different environmental conditions, and from a larger pool of genetic groups and black bear subspecies. Comparing across this type of gradient doesn't effectively improve our ability to compare success levels between individuals in relatively similar situations.

Going through this exercise with other species of animals or plants will yield different results for each species. The point here is not to find an exact magical ratio, but to demonstrate that these types of ratios exist and seem to fall within a certain range. An approximation for that range within a species is about 2 to 1 for typical individuals, 5 to 1 for rather successful versus rather unsuccessful individuals, and 20 to 1 for extremely successful versus extremely unsuccessful individuals.

In nature, what would cause one individual to be more successful than another? It would probably be a combination of things. Foremost might be good fortune—good fortune of timing, placement, climate, etc. Also, superior genetics would play a role in some cases or to some degree. "Individual will" or zest might come into play as well, but of course, for the biologist this is difficult to measure independently of genetics.

> Pattern 9: Of those that reach adult status, individuals of a given species in sustainable natural systems typically have different levels of success. The level of material success between the more successful and less successful varies by species, but a median ratio might be about 2 to 1, or less. The material success ratio between the quite successful and quite unsuccessful averages somewhere around 5 to 1. Extremes can occur also. The material success ratio between the extremely successful and extremely unsuccessful averages somewhere around 20 to 1. The case of the "extremely unsuccessful" individual is likely to be more common than the case of the "extremely successful" individual.

Essential Translation: Individual adults in sustainable complex systems typically have different levels of material success. In a given region or location, the median material success ratio between the more successful and less successful might be about 2 to 1 or less. The success ratio between the quite successful and quite unsuccessful averages somewhere around 5 to 1. The material success ratio between the extremely successful and extremely unsuccessful averages somewhere around 20 to 1. The case of the "extremely unsuccessful" individual is likely to be more common than the case of the "extremely successful" individual.

This pattern can now be translated into the context of human systems:

Economic Translation: In a given area or region, the sustainable annual income of most working adults falls in a range that has a high end less than twice the low end. However, a significant minority fall below or above this range, such that the lowest normal incomes range down to about a fifth of the highest normal incomes. Occasional individuals with even lower incomes, and even higher incomes also exist, but the percentage of people with these extreme incomes is small; these very lowest incomes are roughly one twentieth of these very highest incomes.

For example, a moderately large city with a population of about 600,000 has people working in dozens of fields and occupations, including manual labor, construction, hospitality, teaching, public service, science, medicine, entertainment, consulting, retail business, finance and banking, government, etc. All incomes fall within an annual earnings range of $15,000 to $700,000. But, more than half of the adults who work full time earn between $30,000 and $60,000 per year.

The actual simplified income distribution is as follows: 3% at $15,000-$20,000; 10% at $20,000-$30,000, 20% at $30,000-$40,000; 25% at $40,000-$50,000; 20% at $50,000-$60,000; 10% at $60,000-$70,000; 5% at $70,000-$80,000; 3% at $80,000-$100,000; 2% at $100,000-$200,000; 1.5% at $200,000-$400,000; and 0.5% at $400,000-$700,000.

Above, we defined "wealth" as the accumulation of success measures. As an example, if the number of berries eaten is a success trait for a certain bird species, and if some of those birds not only ate those berries, but gathered and stored or hoarded those berries for their own benefit in the future, we could say that they had obtained material wealth.

Or, imagine a slightly more abstract form of wealth. If certain types of community relationships improved survival for individuals of a particular bird species, and some of those birds were able to accrue those relationships over time—thus improving their success likelihood at a future date—then we could also say that they had obtained a form of wealth. We will view wealth as the gathering of tangible (material) or intangible (non-material) resources in the present, such that there is apt to be an improvement in condition or livelihood in the future.

Let's look at material wealth first. Plants and animals do accrue material wealth, but it is only in direct relation to an immediate or impending survival need. For instance, many mammals need to go all night without eating. For this, they store at least a little energy and fat on their bodies. Most plants also have to manage each night with virtually no sunlight to help them produce energy. They accomplish this with energy reserves made of sugar, starch, etc.

Similarly, many plants and animals survive the winter by building up large reserves of energy in the fall, which they use up through the course of the winter and early spring. For example, many trees and perennial herbs will store sugar and starch in their roots as winter approaches; bears gain dozens to hundreds of pounds in body fat in late summer and fall to help them hibernate; chickadees hide bits of food in tree cracks in the fall to help them survive the upcoming cold; and squirrels hide hundreds of nuts and seeds that they can eat through the winter.

But surviving overnight and preparing for winter aren't the only causes for material accumulation. Some creatures live in places where water and food are sporadic or unpredictable. Cacti store water in their stems to help them survive inevitable dry periods. Camels live where feeding areas might be widely scattered, and they store extra energy in their hump. Cougars will keep returning to a large kill, which might be their only source of food for one or two weeks.

What all these forms of wealth accumulation have in common is that they are designed to provide food or water for about the length of time that they might not be supplied by a harsh or unpredictable environment. These wealth accumulations are limited to the need at hand, and in similar situations where the supply of food and water are more ample or predictable, there aren't significant patterns of material wealth accumulation.

Then we can consider other types of wealth accumulation by individuals. As individual plants and animals grow and mature, their condition, interaction capabilities, and skills often improve. For example, over time, a foraging animal such as a raccoon will learn where the finer feeding spots are and how often to return to those. They remember negative experiences and will often avoid dangerous areas.

Even plants gain leverage or improve their interactions over time, so to speak. A young tree might begin growing among plant species that are not especially complimentary. But as the tree expands, the plants growing in its shadow will shift over time to species that are complimentary with the tree's life cycle. Spring wildflowers will predominate under a shady maple tree. These wildflowers come up early in spring, near the ground where temperatures are warmest, receiving most of their light to grow and flower before the maple unfurls its leaves. The spring wildflowers draw in deer and insects. All the activity aerates and fertilizes the soil. The maple leaves drop in the fall, and after providing some thermal cover over winter, the leaves flatten and disintegrate through the spring as the wildflowers push past them.

These are all examples in which individual-to-individual interactions tend to perfect slightly during the life of each individual. Plants and animals have far less cultural (learned or traditional behavior) than people, but to the extent that animals have culture, it is absorbed by individuals as they mature.

Finally, plants and animals often become more physically fit or attuned through their lives: A spreading vine will grow most rapidly toward sunnier places. Plant roots will extend through the years to reach deeper reserves of water. A hawk becomes a more skilled hunter after a number of successes and failures. Though it isn't quite how the term "wealth" is always used, we can view these individual improvements as a type of intangible wealth accumulation within nature.

> Pattern 10: Individuals often acquire forms of wealth. Material wealth accumulation is limited to situations where it improves the likelihood of surviving challenging physical conditions. A form of improved alignment often occurs between individual plants and animals, yielding wealth that is not as tangible, yet greatly improves survival. Also, as they mature, individuals often learn more advanced behaviors from their own kind, and they often become more cognitively or physically fit.

Essential Translation: Individuals in sustainable complex systems primarily accumulate material wealth in order to address significant and predicable or likely shortfalls in resources. They also accumulate improvements or intangible "wealth" in their community interactions, their cultural knowledge, and their physical/mental adeptness.

This pattern can now be translated into the context of human systems:

Economic Translation: People in sustainable situations accumulate material wealth only in order to meet upcoming shortfalls in resources relative to the demand on those resources.

For example, there is a scenic resort town that employs many hospitality and service workers, such as cooks, waitresses, landscape crews, hotel maids, porters, tour guides, etc. The town is very busy beginning each spring, but becomes far more tranquil by late fall. By winter, crowds are so sparse that most workers are laid off or barely employed for about three months. Those workers who have made a career in this location and who have achieved some situational stability tend to have bank accounts with their normal amount of monthly expense money and about 6 months of additional income. The typical worker tends to dip into this excess cash during the slow winter season, and replenish it during the upcoming spring and summer.

Community Translation: People in sustainable situations tend to form stronger and more appropriate connections (bonds) to other people from childhood through adulthood.

For example, in his late twenties, Jay decided to hang out less frequently with a group of drinking friends in favor of a crowd of people that were focused on their careers and young families. While he enjoyed the partying group, he knew alcoholism ran in his family, and he didn't want to wind up with regrets and in bad shape later. Within five years of this decision, he was fairly happily married, had a son, owned a small house, and had a good reputation at the construction firm where he worked.

Cultural Translation: In sustainable contexts, children spend many years absorbing the culture around them. As they age to and through adulthood, they refine their understanding of that culture in the context of their contemporary conditions, and they become the knowledge bearers who pass the culture on to the next group of young people.

For example, as a child, Kirsten loved to hear family stories from her grandparents, aunts, uncles, and so forth. They would speak of struggles, achievements, and long dead ancestors. As she entered middle age, she decided to begin preserving these family stories through a variety of traditional and modern means so that they would stay accessible to the next several generations. This helped to solidify her immediate and extended family's sense of identity.

Physiological Translation: People in sustainable situations tend to improve their mastery of physical and cognitive activities or challenges as they grow, mature, and age.

For example, Max was a serious football fan for all his life. As a child, he would play ball at the park and follow his favorite pro team on the radio and TV. By high school, he was a starting linebacker. He played in college as well, and even though he didn't always start, his intensity in workouts and his interest in strategies continued. Through his adult life, he remained active and retained much of his strength, coordination, and flexibility. His strong interest for the game continued, bringing him to coach a high school football team, where he was known for being one of the most knowledgeable and successful coaches in the region.

We can also look at the concept of wealth above the individual level, such as with natural communities. Remember that "success" was previously defined as the amount of accomplishment one had made, and "wealth" was the accumulation of success measures. Success for an individual organism was viewed as its resource acquisition, and its ability to continue living and to bear offspring. At the level of a natural community, *success* is more abstract, but otherwise similar: it is some measure of the natural community's ability to continue occurring. *Wealth* for a natural community could be the natural community's storage of assets that are able to improve its future status or continuation (even if those assets don't immediately improve its status).

The next pattern pertains to this type of system-level wealth. It seems that natural systems have a tendency to build up materials that help to secure their own continuation. In other words, they provide themselves with positive support conditions.

One common example of this is in bogs. As bogs form, they build up thick layers of sphagnum moss and partially decomposed "peat," which helps to acidify water and soil. The acid inhibits competing vegetation types and the bog can continue to grow freely and sometimes even expand with each coming year. In other examples of wealth development in natural systems, we can find things such as water table buildup that helps provide a reserve of water; increases in species diversity, which increases system stability; and buildup of woody debris and micro-topography—which help to provide room for more species. Similar concepts of system-level wealth also seem to exist for species, species classes, and ecosystems.

Pattern 11: Species, species classes, natural communities, and ecosystems tend to accrue some forms of wealth that boost their likelihood of persistence far into the future.

Essential Translation: Aggregates, or groups of individuals at different scales in sustainable complex systems, tend to obtain forms of wealth that boost the group's likelihood of persistence far into the future.

This pattern can now be translated into the context of human systems:

Economic Translation: Sustainable groups of people, from extended families to neighborhoods and nations, tend to amass forms of material wealth that increase their likelihood of persistence into the future. The scale of wealth accrual is generally proportional to the size of the group—smallest for families and largest for nations.

For example, the Millers own a cabin on 60 acres of land along a scenic river. This is a favorite holiday and summer getaway destination. It came into their ownership in the 1940s, and they have since rehabbed it and added three bedrooms. Several dozen family members have memories of the cabin, and in respect to this, they view it as a family asset that they are passing forward. Because of this, they managed to transfer the title from just one of them into five heads of family.

In another example, one part of a nation had come from an estate-style land-holding tradition, and in 1800, only 1% of its land was publicly owned and accessible for conservation and recreation. Since that time, residents worked hard to provide money for public land purchase, and through the decades, about 12% of the landscape has become public land for parks, preserves, hunting areas, and greenways.

Occupational Translation: Over time, the sustainable professions, businesses, and business sectors that thrive in a given area diversify to a degree and become more complimentary to each other, which adds efficiency and strength to patterns of economic activity. This in turn reduces the likelihood of interrupting forces from the outside, and so increases the likelihood that the established economy in the area will persist into the future.

For example, three theaters in a city began with much overlap in the types of audience they would attract—opening on the same nights, with similar ticket prices, and trying to attract the same thirties to fifties crowds. But over time, the theatres have become more specialized. One of them opens more frequently during the week and is likely to have cutting-edge attractions. Another is larger and more expensive, trying to draw large traveling shows on weekend nights. The third is more family oriented and has daytime shows, lower prices, and young local actors.

Environmental Translation: Sustainable human communities—from villages to cities to states—refine and improve their land use systems, which improves their likelihood of persistence far into the future.

For example, in one county, plans have been implemented to generate fertilizer from what would have been landfill waste; to minimize nearly all development in disaster-prone areas; and develop zoning patterns that reduce driving requirements. Over time, this has resulted in an actual increase in farmland fertility, a reduction in emergency service costs, and higher net incomes for residents who do not spend as much money on transport. These land use refinements are effectively a type of wealth that the county has accrued for itself.

The last topic in this series is material "waste." All plants and animals generate waste, which is any sort of chemical or matter that is toxic, or will become toxic or obstructive if it is not released into the nearby environment. Among plants and animals, waste can take the form of senescent leaves and twigs, sloughed bark, gasses, scat, skin, hair, feathers, urine, sweat, antlers, etc.

An interesting pattern of waste in nature is that it *does not remain* toxic or burdensome. There are countless examples of this pattern. Leaves and woody material dropped from trees feed innumerable insects, worms, nematodes, bacteria, and fungi. As the leaves and wood are broken down, nutrients, organic matter, and simple elements are released. These released materials enhance growing conditions for other plants, return to the atmosphere, and return to water bodies as nutrients for aquatic life.

Similar patterns apply to all types of animal waste—antlers that are shed provide minerals to rodents that gnaw them down; nutrients in dung and urine are absorbed readily by plants and incorporated so that they can grow larger and faster. At the system level, even whole dead animals never pile up. There are entire species that are adapted to help consume and eliminate carcasses. Vultures, crows, opossums, coyotes, carrion beetles, flies, etc., help break down carcasses to their elemental nutrients and return them to use by plants.

Another example of this pattern—but at the system level—can occur within the hydrologic cycle. Water flowing off a hillside may essentially be a material excess for the natural community inhabiting the hillside (e.g., toxic or damaging if it builds up), but upon release, it enters a stream, which along with its associated wetlands and floodplains, requires that water. Another well-known example of this pattern is the biosphere's oxygen/carbon dioxide cycle. Animals require oxygen and release carbon dioxide, while plants *release* oxygen and *require* carbon dioxide.

> Pattern 12: Waste created at the level of any given individual is a resource to some other individuals, and therefore does not behave as waste at the natural community level. Similarly, excess material at the natural community and ecosystem level is a resource for other natural communities and ecosystems, and therefore, what is released as excess material is generally not received as excess material.

Essential Translation: In sustainable complex systems, material waste exists for a limited duration until useful acquisition or transformation. Thus, material waste does not accrue in sustainable complex systems.

This pattern can now be translated into the context of human systems:

Economic Translation: Material waste generated sustainably at any level, from the individual to neighborhood, and from the local to global economy, is often a valuable resource that is used or transformed to support other individuals or sectors of the economy. In some cases, material waste does not have a recoverable economic value, but is broken down into a neutral elemental form and is released into the environment as a neutral material.

For example, a metropolitan area with about half a million residents has an intricate reuse program that includes regular pickup of excess household items. These go to a warehouse and are available to the public and to charities. This prevents items that have essentially become waste for one individual to become valuable again in the hands of another individual.

Environmental Translation: Excess material sustainably discharged from any local, regional, or national land use is often a valuable resource that is used or transformed to support other local, regional, or national land uses.

For example, a 2,600-acre farming operation has a strategically located conservation area, hayfields, cropland, chickens, and cattle grounds. The several dozen chickens and cattle spend a majority of their time on a few acres near the highest point on the property. Immediately downslope of this area are croplands, then the hayfields, and then the conservation areas along streams and wetlands. Nutrients from animal waste tend to move downslope into cropland, where they can be absorbed and used by the crops. Most of the nutrients that continue to move are trapped in the hayfields, and the small amount that continues further downslope is trapped in wetlands before entering the local stream. This layout turns the animal waste into valuable material for crop and hayfield growth, and keeps it out of the local stream, wherein it would be a pollutant.

That concludes an additional four patterns. Patterns nine through twelve indicate that all individuals experience different levels of material success, but in a given area, most of them fall within a close range, in which the more successful have up to about two-fold the material success relative to the less successful. A significant minority of even more successful and less successful individuals will fit within a five-fold material success ratio, and a small percentile minority will even stretch out to a twenty-fold ratio. Accumulated material wealth at the individual level is limited to the amount that one will need to address probable future resource shortages. However, individuals actually accrue wealth through their community connections, cultural importance and knowledge, and physical and mental abilities and knowledge. Groups of individuals do amass *collective* material wealth that normally outlasts the lifespan of any individual. These groups also amass less tangible wealth via refinements related to collective division of labor and to land uses. Finally, generated waste is temporary, because before it builds up, waste is used or converted to a valuable resource, or it is broken down and released as a neutral material. Similarly, materials that shift out of any given land area are received as a resource by the next land area.

That's twelve different patterns, and more lay ahead. What good are they? At this point, they don't form a full picture. They are showing individual facets of complex systems that might need to be aligned correctly in order to achieve sustainability. The patterns are a bit like puzzle pieces scattered about on a large table. There aren't enough pieces yet to form a whole puzzle, so we can't tell what the big picture(s) might eventually look like. But at least we can see the shape and color of a few pieces, and we can probably start to rule out some possible arrangements of the puzzle pieces.

Eventually, with more puzzle pieces, we might be able to look at alternative system designs quite critically. So for now, let's continue. Next, we will look at how systems mature within the context of competition and cooperation.

Chapter 10 Competition, Cooperation, and Succession

Competition and cooperation both appear to be very important elements at different times and different places within natural systems. Competition in the classic ecological sense can be defined as the case of two or more individuals attempting to utilize the same limited resources. Competition can be for things such as water, space, food, nutrients, sunlight, warmth, a nesting area, a mate, social status, etc.

While two individuals are engaged in competition against each other, one of them, or both of them, will be unable to attain the level of resources that they desire. At least in the short term, having to compete diminishes the survival or success prospects for at least one of the individuals in any competitive match. In nature, competition is often a diffuse event in which numerous individuals compete at different intensities for limited resources. This means that measuring and describing actual levels of competition in nature can be a little messy.

Competition is commonly viewed through an economic lens—for limited food, water, shelter, energy, etc. But here we are also interested in more abstract ideas such as community, occupation, and government, so we will use a slightly broader definition of competition: *Competition is the case of two or more individuals or systems attempting to, or poised to utilize the same limited resources or the same single context.* This broader definition will allow us, for example, to more easily discuss the competition between cultural values or the rise in dominance of a particular cultural value.

Cooperation by any individual toward another individual can commonly be defined as behaviors that avoid competition or that enhance the likelihood of success for the other individual. Just as with competition, cooperation can pertain to a wide variety of resources, such as water, space, food, nutrients, etc. Cooperation can take many different forms, such as divvying up a resource without competing for it, avoiding use of the resource at the same place or same time, using the resource in a way another wouldn't use it anyway, providing an extra resource to another individual, taking a resource from another individual but providing more of a different resource in its place, etc. Just as with competition, cooperation is often a diffuse event in which numerous individuals cooperate at different intensities over different time periods. Measuring and describing actual levels of cooperation in nature can be messy, especially because it can involve the *lack* of an event.

To facilitate an application to a wider variety of situations, we will use a slightly broader definition of cooperation than may be typical: Cooperation is *behaviors or systems that avoid competition, or which enhance the likelihood of success for other individuals or systems.*

At any time or place in natural systems, a scientist could probably make a case for the existence of both competition and cooperation. While they may both be present to some degree, they do not appear to have the same relative importance in all times and places. In sustainable natural systems, there seems to be an emphasis on competition earlier in development, and an emphasis on cooperation later in development. Explaining this might be easiest to do in the context of *succession*, so that is where we will turn our focus. Note, however, that succession relates to other patterns in natural systems (not just competition and cooperation). So this topic will arise later relative to other issues.

In nature, succession is the process by which certain plant and animal types in a certain location will eventually come to be replaced or joined by other plant and animal types. Let's look at an example of succession—suppose a neighbor who lives down the road has a large backyard, and he has paid someone to excavate a shallow pond where his lawn had previously been.

If you visited this new pond right away, it would probably look like a barren pit with some water in it. A few days later, you might see a killdeer running along the muddy edge, and some small bubbles and greenish algae on the bottom. After a few weeks, there might be mosquito larvae near the surface, a couple of mallard ducks, a little bit of floating duckweed, and barnyard grass or witchgrass beginning to colonize the muddy pond edge.

If you revisit the pond, perhaps early in the next year, you will note even more differences. You might see water boatman insects, leopard frogs, red-winged blackbirds, some cattails, water plantain, and very small sandbar willow shrubs. But succession won't stop there. If you revisit ten years later, you might hear green frogs and a song sparrow. You might see a kingfisher, pickerelweed, several types of sedges, asters, dogwood shrubs, a muskrat, backswimmer beetles, and several types of dragonflies.

Not only will more species come to inhabit the pond over time, but some of the first species that showed up, such as killdeer, witchgrass, and water plantain, might be gone. This pond is a typical example of succession and the changes in species and ecological organization that take place.

The process of succession happens with all sorts of areas—lakes, swamps, prairies, forests, deserts, etc. We will run through the process of succession once more with a place that becomes a forest. But this time, we will discuss it within four stages that have different levels of competition and cooperation.

In most upland areas in the eastern half of North America that aren't too cold, rainfall is ample and forests are the prevalent form of vegetation. The type of forest and the species involved differ by region and sometimes by site, but the overall patterns of succession they go through are similar.

Let's imagine that an area of land has recently been cleared of all vegetation, and only soil remains. This can happen naturally when a lake or pond catastrophically drains, when glaciers recede, when a volcano erupts and spews ash, or when searing fires move through downed trees. However, nowadays it is more common for this to happen when an agricultural field is abandoned. With an area of soil opened up, plants and animals begin colonizing. For simplicity, we will focus on the plants as the first stage of succession ensues:

Stage 1: Within months of the clearing event, new plants will spring forth. The exact species will vary by site, but nowadays they might include scatterings of short-lived species such as ragweed, knotweed, dandelion, and foxtail grass. These "first wave" plants are opportunistic. They will probably be sparsely distributed at first, unless their seeds were already present in large numbers due to repeated land-clearing events.

These first-wave species have simple needs and lifecycles, and they don't interact, or don't need to *greatly* interact with each other or with other plants or animals to thrive in the cleared area. This first stage is marked by little competition and little cooperation. It is sort of a "free lunch" stage, where whoever shows up first and runs to the front of the line gets the free meal without having to worry much about interacting with others.

Stage 2: As the plants from Stage 1 mature, they release thousands and thousands of seeds. Many of these seeds will subsequently grow, but now they are encountering a more crowded landscape. Along with them, other species (a second wave) are beginning to move in. These species may take longer to arrive than the Stage 1 weeds, but they are a bit tougher at gaining a piece of ground and holding their turf.

Again, depending upon where and when this event occurs, the second wave species might include old-field goldenrod, wild strawberry, wild rye, broomsedge, dogwood, raspberries, aspen, sassafras, and tuliptree. These species are less independent than the first wave species in that they have slightly more advanced or intricate requirements for reproduction and growth on the site. Not only do the second-wave species compete against the first-wave species, but they also compete against each other.

In fact, some of these second wave species have different ultimate needs, and will begin to pull the community in opposing directions. Broomsedge and wild rye pull toward maintaining open conditions, trying to establish a dense thatch that will help keep out woody plants. Sassafras and tuliptree pull in a different direction, trying to produce enough shade to weaken competing grasses that need full sunlight. This stage is marked by strong competition and limited cooperation. It is sort of a power struggle stage, where the survival of one comes at the expense of another.

Stage 3: As the species established by Stage 2 continue to compete for dominance, some will fare better than others. Maybe this particular site is drier than usual, has more sterile soil, or is prone to a buildup of deep snowdrifts in winter. These local environmental conditions will help to sort out which species will thrive and which will begin to be displaced.

As this happens, the vegetation starts to move in a more coherent and unified direction. It won't have stabilized by this point, but its rate of change will slow. Again, depending on the location, some of the species now present and establishing could be white pine, red pine, red maple, white ash, white oak, hickories, viburnums, hazelnuts, Virginia creeper, hairy wood sedge, bottlebrush grass, arrow-leaved aster, rough goldenrod, violets, wild geranium, tick trefoil, and agrimony.

Whatever species do begin to dominate the site will generally be ones that are adapted to grow well near each other. So trees won't crowd out all the shrubs, and flowers will bloom through the growing season, giving a stable food supply to pollinators. Plants begin to settle into locations where there is just enough sunlight to fully support them. This stage is marked by a gradual decline in competition, and a gradual increase in cooperation. It is sort of a "decision" stage, where the general direction of the community is clarified.

Stage 4: As succession continues, the species that are a bit more pioneering often have a bit less staying power in Stage 3, and they grow a bit sparser—perhaps the white ash and agrimony are examples of this group. At the same time, additional species move in for the longer haul. These latter entrants into the system are not found in earlier successional stages. They can be rather selective about where they grow and what other species they will grow near.

Some of the new species could include beech, red oak, hemlock, flowering dogwood, bladdernut, trilliums, lady's slipper orchids, bellwort, plantain-leaved sedge, Indian cucumber root, etc. Although the vegetation in Stage 4 will appear similar and fairly stable year on year, there will be continual activity with growth, metabolism, etc., according to Pattern 1 (see Chapter 7).

The vegetation will also continue to slowly shift and ebb in order to perfect and continually refine its internal balance in response to changes within and without, according to Pattern 2 (see Chapter 7). While competition is still present in Stage 4, it is deemphasized relative to cooperation. This is sort of the "harmony" stage, where the community type is fully present and yet continually refining itself.

Of course, there's no abrupt separation between these stages, so the changes from the beginning of Stage 1 through Stage 4 are incremental. The successional stages happen often enough in nature that in many environments, different species are specifically adapted for the different stages.

Patterns of competition and cooperation are perhaps easier to observe and imagine in a natural community or ecosystem context where something like an empty field eventually transitions to mature forest. But the same sort of processes seems to happen at other scales as well.

For example, on a shorter and more local timescale, a tree in a forest will release many thousands of seeds. Hundreds and hundreds of those seeds will often establish in the soil, growing into small tree seedlings. These seedlings don't compete much, and they have little interaction; at this point they represent Stage 1 of succession. As the seedlings grow, they enter Stage 2, where they use more resources and begin to compete with one another. This competition intensifies so much that most of the seedlings and saplings will die, but some will survive.

The death of most, and the survival of some, marks the entry into Stage 3. This stage produces a level of order in which there is a survivable, cooperative spacing between the remaining individuals, and in which the remaining individuals tend to be the best adapted for the immediate situation.

In Stage 4, the young trees hold their position or grow very slowly, maintaining a survivable spacing and growing or awaiting an opportunity to grow into canopy position, which they can only do in response to a new gap in the forest canopy. The coordinated growth delays and surges of Stage 4 will be according to how much light is available, and are again in line with a cooperative pattern.

This same sort of four-stage pattern can even be seen in the growth of one tree. In its lifetime, a tree will normally form hundreds of very short twigs of just a few leaves each. All of these twigs are potential new branches (Stage 1). However, the tree will invest more in the twigs that receive more sunlight as it deemphasizes those that receive less than optimal lighting (Stage 2). As this process of investment and de-emphasis continues, the tree begins to take its own unique form, creating branches based upon good light reception (Stage 3). Finally, in Stage 4 the tree will continue adding growth upon these branches, which are now its main limbs, as it works to maximize the incidental sunlight and minimize the risk of branch fracture or tipping.

We will separate these four stages and discuss each separately as its own pattern in reference mainly to competition and cooperation. Each of these four stages is most meaningful in the context of the other three stages. So unlike the translations and examples provided for previous patterns, translations and examples under a given category, such as culture, will occur together.

Pattern 13: Ecosystems, natural communities, species classes, species, individuals, and offspring have developmental stages. The first of these is an initiation or free-lunch stage marked by little competition, little cooperation, and overall relative disorganization.

Essential Translation: The first stage of four stages in the development of a still unformed sustainable complex system is often marked by relatively little interaction, competition, or cooperation between the first components to become established in that system.

> Pattern 14: Ecosystems, natural communities, species classes, species, individuals, and offspring have developmental stages. The second of these is a testing or power-struggle stage marked by high competition and minimal cooperation over available resources, and also marked by uncertainty over eventual form.

Essential Translation: The second of four stages in the development of a still rudimentary sustainable complex system is often marked by increasing competition and little cooperation between the components that have become established in that system. Also, there is often an uncertainty as to what form the complex system will eventually take.

> Pattern 15: Ecosystems, natural communities, species classes, species, individuals, and offspring have developmental stages. The third of these is an aligning or decision stage marked by decreasing competition, increasing cooperation, and the emergence of patterns or form.

Essential Translation: The third of four stages in the development of a still forming sustainable complex system is often marked by decreasing competition and increasing cooperation between the components that are still established in that system. In this stage, patterns and form also tend to become apparent.

> Pattern 16: Ecosystems, natural communities, species classes, species, individuals, and offspring have developmental stages. The fourth of these is a continually balancing or harmony stage marked by little competition, significant cooperation and coordination, and repeating patterns and forms. This fourth stage is also a merging into Patterns 1 and 2.

Essential Translation: The fourth and final stage in the development of a sustainable complex system is often marked by little competition, but significant coordination and cooperation between the components that occur within that system. This stage includes patterns and forms and the non-static continual activity and adjustment expected in all complex systems.

These four patterns can now be translated into the context of human systems:

Economic Translation of Pattern 13 through 16: A newly forming economy that can eventually become sustainable is often marked by relatively little interaction, competition, or cooperation between the first economic units that become established. As that economy continues forming, it enters a second stage marked by increasing competition, yet little cooperation between the economic units that have become established and are establishing. The third stage in the development of the economy has decreasing competition and increasing coordination, alignment, or efficiency between the economic units that are still established. The fourth and final stage in the development of an economy is often marked by little competition, but significant coordination, alignment, and efficiency between the system's economic units. This final economic stage is not assumed to be static, but rather will have units that ebb, wane, adjust, and improve in the face of changing internal or external conditions.

The backdrop for the next seven modal examples will be a territory that opens up to human settlement after being uninhabitable for centuries due to military hostilities. Though lacking an exact time in history, it provides a convenient setting to discuss the stages of system establishment.

Within months after settlement in the new territory has started, economic activities have begun. Though these activities are interesting, they appear somewhat haphazard. For example, sometimes one can buy food at a store (usually more of a shack), but in some locations there are few stores. When stores do exist, their inventories are like a hodge-podge that varies almost daily. One store might carry all canned goods, and another might carry liquor, eggs, and fresh chickens. One can order foods or items from outside the territory, but delivery has to be arranged through individual truck or mule team drivers who seem very busy and have inconsistent routes. On the whole, choices are quite limited, and customers must patronize the stores that are available, or go with a busy delivery man who may or may not deliver on time.

Through the next few years word gets around that there is money to be made. Additional economic enterprises and stores spring up as new vendors, who hope to get a piece of the action, arrive. Some of the new stores are larger, or have more selection than the first stores that had established. Other new stores try to carry specialty items or are located at points of convenience. There are also more delivery routes, and at the same time, many individuals have purchased equipment to make some items at home, such as furniture.

Through this time, many storeowners are finding that competition is more intense, and what was at first very lucrative no longer provides easy income. After a few decades, some stores have gone out of business, and many vendors have specialized in order to keep a customer base. One storeowner finds that he can make a living selling roofing materials, but to concentrate on that, he must forego his bike and car sales. Several individuals who originally sold tanks of potable water have gone out of business, but there is now a provider of home water filtration systems, and several people make a living repairing those systems.

After several decades pass, there isn't a great deal of annual change in the types of goods available and the amount of economic activity in different parts of the territory. Stores will carry an occasional new item, or they may stop carrying an unpopular item, but by and large, people know which stores are in their area, including what they carry, and their prices. Each storeowner tends to know what the other stores in their region carry, and beyond a few staples, they generally choose not to compete with the specialty items found at other regional stores. People's choices regarding shopping, ordering for delivery, or going with homemade items are regular and planned through a combination of convenience, cost, and reliability. At this stage, it is rare for entirely new stores to open, and similarly, very few close in a typical year. Very few entrepreneurs in the territory have gotten "rich quick" since the first few years of settlement, but people in business have now settled into more reliable, moderate incomes.

Community Translation of Pattern 13 through 16: A newly forming community that can eventually become sustainable is often marked at first by relatively little interaction, and the interaction that does occur is somewhat haphazard or inconsistent, and lacks strong bonds between individuals or groups. As that community continues forming, it enters a second stage marked by increasing interactions and relationships, but the forms that these take still vary greatly, such that coherent relational patterns still aren't clear. The third stage in the development of a community is marked by increasing coordination and patterns (predictability) in community relationships. The fourth and final stage in the development of a sustainable community is marked by significant coordination and pattern. This final stage is not assumed to be static, but rather will have community relations and patterns that ebb, wane, adjust, and improve in the face of changing internal or external conditions.

For example, staying with the prior scenario in which an uninhabited territory was opened up to human settlement, sustainable community characteristics form through four stages. People have left their original communities behind and entered the territory as individuals, couples, or small families. At first, the only significant relationships they have are with those they arrived with (partner, spouse, family, etc.). The new relationships that people make over the next few months are very rudimentary—between people of any age or group, and based on little mutual understanding other than the need to complete a transaction, provide a repair or service, or resolve a conflict.

As people begin to settle into locations, occupations, and lifestyles within the territory, many new relations are initiated between individuals. These are often associated with their housing location (neighbors), their work (coworkers), or mutual acquaintances. Even though these new relations are being sparked by the multitudes, they still have a bit of a random quality—they lack much clear grouping or structure by common interest, life path, or long-term bond.

Five or ten years after initial settlement, definite patterns have developed. Some people have met and married; others have become close friends or formed close-knit groups. Informal activity groups have become common—the boys that like to hunt and fish together, the adults that play cards on weekends, religious devotees that read together, etc.

After another 30 years has elapsed, the final stage of community development is present. This stage has much more nuanced community structure than could have existed right after settlement. Not only do individual people have kin and circles of friends or comrades with which they pursue their interests, but there are also patterns of formation across the territory. Multi-generational families have begun to form, and especially in the countryside, these families often spend a great deal of time together. Different parts of the territory each have their own "leader families" that tend to plan social activities. The social groupings that have formed have done so in conjunction with a diversity of population densities and developing customs across the territory, which is reflected in things such as the types of social groupings that children create. As the decades go by, the community structure often undergoes gradual change in accordance with a variety of other factors such as the economy, technologies, and culture.

Occupational Translation of Pattern 13 through 16: The first stage of a newly forming occupational force of business sectors and professions that can eventually become sustainable is often marked by skill generalization and relatively limited coordination or synchronicity between the members of that force and the economy. As that occupational force continues forming, it enters a second stage with an increasing diversity of skill sets, and increasing competition within the occupational force, but there is still poor coordination within the force as a whole. The third stage in the development of the occupational force is marked by decreasing competition between its members and increasing coordination (niche filling). The fourth and final stage in the development of a sustainable occupational force is often marked by little competition, but significant coordination, alignment, and niche efficiency among the force members. This final stage is not assumed to be static, but rather will have occupations that ebb, wane, adjust, and improve in the face of changing internal or external conditions.

For example, staying again with the uninhabited territory that is opened up to human settlement, sustainable occupational characteristics will form through four stages. The first arrivals to the territory tend to be people without highly advanced skills. Out of necessity, even those that have an occupation or skill must be a jack-of-all-trades, and work on whatever needs doing. This doesn't mean that they are financially unsuccessful, just that there is much more activity occurring in bartering, logging, fishing, and food delivery than in music, urban planning, or medicine.

Over the next decade or so, many more individuals enter the territory, bringing their own businesses and skill sets—architecture, plumbing, teaching, legal, and administrative backgrounds, along with people arriving to sell furniture, rugs, and the latest gadgets. Many of these new arrivals have a speculative quality, hoping that the situation will be good from their business angle. This is a period of good fortune for some, but a growing percentage face disappointment relative to their expectations. In any case, it is a period of rapid flux in the occupational sector.

By the time twenty or thirty years have elapsed, the occupational field appears more clarified and organized. There tend to be professionals of various types spread through the territory. For example, one no longer needs to travel to the other side of the territory to find a dentist. Also, most businesses that originally competed head to head no longer do; some of them simply went bankrupt, but in many cases they were able to adjust their goods or practices to serve different needs, eliminating their areas of redundancy that led to extreme competition.

Within fifty years, the occupational force is generally stable; the number of people in each profession does not tend to change dramatically from year to year, and new businesses do occur, but they rarely do much to shake up the scene. Competition between professionals and businesses still exists, but the pressure of that competition tends to be channeled toward specialization and self-improvement rather than toward undercutting or overtopping.

Governmental Translation of Pattern 13 through 16: A newly forming governmental structure that can eventually become sustainable is often marked by very little development of rules, agreements, and behavioral expectations—or codes. As that governmental structure continues forming, it enters a second stage marked by increasing numbers of codes. However, they are often in conflict with each other, are not highly refined, or are somewhat misaligned with the real situations they are intended to pertain to. The third stage in the development of governmental structure is marked by an abandonment or significant modification of some codes, and a strengthening of others that form a coherent group or that are especially germane to existing circumstances. The fourth and final stage in the development of a sustainable governmental structure is marked by significant coordination, alignment, and efficiency between codes and the circumstances that they target. This final stage is not static, but rather allows continual review or refinement of the rules, agreements, and behavioral expectations to remain synchronized with internal and external conditions.

Examples of the four stages of sustainable governmental development are illustrated through the territory opening up to human settlement. As settlement begins, people arrive singly or in small groups. These individuals and small groups have many of their own behavioral codes, but many individuals behave in ways that provide the highest level of personal benefit. This is a somewhat chaotic period in which many people believe that others are committing crimes or trespasses.

Once several years have elapsed, there is a nascent formation of rules or behavioral expectations throughout the territory. This, for example, takes the form of general rules and laws that bar the most disruptive or self-centered behaviors, such as burglary or arson. These rules are somewhat local throughout the territory; that is, they apply to one enclave but not to others. Often these rules are based more on what has already gone wrong than on what should go right, and thus the rules in any area may leave significant room for more to go wrong. Also, many of these initial rules fail to address the implications of their own existence, such that they sometimes cause more disruption than they prevent.

Through the passage of a few more decades, a more organized overall set of rules and expectations takes shape. For one thing, rules in each enclave have become slightly more aligned with adjacent enclaves, so that they do not disrupt conditions nearby. Also, there is some formation of overall territorial laws, and local rules also align somewhat with these overall laws. Furthermore, the laws and rules themselves have come to a higher point of balance by closing loopholes while adjusting for their own disruptive potential.

Within a century after settlement, the territorial laws and local behavioral expectations are relatively stable. Details of laws are changed sometimes, but the central theme of local rules or territorial laws rarely change. The laws and rules are intuitively and carefully intertwined (coordinated) with contemporary conditions, including the economy, community, culture, etc. This intertwining means that for the most part, laws and rules are consistent with conditions in the economy, community, and culture, and people readily choose to follow laws and expectations rather than violate them.

Environmental Translation of Pattern 13 through 16: A newly forming environment (land use system) that can eventually become sustainable is often marked by a relative lack of interaction within and between land use areas. As that environment continues forming, it enters a second stage marked by increased diversity of primordial land uses in each area, and increased tension over appropriate land use per each area. In this second stage, there is interaction between different land use areas, but this interaction is still rudimentary. The third stage in the development of an environment includes a favoring of certain land use patterns and an abandonment (at least locally) of others. Coordination of processes and patterns within land use areas and between land use areas greatly increases during this third stage. The fourth and final stage of sustainable environment development includes rather stable, repeating or patterned land uses that are coordinated, efficient, and compatible internally and with neighboring land uses.

Examples of this environment development pattern are evident in the territory that is opened up to human settlement. The land uses in the first few years are somewhat simple, not requiring complicated coordination, and they are also distributed somewhat haphazardly on the landscape. Many people clear the most convenient or accessible patches of land for simple agricultural activities such as bean or corn production, or raising steer, dairy cows, goats, or pigs. Homes and roads are built, though their designs tend to be simple. Land uses at this point do not appear to be very coherent—there are hog farms right beside town centers, and roads that travel through floodplains and are washed out frequently.

Stage two unfolds over the next twenty years or so as many new people enter the territory. The new arrivals bring new ideas and consider them relative to what others are already trying. This gives a slightly more complex quality to new land uses. Some settlers try innovative ideas such as pasturing goats in fruit orchards. Others try crop rotations, or experimental crop types. In some cases though, land uses begin to clash or compete. Some expanding towns are in conflict with existing livestock operations. Some of the intensive agricultural operations cause water-quality problems for nearby homes or towns. In other cases, businesses compete to acquire the same parcel of land to support their expansion.

The third stage of environment formation occurs over the next 75 years. As this stage unfolds, patterns in land use begin to clarify. For example, large cattle operations have shifted to the south end of the territory where the climate is drier, and several agricultural crops that were prevalent early in settlement have virtually disappeared from the landscape due to their mediocre performance in local soils. Certain crop rotations, discovered by trial and error, have become popular. Towns have settled most of their expansion and public works conflicts with livestock operators, and water-quality improvement systems have been created downslope of some of the larger farms. Most of the roads that were hastily built in the first few years of settlement have been improved, although some have simply been abandoned.

The final stage of environment development includes quite stable land uses that are each nuanced according to their specific context and earth location. Farms are operated with a careful eye on soil, climate, and pest conditions relative to crop type, timing, and land management. Farmers are careful to make choices that will not decrease their future livelihood or leverage. In other examples of this coordination, homeowners have landscaping that is synchronized with their particular lot. As a point of efficiency and practicality, livestock operations that still occur near towns generally supply the meat for that town. Finally, water availability and quality is sufficient throughout the territory for the land uses that have taken hold.

Cultural Translation of Pattern 13 through 16: A newly forming culture that can eventually become sustainable may at first be weakly developed, without many belief and behavior patterns. In this first stage, belief and behavior patterns that do exist may be unsynchronized with each other and with the local economy, community, environment, etc. As a culture continues forming, it enters a second stage marked by an increasing number of beliefs and practices. These beliefs and practices may occur in pockets, interspersed with—and somewhat incompatible with—other beliefs and practices. The overall form of the culture may not be very clear in the second stage. The third stage in culture development is marked by increasing coherence. Some beliefs and behaviors are abandoned, while others are strengthened, such that most people's beliefs and practices move closer together. The emerging culture will likely be somewhat aligned with a variety of contemporary factors (economic, environmental, etc.). The fourth and final stage in the development of a sustainable culture is marked by significant coordination and pattern. That is, the culture between individuals and between regions is compatible, and the culture is generally synchronized with other factors such as the economy, the government, etc. This final stage of a sustainable culture is not static; rather it will shift to continually improve its alignment in the face of changing internal or external conditions.

Considering the uninhabited territory opens up to human settlement, a sustainable culture will form through four stages. People leave their home communities behind and enter the territory as individuals or in small groups. In leaving their people, their homeland, and many possessions, they carry a reduced or simplified version of their culture into the new territory. In addition, people have difficulty carrying out cultural activities while moving. Finally, the limited social organization present as the first settlers arrive creates a somewhat unsorted mix of different cultural fragments.

Within a few years, people have been able to return to a more comfortable rhythm and recover more of their original cultural patterns. However, these previous patterns are now alongside new habits and beliefs that are developing in order to cope with new life circumstances. Furthermore, each person is affected by the cultural practices and beliefs of their new neighbors. These three cultural influences (old culture, new culture, and neighbors' culture) are not fully compatible. At this point, there is still not a strong overall pattern to the culture of the territory.

The third stage occurs over the next thirty or forty years. What had been a large number of inconsistent cultural patterns is now shifting to take on a clearer form. Some older cultural patterns have been dropped, but others have been strengthened. Completely new elements of culture native only to the new territory are also evident. These new elements seem to fit with the new economic and social structures that that have emerged as the territory has been settled. The larger-scale pattern of culture within the territory has also been influenced by new arrivals, who have been attracted to regions of the territory where their own culture is practiced, resulting essentially in a number of different enclaves.

In the final stage of cultural development, cultures are relatively stable—that is, not changing tremendously within the lifetime of any individual. Changes that do occur are often evidently a response to shifts in the economy or environment. Across the territory, culture does vary from community to community, but the variance is such that there isn't significant conflict between neighboring cultures.

Physiological Translation of Pattern 13 through 16: Individual people can develop sustainable skills and interests through four stages. The first stage is exposure to basic aspects of a wide number of mentally and/or physically based subjects (i.e., potential interests). In the second stage, exposure to these subjects continues in greater detail, and at the same time, additional subjects are introduced. At this point, more subject matter has been introduced than can be pursued in a lifetime. The third stage is a focusing or pursuit of the subset of interests that are most attractive to that individual, and a simultaneous dismissal of other interests. The fourth and final stage sees the individual practicing and developing their selected skills and interests. While the individual may not reach a level of perfection with them, they can continue to improve their skills and abilities from their own level, and can continually adjust their focus within an overall topic as they learn more, and as conditions around them change.

For example, in the scenario of an uninhabited territory opened up to human settlement, the selected interests of individual people sustainably form through four stages. These stages occur in the context of the existing culture, laws, economy, etc. In this territory, most children do attend school from about age five to fourteen. During this time, they are taught reading, math, basic science, and history, along with a variety of sports.

The work in these subjects expands each year, and by the end, includes an introduction to other topics such as foreign languages, story writing, economics, etc. Meanwhile, their local town or neighborhood life includes all sorts of other pursuits, including family activities, religious ceremonies, raising food, and interactions with local crafts and trades. By about age fourteen, each of the children has been introduced to far more topics than they could pursue, and for most of them, this marks the approximate end of Stage 2 and the beginning of Stage 3.

At this point each child begins to focus based on their aptitudes, interests, and other economic or cultural considerations. Over the next ten years or so, most embark on their own path. Stage 3 for these individuals can include direct local apprenticeships, trade schools, business pursuits, religious training, musical practice, family or village caretaking, starting their own families, or advanced schooling at a university.

The fourth stage occurs through the remainder of each individual's lifetime, in which each acquires more skill and insight as their experience grows. Oftentimes, they are able to specialize more and more in a particular aspect of a field. This does not mean that they spend all of their time in one single pursuit, but they do have a limited number of strong interests and skills.

Overall, Patterns 13 through 16 suggest that competition and cooperation both play important roles in four stages of sustainable complex system development. Within a complex system that is undergoing initial development, there is little competition or cooperation. At this early stage, organized interaction is weak overall. Soon, the system shifts into a mode that includes high levels of competition between the system members. As the competition produces winning and losing outcome(s), a more organized system begins to take form, and cooperation begins to become more important within that system. In the final stage of development, the complex system displays significant cooperation and patterning, as it continually shifts in a direction of fine-tuning. Next, we look at patterns under the theme of *time*.

Chapter 11 Time

In the previous chapter, we looked at natural system succession relative to competition and cooperation. Now we can look at succession relative to *time*.

Even though the four stages of succession might be equally important from a developmental perspective, they don't usually exist for the same lengths of time. In nature, there is relatively less time spent early in succession, and a relatively more time late in succession. To see this, we can look back at our example of a cleared field that eventually succeeds to a forest.

In Stage 1, new plants sprung forth in a sort of free-lunch stage with little competition, cooperation, or organization. A typical length of time for Stage 1 from beginning to end might be about three years. In Stage 2, increasing populations create a power-struggle stage, where competition is severe and cooperation is minimal. This stage might go from approximate start to finish in about fifteen years. Stage 3 is the decision stage, in which the best adapted vegetation gathers momentum. During Stage 3, competition slowly declines relative to cooperation. This stage might go from approximate start to finish in fifty years. In Stage 4, there is continual refinement and adaptation of an established system in a harmony stage. Stage 4 will continue until a major disruption occurs. The amount of time before a major disruption occurs can vary greatly, but perhaps 1,500 years is a reasonable, typical value under natural conditions.

If we convert these times into a ratio between Stages 1 through 4, we get about 1 to 5 to 50 to 500. In other words, in this example about 0.2% of the time in Stage 1; 1% of the time is spent in Stage 2; about 9% of the time is spent in Stage 3; and 90% of the time is spent in Stage 4.

We could work up a similar example with a cohort of young trees, in which Stage 1 is the open germination period; Stage 2 is the intense competition and die-off period; Stage 3 is the spacing and waiting stage, and Stage 4 is the canopy accession stage and mature stage. If the cohort of trees in our example were sugar maples, Stages 1 through 4 might run for about 2, 25, 60, and 300 years, respectively. If we convert these times into a ratio for Stages 1 through 4, we get about 1 to 13 to 30 to 150 (0.5%, 6%, 16%, and 78% of the time).

We could perform a similar calculation regarding an individual tree. We had an example where a tree sent out numerous young shoots; then it experienced competition for light between the shoots; next it chose which shoots and branches to favor as it developed its overall form; and finally, it maintained and refined its form according the shifts in conditions around the tree. If we consider a white pine through these stages, it might spend about 15 years in Stage 1, 30 years in Stage 2, 60 years in Stage 3, and 150 years in Stage 4 (6%, 12%, 24%, and 59% of the time).

Notice that as we went down in hierarchical level from a natural community to an individual tree, our ratios became less skewed toward the latter stages of succession.

> Pattern 17: Ecosystems, natural communities, species classes, species, individuals, and offspring each spend unequal amounts of time moving through four successional or developmental stages. The least time is spent in Stage 1; and progressively, more time is spent in each successive stage. Time spent in latter stages is most heavily weighted at higher hierarchical levels.

Essential Translation: Sustainable complex systems develop through a four-stage successional pathway in which progressively more time is spent in each stage. The higher the system level, the more skewed the system will be toward latter stages.

As with Patterns 13 through 16, the backdrop for the next seven examples will be the new territory that opens up to human settlement after being uninhabitable due to military hostilities.

Economic Translation: A newly forming sustainable or potentially sustainable economy will proceed through four successional stages. The shortest stage is the first one—a stage characterized by limited interaction. The next shortest stage is the second one—a stage characterized by increasing competition. A still longer stage is the third one—a stage characterized by decreasing competition and increasing coordination or efficiency. The longest stage is the fourth, which is characterized by refinement of efficiency in a relatively low-competition atmosphere.

For example, within the first few months after settlement has started in the new territory, Stage 1 of economic activities begins. This somewhat haphazard period coincides closely with the rapid settlement period, and lasts about 6 years. Stage 2 is less chaotic, but on the whole, more competitive. Stage 2 might last from about the 6th year until the 25th year. Stage 3, marked by loss of some business ventures and increased specialization and efficiency of those that remain, lasts from about the 25th year through the 90th year. The fourth and final stage is characterized by relative stability and continual improvements in alignment and efficiency. This stage might begin in about year 90. The endpoint of this stage could vary, and would probably coincide with a major social, political, environmental, or other shift or disruption within the territory. One of many potential approximate endpoints of Stage 4 could be at 1000 years.

Community Translation: A newly forming sustainable or potentially sustainable community will proceed through four successional stages. The shortest stage is the first one—a stage characterized by limited interaction or fidelity of interaction. The next shortest stage is the second one—a stage characterized by increasing relationships that lack a clear overall pattern. A still longer stage is the third one—a stage characterized by increased coordination and pattern development (predictability). The longest stage is the fourth, which is characterized by continual adjustment of relatively coordinated and pattern-forming relationships.

For example, as settlement in the new territory has begins, so does Stage 1 of community development. This period in which new relationships are rather utilitarian lasts about 1 year, though this stage lingers longer wherever settlement is ongoing. In Stage 2, many new relationships grow, yet there is still an element of uncertainty over which relationship will strengthen and which will weaken. Stage 2 lasts from about the 1st year until the 5th year. Stage 3, marked by strengthening of certain bonds and formation of patterns, lasts from about the 5th year through the 25th year. The fourth and final stage is characterized by relative relational stability and patterning, and continual gradual adjustments. This stage begins in about year 25. Individuals experience the end of Stage 4 at the end of their lifetime, but the greater community endpoint could vary, and would probably coincide with a major social, political, occupational, or other shift or disruption within the territory. One of many potential approximate endpoints of Stage 4 could be at 1000 years.

Occupational Translation: A newly forming sustainable or potentially sustainable occupational force will proceed through four successional stages. The shortest stage is the first one—a stage characterized by skill generalization. The next shortest stage is the second one—a stage characterized by increasing specialization and competition. A still longer stage is the third one—a stage characterized by decreasing competition and increasing niche filling and efficiency (non-overlap, etc.) between occupational groups. The longest stage is the fourth, which is characterized by continual refinement of niches and specialization in a relatively low-competition atmosphere.

For example, Stage 1 of occupational development begins in tandem with the first settlers' arrival in the new territory. This period, in which most people act as generalists from an occupational perspective, lasts about 3 years. Stage 2 sees the influx of many skilled tradesmen. Stage 2 lasts from about the 3rd year until the 15th year. Stage 3 is marked by a disappearance or modification of some of the occupations that were present in Stage 2, though the occupation types (including business types) that remain are becoming more specialized. Stage 3 lasts from about the 15th year through the 50th year. The fourth and final stage is characterized by relative stability and continual improvements in alignment and efficiency. This stage begins in about year 50. The endpoint of this stage could vary, and would probably coincide with a major social, political, environmental, or other shift or disruption within the territory. One of many potential approximate endpoints of Stage 4 could be at 1000 years.

Government Translation: A newly forming sustainable or potentially sustainable governmental structure will proceed through four successional stages. The shortest stage is the first one—a stage characterized by a general lack of agreed behavioral codes. The next shortest stage is the second one—a stage characterized by increasing behavioral codes that are not tightly aligned with real-world conditions. A still longer stage is the third one—a stage characterized by a honing and coordination of behavioral codes. The longest stage is the fourth, which is characterized by continual refinement of behavioral expectations in order to align them as closely as possible with their intended purpose.

For example, as the first settlers arrive in the new territory, Stage 1 of government begins. This period of minimally agreed-upon expectations lasts about 5 years. Stage 2 witnesses the rapid but relatively uncoordinated growth of various rules. Stage 2 lasts from about the 6th year until the 25th year. Stage 3 is marked by somewhat improved alignment between behavioral codes and practical conditions. Stage 3 lasts from about the 25th year through the 75th year. The fourth and final stage is characterized by the relative stability of behavioral codes and expectations, though still with continual refinements in their form and execution. This stage might begin in about year 75. The endpoint of this stage could vary, and would probably coincide with a major social, political, environmental, or other shift or disruption within the territory. One of many potential approximate endpoints of Stage 4 could be at 1000 years.

Environmental Translation: A newly forming sustainable or potentially sustainable environment will proceed through four successional stages. The shortest stage is the first one—a stage characterized by relatively simplified land-use units that lack great interaction. The next shortest stage is the second one—a stage characterized by increasing land-use diversity. A still longer stage is the third one—a stage characterized by increasing coordination or juxtaposition of land uses, according to the inputs and outputs of each land-use type. The longest stage is the fourth, which is characterized by a coordinated set of efficient land uses that undergo continual adjustments in their internal and external alignments.

For example, Stage 1 of environmental development begins in tandem with settlement. This somewhat disorganized period of relatively simple land uses lasts about 10 years. Stage 2 witnesses an increased diversity of land-use types, but still within an overall atmosphere of disorganization. Stage 2 lasts from about the 10th year until the 30th year. Stage 3, marked by loss of some land uses, but increased coordination of those that remain, lasts from about the 30th year through the 100th year. The fourth and final stage is characterized by relative land-use stability and alignment or coordination within and between land uses. This stage begins in about year 100. The endpoint of this stage could vary, and would probably coincide with a major social, political, environmental, or other shift or disruption within the territory. One of many potential approximate endpoints of Stage 4 could be at 1000 years.

Cultural Translation: A newly forming sustainable or potentially sustainable culture will proceed through four successional stages. The shortest stage is the first one—a stage characterized by limited cultural development and limited alignment between the culture and other complex systems. The next shortest stage is the second one—a stage characterized by increasing cultural patterns, though these may exist as pockets rather than as a coherent overall culture. A still longer stage is the third one—a stage characterized by increased unification within the culture and increased cultural alignment with outer conditions. The longest stage is the fourth, which is characterized by refinement of structured culture as it aligns continually with internal and external conditions.

For example, Stage 1 of culture development would occur as settlers arrive in the new territory. This period of weak and disjunct culture lasts about 3 years. In Stage 2, numerous identifiable cultural elements will exist, yet they still won't form a coherent overall culture. Stage 2 lasts from about the 3rd year until the 15th year. Stage 3, marked by a strengthening of some cultural patterns and a weakening or loss of others, lasts from about the 15th year through the 50th year. The fourth and final stage is characterized by a relative cultural stability and patterning such that there isn't significant internal or external conflict being caused by the culture. This stage begins in about year 50. The endpoint of this stage could vary, and would probably coincide with a major social, political, environmental, or other shift or disruption within the territory. One of many potential approximate endpoints of Stage 4 could be at 1000 years.

Physiological Translation: Individuals can develop sustainable skills and interests through four successional stages. The shortest stage is the first one—a stage characterized by introduction to several basic subjects. The next shortest stage is the second one—a stage characterized by an increasing number of subjects and level of detail. A still longer stage is the third one—which is characterized by a focusing toward a smaller set of interests. The longest stage is the fourth, which is characterized by continual practice and skill refinement.

For example, Stage 1 of physiological development for children in the territory begins at about age 3, when the toddler years are done, and ends at approximately age 8. This period includes a basic introduction to physical and cognitive topics, such as swimming and reading. In Stage 2, the subject matter that children are taught begins to expand and deepen. Stage 2 lasts until about age 14, when most public schools in the territory terminate. Stage 3, longer yet, is a period when children branch off to pursue a few more focused interests. The last and longest is Stage 4, which runs from about age 26 until death. In Stage 4, each individual practices their skill(s), which includes a continual attempt toward mastery.

Pattern 17 describes the extended time spent in Stage 3, and especially Stage 4 of succession. This means that individuals or groups that are sustainable will typically experience Stage 4 as the dominant, or longest-lasting stage of their life (or lives). Recall that Stage 4 is a sort of balancing or harmony stage. This doesn't mean that Stage 4 is better or more perfect than other stages, but it is the stage with the most physical and energetic coordination, refinement, balance, and continuity over time. Another way of expressing this is that per unit time, Stage 4 witnesses the least amount of dramatic change.

Presumably, Stage 4 in each complex system does come to an end. For example, the end of Stage 4 for an individual is death, and for a species it is its extinction or evolution/transformation into a new species. We can define a system's *lifecycle* as the total time it takes for the system to go from the start of Stage 1 until the end of Stage 4. The next topic pertains to these lifecycles.

In nature, the *lifecycle*, or time it takes for a system to start at Stage 1 and move to the end of Stage 4, varies greatly. An annual plant species can live just a few months before it sets seeds and dies. A bird of prey can live a few decades. A forest stand might survive several hundred years, and a forest ecosystem thousands and thousand of years. It seems that in nature, there is a prevailing pattern in which lifecycles are progressively longer from the offspring level to the biosphere level. The word *prevailing* is used to indicate that while lifecycles are not *always* longer at each hierarchical level, they usually are (or at least they aren't shorter very often). Let's go through a hypothetical example of this in a natural setting by looking at lifecycles starting with the lowest hierarchical level (offspring) and ending with the highest (biosphere). After that the importance of this pattern will be discussed.

Imagine a large forested swamp—maybe a 3,000-acre patch of ash, maple, and oak, or maybe the dominant trees are spruce and birch. This forest is in a temperate climate with many large animals such as deer, elk, wolves, bears, etc. A progressive lifecycle pattern can be demonstrated starting with the deer that live in this forest:

Beginning at the lowest hierarchical level with a fawn (offspring), fetal development through birth and adolescence will take about 2 years. The parents (buck and doe individuals) of that fawn might live an average of 4 or 5 years. The end of the lifecycle of the group or herd of deer (species) inhabiting that swamp will be defined by a dramatic change rather than a complete die-off. Perhaps the lifecycle for deer in this swamp averages about 60 years, when an infrequent harsh, snowy winter leads to starvation of over half the herd. In this case, the deer, elk, and moose of this swamp form the grazer and browser species class. The lifecycle for this group might be longer still, maybe 250 years—marked by a dramatic and simultaneous reduction in deer, elk, and moose after a dry summer, a harsh winter, and a peak in the wolf population.

The 3,000-acre swamp forest might have a lifecycle of about 700 years, which could be the frequency of stand-replacing fires or windthrow events. The forest will grow back after these events, but they still mark a temporary endpoint of a Stage 4 condition. Longer still might be the lifecycle of the forest cluster ecosystem within which the swamp forest is nested. Maybe the greater ecosystem is spread across 100,000 acres, and has a lifecycle of about 4,000 years, at which time a dry period allows grassland vegetation to invade and displace most of the forest.

Finally, the regional landscape (here interchangeable with the biosphere level) might have a lifecycle of about 20,000 years, which is a potential amount of time between ice ages. From offspring to biosphere, we've seen lifecycles rise progressively from just 2 years to 20,000 years.

This example focused upon a single site. But would a similar pattern emerge if we try to look at lifecycle patterns from an evolutionary perspective by looking at types, rather than a particular site? For example, if we look at the lifecycle of a species as a whole through its entire geological existence, do we see a shorter lifecycle than that of the species class to which it belongs?

It appears that the lifecycle analysis becomes more difficult at an evolutionary level, because the fossil records are probably insufficient to determine most lifecycles, and because the way we categorize biological groupings that existed before recorded history becomes confusing or subjective (see Chapter 6). However, here's one attempt to accomplish this abstract-type of analysis beginning at the species level with white-tailed deer, which is the most common ungulate in eastern North America:

Apparently there is fossil evidence for the existence of white-tailed deer, as a single species, going back roughly 4 million years. If we are still thinking of our species class as deer, elk, and moose, then its lifecycle is a bit subjective. Technically this species class has only existed in North America for about the last 120,000 years, when moose migrated from Asia or Europe. Though moose residency is shorter, certain elk species have been in North America for over 20 million years. Furthermore, the deer family group (which includes deer, elk, and moose) has been on Earth for maybe 30 million years. For choosing the lifecycle of our species class, this puts us somewhere in the range of 120,000 years to 30 million years, depending upon how we want to define that species class.

The next level would be the swamp forest natural community type. It will be impossible to determine how long a semblance of this type has been present on Earth or in North America, but one common forest type, the "beech-maple" type, is thought to have been in North America for perhaps 5 million years, so maybe we can make a large assumption that our swamp forest has had the capability of existing for about the same duration.

For the ecosystem-level lifecycle, we might be able to use the *temperate deciduous forest* as our concept type. Temperate deciduous forest types are thought to have spread across North America 30 million to 60 million years ago, so these can represent the ecosystem lifecycle range.

Finally, what is the geological age of our biosphere? One way to look at biosphere lifecycles is that they end with mass extinction events. The last mass extinction event was about 65 million years ago, when dinosaurs died out. Following this, mammals increased, and many modern plant groups developed. Thus, our evolutionary-level lifecycle analysis provides results from the *species level* to the *biosphere level* of roughly: 4 million years; 120,000 to 30 million years; 5 million years; 30 to 60 million years; and 65 million years.

Even though this evolutionary group-level analysis is difficult to do (or difficult to do well), it still yields an approximate result of increasing lifecycles at higher hierarchical levels. Notice that we can't include offspring and individual levels in this comparison, since they represent single animals, and in this case, we are conducting a group comparison.

Why would these lifecycle patterns matter? It means that under natural conditions, individuals and their groups typically experience a degree of contextual stability or predictability. So, a baby born in a herd normally comes of age while certain adults, such as its parents, are alive and present. The individual then lives out its entire lifecycle while its species does not undergo dramatic changes in conditions—at least none that are significant enough to jar the population. This happens while its species survives in a virtually unchanged species class. The species class then lives in a highly predicable natural community setting, and the natural community is situated in a rather stable ecosystem context. Finally, all of these are happening while set in a stabilized biosphere.

This contextual stability would presumably allow each lower level in the hierarchy to reach its stability phase (Stage 4), more rapidly. Perhaps this helps explain why Pattern 17 exists; by reaching Stage 4 more rapidly, individuals and groups are left with more time to spend in that stage. It also means that individuals and groups are in a context or setting where organizational complexity and refinement are high, and cooperative patterns are maximized.

> Pattern 18: Offspring, individual, species, species class, natural community, ecosystem, and biosphere lifecycles are generally shortest at the lowest hierarchical level (offspring), and become progressively longer up to the biosphere level. Because Stage 4 of development at each level is usually longer than the other three stages combined, this pattern usually enables the biota of each level to experience their entire existence and reach their own Stage 4 while living within a Stage 4 context, meaning that they experience a context of relative stability, refinement, organizational complexity, and cooperation.

Essential Translation: Sustainable complex systems go through their entire lifecycles, including their own Stage 4 level of development, in a relatively stable, refined, organizationally complex, and cooperative context.

This pattern can now be translated into the context of human systems:

Economic Translation: Sustainable economies go through their entire lifecycles, including their own Stage 4 level of development, in a relatively stable, refined, organizationally aligned, and low-competition context.

For example, Frank and Jan grew up within a community that had been founded about 300 years prior. Their local economy had long been stabilized, and had relatively refined organization. Frank and Jan married in their early twenties, and had three children by their late twenties. The next six to eight years brought them numerous financial transitions and strains, including the cost of raising young children, health care, a career change, a loan, house payments, two job changes, a promotion, and a small inheritance. This was a period of rapid financial changes for them, but by their late thirties, they had entered a smoother Stage 4-like period, which essentially lasted the rest of their lives. Their ability to reach economic stability (i.e. a degree of harmony) was aided by the fact that their local economy was fairly stable during their lifetimes. The stable local economy helped them to plan their future, and to make strategic choices whose risks and rewards were reasonably predictable. It also provided a context that was somewhat assistive should they slip in trying to establish themselves within that economy.

Community Translation: Sustainable communities go through their entire lifecycles, including their own Stage 4 level of development, in the context of a relatively stable, coordinated, and organizationally patterned greater community.

For example, a single ethnic group has inhabited a region for several thousand years. One of the towns within that region has been in Stage 4 with regard to community development for about 500 years. When a new cluster of homes was built at the edge of that town, it only took about 12 years for the residents within the new cluster to reach a more stable, organized level of community development (i.e., Stage 4). The context of a long-standing community helped that subcommunity to stabilize as rapidly as it did.

Occupational Translation: Sustainable occupational groups go through their lifecycles, including their own Stage 4 level of development, in the context of a relatively stable, organizationally specialized, and efficient greater occupational setting.

For example, a 200-year old publishing company that has been stable for many decades creates fourteen new job openings to address changes in business practices associated with an advance in technology. Within six months of creating the positions, there are numerous adjustments that take place as the new employees negotiate their new situations—the company moves the new staff into a sales division, adds two more employees to this group, and moves three of the employees to a different office. After about three years, the new positions undergo relatively few changes in their duties, expectations, locations, etc. The positions within the company stabilized as quickly as they did with the help of the company's preexisting, stable occupational structure.

Government Translation: Sustainable governments and behavioral expectations go through their lifecycles, including their own Stage 4 level of development, in the context of a relatively stable, consistent, and coordinated greater government and set of behavioral expectations.

In a hypothetical example, international laws have been in place for many decades. Most of these laws apply to national behaviors rather than to a nation's internal affairs, and they are rather strictly enforced. When an ongoing conflict between two religious groups inside one nation leads to a negotiated division of land and authority, a new nation is born. The new nation bases its external law upon international law, and its internal law upon centuries of religious practice. Within just twenty-five years, the government within the new nation has stabilized (i.e., entered a Stage 4 period). This stabilization was able to occur as quickly as it did because of the anchoring effects of international law and religious edicts and customs.

Environmental Translation: Sustainable environments go through their entire lifecycles, including their own Stage 4 level of development, in the context of environments and land uses that are relatively stable, coordinated, and efficient.

As part of a conservation plan, homesteaders are resettled from hillsides into a fertile valley. The valley was historically occupied, with a history of use for pasture, hay, and orchards, but its inhabitants had left in pursuit of urban economic opportunities. As the homesteaders arrive onto land that had become unoccupied, they are provided with farmsteads that need repair, but that had been worked for centuries. Their uncertainty of how to operate in this new landscape only lasts a few years. The long-established preexisting land uses and infrastructure in the valley enable the resettled families to stabilize their own land-use patterns and activities with relative ease.

Cultural Translation: Sustainable cultures go through their entire lifecycles, including their own Stage 4 level of development, in the context of a relatively stable, coordinated, and organizationally patterned greater culture.

For example, Marie was born in a large coastal fishing village that had been in place for about 600 years. The culture of the village had formed around the sea—its steady seasonal rhythm, the village's reliance upon it, and their vulnerability of villagers to its changes. Growing up, Marie felt like a black sheep in this setting—she was artistic and creative, and she felt like the village culture was stubborn and predictable. As a teen, her father died in a tragic accident, setting off several years of psychological difficulty for her. During these difficult years, she came close to engaging in self-destructive behaviors and she nearly left home; but by her early twenties, she was stabilizing. She eventually came to engage with her native culture by creating elaborate funeral arrangements for deceased villagers. This allowed her to express her personality in the context of the culture she had once considered stubborn and dull.

The fact that her native culture was stabilized during her entire life essentially provided her with the support she needed during her years of difficulty, and allowed her to reach personal stability and contentment much more rapidly than she otherwise might have.

Physiological Translation: Groups and individuals sustainably go through their lives, including their own Stage 4 levels of development, in an atmosphere in which many skills, trades, and hobbies are well developed and available to newly interested people.

For example, as Chet grew up, he was introduced to many possible areas of interest. His interests were diverse, but he especially loved machinery and design, and he seemed to be "tuned in" to whether or not protocols were being followed. His skills and interests came together well in the field of automotive testing. The fact that he lived in a society that had technological development over hundreds of years, and had many well-established skilled trades associated with those technologies, gave Chet a much better opportunity to train into a career that challenged him and matched his unique interests.

The next pattern also includes the element of time, and how long it can take for a balance to be reached. To discuss this, we will look at a concept often important in physics and chemistry—*equilibrium*. Equilibrium is a condition, or state of being, in which stability is reached—that is, when some type of motion comes to a stop or net changes cease to occur. We all intuitively understand what equilibrium is in our everyday lives. The mug full of hot tea eventually cools down to the temperature of the room it's in, reaching a sort of thermal equilibrium. When riding your bike on the sidewalk as a child, maybe you've stopped pedaling to see how far you would coast. When you coast for a while but then can't go any further, you come to a stop, which is a point of energetic equilibrium.

All around us, the world is full of similar equilibrium processes, in which there is transition toward a point of stability or balance. If someone walks out to their lawn on a dry day and pours a gallon of water on the grass, they have just created an unstable condition that will immediately begin moving toward equilibrium.

Even though they essentially poured the water into a "pile" or puddle, the pile is unstable because gravity and chemical attraction immediately pull on the water. These forces suck the water down into the soil, and eventually into the grass. As that water moves into the soil and then into the grass, it is seeking a form of equilibrium. Nature is full of these equilibrium processes, in which there is movement to right any imbalances.

As an example, imagine a small lake ringed by wetland vegetation. Maybe there have been three years of fairly dry weather where this lake occurs. With less water coming in, the lake level begins to be out of equilibrium with the existing conditions, and it might drop by an elevation of two feet.

We could assume equilibrium had been reached and stop the example there, but it would be unrealistic to do so. Because natural systems are complex, it is rare for only one equilibrium change to occur. For example, with a lowered lake, the wetland vegetation along the lake edge would then be out of equilibrium, and would begin following the lakeshore, growing downward (lakeward) of where it did previously. That vegetation shift would in turn create another disequilibrium—the upland vegetation growing upslope of the wetland vegetation would then be well adapted to conditions at its lower edge, where it meets the wetland vegetation. The upland vegetation would then begin to migrate downward toward a new equilibrium, outcompeting and replacing some of the wetland vegetation that is left on dry ground.

For another example, imagine a spruce and fir forest along a mountainside. This forest sits in an avalanche pathway, and one day, the snow mantle, seeking equilibrium, avalanches and wipes out the forest in its path. While the avalanche clears the vegetation, it leaves behind much of the soil and space for plant growth. So vegetation—attempting to establish equilibrium—will begin to fill the mountainside again. The initial re-vegetation growth might include mosses, herbs, and grasses, which might transition over time to shrubs, and then once again, spruce and fir trees.

We can even see this at work on a very small scale. Imagine a forest where a squirrel is building a nest of small branches, and a leaf is knocked loose from high in a tree. The leaf is immediately subject to gravity, and to reach equilibrium, it must fall to the forest floor. Once on the forest floor, it is still in disequilibrium because it is a favorite food for certain beetles, slugs, worms, etc. These creatures begin to consume the leaf, eventually eating all but the larger leaf ribs. But the situation is still out of equilibrium, because the leaf ribs will then be consumed by bacteria and fungi.

These three examples show how an initial event causes, or leads to, other events. We can call these initial events, such as the drought, the landslide, and the fallen leaf, *primary events*. We can then refer to the proceeding events, such as the wetland vegetation shift, the forest regrowth, and the leaf decomposition as *response events*. After a primary event has moved the natural system away from equilibrium, the response events occur in order to swing it back toward equilibrium. When we look at any area of nature, or any hierarchical level such as an individual or a natural community, we find examples of these primary events and response events.

What is so often interesting to the ecologist is that these response events *take time*. Well, you might think, "Of course they take time." Everything that happens takes time, right? That is true, but in nature, the *amount* of time these events take is highly variable, and this variability has a significant influence on the condition of nature at any point in time and space.

In our first example above, when the three-year period of dry weather began to cause lower lake levels, the lake might be unable to come to its new level for about three-and-a-half years. Why is this? Because it takes time for groundwater levels to drop, which then cause the lake to drop. And while the lake drops over a period of three-and-a-half years, the wetland vegetation might follow in its wake over a five-year period. This is because it takes time for the plants to send out new shoots and suckers, and for new seeds to germinate and grow. And then, while the wetland vegetation line may have shifted downslope over five years, it may not have been fully displaced by upland vegetation for ten years. Again, this is because it takes time for the upland vegetation to out-compete the wetland vegetation.

We have a similar situation with the avalanche. The avalanche and loss of the spruce-fir forest could happen in a minute or two. The time required for the return of vegetation would depend a lot on local conditions, but the widespread growth of mosses, herbs, and grasses might take ten years; the increase in shrubs might take twenty-five years; and the resurgence of mature trees might take a total of 175 years.

Even the small-scale example shows these time issues. The leaf falls quickly from the tree, but its flat blade might not be eaten by slugs and insects for a few weeks, and breakdown of the leaf ribs by bacteria and fungi might take another nine months.

At any given moment after a primary event, there are often one or more response events in different stages of progress. The drought in our example occurred over three years. At the end of those three dry years, the lake had accomplished most, but not all, of its level drop; the downslope side of wetland vegetation might have been about halfway through adjusting; and the upslope side of the wetland vegetation might have been in early stages of decline and replacement by upland vegetation. The upland vegetation might not finish claiming new space for ten years.

The three-year drought primary event is causing three response events, including a ten-year vegetation response event. But that's not the end of the story, because oftentimes, nature being as it is, a new primary event will occur before the response events to the prior primary event are completed. When the three-year drought ends, rainfall might return to normal, and the lake level will slowly rebound. So there's the wetland vegetation, in the middle of moving downward toward the new lake level, and the lake is already rising again. On the mountainside, our forest regrowth was projected to take 175 years. But maybe after only 75 years, another massive avalanche occurs, resetting everything before our forest can become old. Before the leaf has completely disintegrated, many other leaves will have fallen and will be partway through their disintegration processes.

What happens in all these cases is that before natural systems can fully establish equilibrium, the target for equilibrium is moved, and/or the condition of disequilibrium is reestablished. This is why, in nature, there is never really a state of perfect and final balance. It is more like there is a *process of developing* balance, or moving toward balance.

If natural systems are not in a state of perfect equilibrium or balance, then where are they? They are *moving toward an equilibrium target*, but doing so in reference from a prior equilibrium target. Perhaps an analogy would help convey this: suppose that during track practice a young man is alternately sprinting quarter miles and then jogging quarter miles. As he ends a quarter mile sprint, he slows to a jog. In the first ten or twenty seconds of jogging, he will be extremely out of breath. Then gradually, while jogging, his breathing will slow down to match the jogging pace. His body is shifting to match current conditions (jogging), but it is doing so from a reference point of prior conditions (sprinting), and that shift takes time.

Because nature is so full of primary events, its systems are always trying to do the same thing as that track runner's body—approaching equilibrium from an orientation of previous, or historic conditions. If given enough time, natural systems could find a final equilibrium point, but in the practical world, new primary events will occur first. This resets the equilibrium target, which requires new adjustments, and keeps natural systems in continual motion.

This doesn't mean that natural systems are inherently "imbalanced" or "out of balance," because those terms suggest something that is substantially off-kilter, or unable to be where it needs to be in order to function. Certainly, nothing like this type of imbalance is the case in undisturbed nature. At any given point, nature's dynamic systems are simply using their tools to approach a new equilibrium target, and that takes time.

Remember Pattern 1, which noted that natural systems were full of internal activity? Add to this the fact that every event, including each response event, is its own little primary event (engendering its own little response events), and you can imagine the grand and continual movement occurring in the natural world.

Pattern 19: Offspring, individuals, species, species classes, natural communities, ecosystems, and the biosphere are in a continual pursuit of equilibrium with existing conditions, shifting from a reference point of prior conditions.

Essential Translation: Shifting from a reference point of prior conditions, sustainable complex systems are in a continual pursuit of equilibrium with existing conditions.

This pattern can now be translated into the context of human systems:

Economic Translation: Shifting from a reference point of prior conditions, sustainable economies are in continual motion toward equilibrium with existing conditions.

For example, a hilly province has an economy centered upon small, scattered farms, small towns, and limited tourism. Incomes in the area are modest, as are government expenditures and revenues, which are based on sales taxes and property taxes. Institutions such as a local senior center have wish lists of what they could accomplish, if they only had more government funding.

One year, mineral deposits are discovered in this province, and not long afterward a mine is established. There is an expectation that the new mine will enrich the region and enable more economic support for institutions such as the senior center. This does occur, but it takes some time. After the mine opens, operations expand gradually, and it takes three years to reach full mining capacity and employment levels. Local incomes and consumer spending increase over ten years, as miners' earnings are spent and ripple through the economy. Home prices go up slightly as the first a few dozen mineworkers move to the area, but they go up more over the next fifteen years as longtime residents become wealthier and are able to buy first or second homes. In addition, even though government revenue is slightly higher in the year that follows the mine opening, the first inclination of officials is to pay for emergency repairs at local grade schools. The senior center sees budget increases beginning within three years after the mine opens, and leveling off twelve years after that. This funding time lag was because funding required a series of precursor events after the mine opened.

Community Translation: Shifting from a reference point of prior conditions, sustainable communities are in continual motion toward equilibrium with existing conditions.

For example, a family of eight immigrates to a new, more fertile land. In doing so, they leave behind very strong bonds with the rest of their clan. The family responds in several ways to this dramatic change. They turn inward to some degree to form their own community. They also begin to develop relations with people in their new land. Through this though, they still find ways to maintain some correspondence with their clan. It isn't until about forty years pass that the members of this migrant family have a more stable sense of community, which includes lots of self-reliance, improving relations with their new neighbors, and periodic correspondence with some members of their own original clan.

Occupational Translation: Shifting from a reference point of prior conditions, sustainable occupational systems are in continual motion toward equilibrium with existing conditions.

For example, a new transport vehicle is created, and people are needed to operate the vehicles almost immediately. At first, it is difficult to find enough people with the mental alertness and dexterity for the job. Furthermore, there is significant variability in the skill level and behavior of the first wave of operators. Within two years, training centers are established to prepare more operators. Training takes about nine months, and during the next decade as these new trainees enter the workforce, the extreme demand for new operators relaxes somewhat. It ultimately takes about eight years from the time the vehicle was first developed before skill types and skill levels for these workers become very predictable and uniform.

Government Translation: Shifting from a reference point of prior conditions, sustainable rules and behavioral expectations are in continual motion toward equilibrium with existing conditions.

For example, technical changes in the way banks handle money enables them—or causes them—to obtain a portion of the interest that would have gone into individual bank accounts. While technically legal, many people believe this activity goes against the intention of at least two existing banking regulations. After a year of debate and negotiation, new regulations are passed to prevent the bank's interest-gathering activities. However, the combination of the new regulations and preexisting regulations causes the banks to have to choose between losing money or breaking at least one regulation. After another year, more negotiation result in elimination of one older regulation, and a fine-tuning of three others. About five years after the issue first arose, banks can operate within the law, distribute all interest, and make a profit. Despite this achievement, several of the banks find themselves still fending off demands for even stricter controls on their behavior—an outfall of the negative publicity that the incident generated.

Environmental Translation: Shifting from a reference point of prior conditions, sustainable environments and land uses are in continual motion toward equilibrium with existing conditions.

For example, to reduce congestion, a mid-sized city designates a fifteen-block zone of its downtown as automobile-free. After cars are removed, congestion becomes much less of a problem, but numerous other unplanned changes begin to take effect shortly thereafter. Without the use of their autos, more people must access the fifteen-block area on foot. These pedestrians immediately attract more sidewalk vendors. Some of the successful vendors then open up storefront shops. As the number of shops increases, people begin to go downtown just for shopping, and restaurants and stores become even more numerous. This attracts more people to live on the outskirts of the downtown area, and all the activity attracts a small art museum, a playhouse, etc. After about thirty years, the automobile-free zone is much more vibrant with economic and cultural activity than was expected, and a two square-mile ring around this zone has responded with an increase in permanent residents, many of whom have brought their own innovative business ventures.

Cultural Translation: Shifting from a reference point of prior conditions, sustainable cultures are in continual motion toward equilibrium with existing conditions.

For example, Clarence and Olivia have been married for about a decade. They have three children in grade school, and a relatively stable family life. But this changes when Olivia goes to work full time, and two months later, Clarence is assigned to rotating shifts at his job. They manage logistics fairly well—getting to work and school on time, paying the bills, and getting the kids to bed. Yet they both gradually feel troubled, because six months after their job changes, the children are often cranky, and they themselves feel disoriented. They realize that two cornerstones for their children and themselves had been their family dinners and periodic weekend outings. With these now mostly gone, everyone in the family was being affected.

After recognizing this, they devise a plan to reinstate their family time, which requires both of them to adjust their work hours slightly. It allows them to have breakfast as a family four mornings a week, and to take an outing every other Sunday. They even assign the children to put these events on the calendar to bring back their sense of family predictability.

This is a big improvement, but after another year, they feel they still need more downtime as a family. In response, Clarence is able to drop every other Friday shift so the family can enjoy one more free night together on those days. Finally, about two years after Olivia went back to work, they feel a sense of overall family balance. However, a sense of *perfect* balance is elusive, as the couple has to make additional periodic scheduling adjustments to accommodate their growing children's sports participation.

Physiological Translation: Shifting from a reference point of prior conditions, sustainable mental and physical status and skills are in continual motion toward equilibrium with existing conditions.

For example, Amanda received an arm fracture and a badly broken leg in a horse riding accident. Despite her plaster casts, she became quite functional after a few days by relying heavily upon her left arm and left leg. During the next six months, her bones healed, and she expected to have her "normal self" back when her casts were removed. But by this time, her left side had become quite strong, and her right had become quite weak. She had to work with her body to manage this new imbalance, and gradually her body could function quite normally. By the time two years had passed since her accident, she only thought about it occasionally. Yet some of the injury sensations still lingered in her body, and for several years she would feel dull or fleeting pains when she had to lift heavy boxes or climb several flights of stairs.

The next natural system pattern pertains to the *rates* at which events occur. The focus will be upon the rates of the most impactful events, or events associated with the most significant transformation.

We can refer to events that are relatively impactful or transformative as *major events*. Let's define a *major event* as, *a large and long-lasting or permanent change in the internal condition or status of a natural system or individual.*

By this definition, a tree does not experience a major event when the wind shakes its leaves, because it is not greatly changed or permanently changed as a result. But it does experience a major event when it loses a thick limb in a windstorm, because its shape will be permanently changed, and its health status could be lowered, at least for a while. By similar reasoning, a turtle does not experience a major event when it eats one insect. One insect is just a small part of the turtle's weekly diet, and it will have chances to eat other insects again and again. But, when a turtle grows from a weight of three ounces to five pounds, its status is greatly affected. That difference in size and weight affects the foods it is able to eat, its strength and safety, its breeding status, etc. Thus, growing from three ounces to five pounds is a major event for the turtle.

Bearing that in mind, let's start by looking at a few examples of major event rates in nature through the lifetime of a goldfinch, a leaf on a tree, and a mountaintop forest.

A goldfinch begins its lifetime in a minute embryonic form inside a shell. It grows inside the shell until it begins to hatch. Once hatched, the baby bird is fed regularly by its parents as it grows toward adult size. As it grows in the nest, it stretches and flutters its wings in preparation for its first flight. Eventually, the bird might take its first short or clumsy flight to a nearby branch. When first out of the nest, the young adult bird will still be fed by its parents as it gains coordination and experience. After gaining independence, depending perhaps upon weather and food availability, the finches will often migrate in groups. Migration distance and times might vary by location, but trips in the range of 1,000 miles might be typical. After wintering, the birds will fly back north, and courtship will begin. After forming a pair, a female will build a nest and the male will assist. The female will begin laying eggs in the completed nest. And this picks up where our cycle began, with newly laid eggs. Death for an individual bird can come at almost any time and in a variety of ways, such as being snared by a kestrel, being killed in a storm during migration, dying from a disease, or starving after an injury.

For the goldfinch, these are all major events—being created as an egg, hatching, growing, fledging from the nest, gaining full independence, migrating once, migrating again, courting, nest-building, laying new eggs, and dying. How long does each major event take? Is it quick, and nearly instant? Is it slow and gradual? Consider each of these events against the time they take. The egg is laid rapidly by the female finch, though it has taken about one day to form inside her body. The embryo grows progressively toward the chick stage, only becoming mature enough to hatch in two weeks. After breaking its air sac, the hatching baby bird will usually require a few hours to get out of its shell. The next two weeks involve gradual growth in which feathers form. The first flight of the maturing bird happens in just seconds, although its training for flight began several days prior with stretches and flutters in the nest, and the goldfinch will continue to gain flying skill for several weeks after its first flight. The bird's migratory movements literally occur one flap at a time, covering one mile, and then another, and then another, etc., over a number of days. Courtship is likely to last several days, which will presumably result in a pairing. Building the nest takes patient placement of fluff, grass, bark, etc.—one piece at a time—over the course of about five days. The female will lay eggs. She lays only one per day, and might lay them for up to a week. The death of a finch can happen quickly or gradually. She or he could be picked off in the blink of an eye by a falcon or a cat. She or he could get knocked into a tree during a nighttime storm, dying rapidly. Or the finch could die more slowly on a cold night with low food availability.

It appears that most of these major events in the life of the goldfinch are *gradual*, or progressive—gradual growth inside the egg, gradual growth in the nest, gradual gain in flight skill, gradual independence, gradual progress in migration, etc. But none of these things just "happen" with the snap of a finger. They are major transformations in the life of the animal, and they progress, form, or expand over hours, days, or weeks.

Even the major events that we might think of as instant—the laying of an egg—is the last stage in a process that took about 24 hours for the mother's body to complete. Or the first flight out of the nest—this is really just one amateur flight. The more transformational thing for the bird is to shift from being completely flightless to being a skilled aviator. This takes many weeks from the first stretches through many practice flights. The only events here that seem fully abrupt and non-gradual are the death in the talons of a falcon, the claws of a cat, or against the trunk of a tree. The rest are either completely gradual or gradual and leading up to an abrupt finish (such as laying an egg).

We can see this same tendency all around in nature, where most of the major events in the life of a system or individual seem to happen gradually. A primordial bud forms on a tree in the summer. The bud sits through winter, then slowly swells in spring. Eventually, a swelling leaf inside the bud begins to push open the bud scales, and emerges. The young leaf grows over a couple of weeks, and then begins to harden. The leaf is held through the sun and wind for weeks and weeks, and as it does so, it gradually ages—often obtaining little dents or cuts, some surface mold, or an insect hole. As fall settles, the tree pulls nutrients back out of the leaf, and over a few days or weeks, it changes colors and dries. Then, as its connection to the tree weakens, a breeze will kick up and it will suddenly fall.

Again, these major events are largely gradual, progressive. It is true that the leaf will fall in an instant, although even this was the endpoint of a much more gradual deciduous process.

Even at the higher system levels, most major events seem to be gradual. Imagine a pine forest on a mountaintop. The forest will experience major events with each new season. In spring, it will gradually thaw out from a winter grip, migrant birds will pass through, and plant growth will resume. In summer, the forest will be full of insect life, birds, flowers, and perhaps grazing or browsing animals. As fall arrives, the forest changes again—migrant birds pass back through, trees drop seeds or nuts, and wildflowers start to pull their nutrients back down underground. When winter arrives, cold temperatures and snow will keep life in more of a dormant state as the new snowpack builds up again. Over the years, the density of certain plant species could expand while certain others contract in response to gradual changes in climate or animal activity. For example, a decade-long drought could occur, enabling more xeric species to expand relative to other species.

A change in temperature during the day, the arrival of a few migratory birds, and the blooming of a few flowers don't really constitute major events for the forest because they don't strongly affect its status, and they are rather transient events. But the overall seasonal changes lasting a few months each could be considered major events. The permanent or semi-permanent changes in plant composition due to climate fluctuation can also be considered major events. Clearly, these seasonal and semi-permanent major events are *gradual*.

It is true, however, that the forest could experience rapid major events. Over the course of several hundred years, maybe three fires will move across the mountain, essentially resetting tree growth. Or, maybe a great windstorm causes a throw event, knocking over nearly every tree in the forest. These would be rapid, or relatively instantaneous events. But in any case, the *majority* of major events—such as the seasonal changes—are gradual, and instantaneous major events are relatively infrequent by comparison.

If most major events over the course of an individual or system lifetime are gradual, then it seems plausible that gradual events play a more influential role in their existence. It doesn't mean that rapid events don't play a role in nature, but it does suggest that individuals and natural systems have more opportunity and need to cope with, and prepare for gradual events.

> Pattern 20: Offspring, individuals, species, species classes, natural communities, ecosystems, and the biosphere experience gradual or cumulative major events much more often than they experience instantaneous or concentrated major events. Altogether then, gradual major events can be seen as more impactful than instantaneous major events. However, a given gradual major event is not necessarily more important or impactful than a similar instantaneous major event.

Essential Translation: Sustainable complex systems undergo gradual or cumulative major events more often than they undergo instantaneous or concentrated major events, and thus gradual or cumulative major events tend to have more overall impact on those systems than instantaneous or concentrated major events. Here, *major event* refers to a large and long-lasting change in the condition or status of a system.

This pattern can now be translated into the context of human systems:

Economic Translation: Sustainable economies undergo gradual or cumulative major events more often than they undergo instantaneous or concentrated major events, and thus the gradual or cumulative major events tend to have more overall impact on sustainable economies. Here, *major event* refers to any large and long-lasting change in the condition of an economy.

For example, in addition to its everyday internal economy, a small nation has three major economic sectors. These are tourism, fishing/seafood, and export of a rare tree oil. It so happens that nearly all the income provided from these sectors occurs from about February through July of each year. But the overall economy does not jump up or down abruptly in response to these cycles; rather, it gradually accelerates each year from about February to April, and it gradually shifts back down from August to September.

Community Translation: Sustainable communities undergo gradual or cumulative major events more often than they undergo instantaneous or concentrated major events, and thus the gradual or cumulative major events tend to have more overall impact on sustainable communities. Here, *major event* refers to any large and long-lasting change in the condition of a community.

For example, over the course of about 170 years, one branch of a family undergoes many gradual transitions as infants arrive, children grow and mature, adults go through life changes, and the elderly pass away. Through many gradual changes, this particular extended family also encounters two sudden events—one of these is when five family members leave the home region, and another is when a middle-aged parent of four children is lost in an untimely accident.

Occupational Translation: Sustainable occupational sectors or forces undergo gradual or cumulative major events more often than they undergo instantaneous or concentrated major events, and thus the gradual or cumulative major events tend to have more overall impact on sustainable occupational groups. Here, *major event* refers to any large and long-lasting change in the condition of an occupational group.

For example, a data company tracks twenty-eight career types in the workforce of a metropolitan county. They find that over a sixty-year period, eighteen of the twenty-eight career types underwent large (i.e., greater than 30 percent) net changes in workforce level. However, they cannot find more than three instances when any of those career types gained or lost more than 2 percent in a single year.

Government Translation: Sustainable government and behavioral codes undergo gradual or cumulative major events more often than they undergo instantaneous or concentrated major events, and thus the gradual or cumulative major events tend to have more overall impact on sustainable government and behavioral codes. Here, *major event* refers to any large and long-lasting change in the condition of a government/behavioral codes.

For example, there are a series of international laws that provide cooperative military protection to nations that meet certain minimum behavior thresholds. Whenever it is agreed that one of those thresholds needs to be raised, it is raised incrementally over five years rather than in one day.

Environmental Translation: Sustainable environments and land uses experience gradual or cumulative major events more often than they experience instantaneous or concentrated major events, and thus the gradual or cumulative major events tend to have more overall impact on sustainable environments and land uses. Here, *major event* refers to any large and long-lasting change in the condition of an environment or land area.

For example, a kingdom has carefully designed a stormwater system for its capitol city. This system of drainageways, ponds, empty fields, sand filters, and flow restrictors causes water runoff from even the biggest storms to be released at moderate rates and with low pollutant levels into a swamp wetland and a local river.

Cultural Translation: Sustainable cultures undergo gradual or cumulative major events more often than they undergo instantaneous or concentrated major events, and thus the gradual or cumulative major events tend to have more overall impact on sustainable cultures. Here, *major event* refers to any large and long-lasting change in the condition of a culture.

For example, within one tribe there is an ancient storytelling tradition. There are stories about battles, orphaned children, animals, spirits, stars, and sea voyages. Because it is an oral tradition, the stories change a bit with the tellers and each new generation of children. Though it is common for stories to shift according to people and times, it is much less common for any to be dropped, or for entirely new stories to be introduced.

Physiological Translation: Sustainable human bodies and minds experience gradual or cumulative major events more often than instantaneous or concentrated major events, and thus the gradual or cumulative major events tend to have more overall impact on sustainable human bodies and minds. Here, *major event* refers to any large and long-lasting change in the condition of a person's body, skills, or knowledge.

For example, Paul and his younger sister Sarah were born to active parents who expected their children to also be active and athletic. Along with other sports they were taught how to swim, ice skate, ski, ride bicycles, and ride horses. By adulthood they were both rather proficient at all of these things. Although both of them attained these proficiencies starting from a beginner level, at no point in time were either of them suddenly exposed to an advanced skill, nor did either suddenly obtain those skills. Instead, they were gradually introduced to each activity, and as they practiced and gained confidence, their skill levels grew.

Patterns 17 through 20 all have an important theme of time. Overall, these patterns suggest that sustainable complex systems spend a great majority of their lifecycles in their most developed, balanced, and stable stages, and that their lifecycles also occur in a relatively developed, balanced, and stable context. But, those "developed," "balanced," and "stable" stages are never *fully* developed, *perfectly* balanced, or *completely* stable, because they are always dynamically approaching these conditions from a reference point of prior conditions. Finally, the majority of important changes that a complex system experiences through its lifecycle are relatively gradual or incremental, rather than abrupt. Next, we look at the topics of grouping and group decision-making.

Chapter 12 Group Decision Making

In nature, animals often gather into groups. You may even see this near your own neighborhood—a flock of doves on a telephone wire, a small herd of deer, a school of baby catfish near the shore of a pond, a gathering of turkeys near the edge of a cornfield, a pair of ducks in a marsh, a large flock of geese in flight, or sea lions gathered along a beach. Some animal groups occur in the wilderness, far from most people—packs of wolves, drifting herds of buffalo, traveling troops of monkeys, or families of elephants looking for food or water. It is certain that these animals are purposely grouping together, because if they weren't, we'd instead see them scattered more individually across the landscape.

Biologists believe that animals form groups for numerous reasons. A group of animals has many more eyes, ears, and noses than an individual. So compared to one turkey, a flock of turkeys probably has an enhanced ability to see or smell a predator. A group of animals can effectively defend a territory from intruders of their own kind. For example, a pair of eagles can team up and chase another eagle away from their nesting area.

Groups of animals can divide up a task and accomplish things they could never accomplish individually (see in part Patterns 6 and 7). Several members from a pride of lions can go out and hunt large, dangerous prey such as water buffalo. During the hunt, some lions can chase while others wait to pounce. Meanwhile, other pride members remain back with the family, protecting the cubs from hyenas and other lions.

By congregating, animals can pass on knowledge through generations. Some fish species learn when or where to travel by following the experienced group members. Similarly, bobcat cubs will learn how to hunt from their mother before going off on their own.

Forming groups often gives animals quite an advantage. But being a member in a group has its downsides as well. It can hinder the acquisition of resources—one buffalo inside a large herd might be insulated from predators, but the grass in the middle of the herd isn't as good as the ungrazed grass on the edge. A pride of lions might cooperate to take down an animal, but if the pride is large and exceedingly hungry, there might not be enough food to go around, and some of the lions might not get a share of the meal.

In other cases, the needs of the group members don't match up. One bird in a migrating flock might be particularly tired and hungry, and ready to descend, but the other members of the flock might wish to fly many more miles before the next rest. Staying with the group too long could mean certain death for that individual. Some members of the monkey troop might want to feast on figs in the tree canopy, while others are in need of protein and would rather pick for ants and beetles off the ground.

Animals indeed form groups for good reasons, but in doing so, they face a conundrum. How can they enjoy the benefits of their group while minimizing the downside of being in that group?

Animals are keen, both to the positives and the negatives that come with grouping, so they often limit their group size. They may also choose to group and then divide based upon changing conditions. This occurs, for example, with Canada geese. These birds often migrate together in large numbers. This probably helps them navigate and stay safe during rest stops. But when it comes time to breed, they separate off into pairs and generally create their own nesting territories—which probably helps them to obtain food to raise chicks.

African elephants also form groups that later disband. Their matriarchal family groups might contain about a dozen individuals. But sometimes a family will socialize and bond with another family. A handful of families can congregate, forming a clan. Under certain conditions or stressful situations, these elephant clans coalesce even further and form herds of hundreds. The main foundation of all these groupings is the small matriarchal family unit. The small family unit never breaks up, but the other groups break up after they form.

This pattern of congregating and then separating has been termed "fission-fusion." It means that animals form groups, but they also disband. This fission-fusion behavior is present in many species, from pronghorn antelope to killer whales to baboons.

Further, the predominant pattern in this fission-fusion behavior is that the larger the grouping, the less time individuals spend within that grouping. Why would this be?

One primary factor seems to be that, as groups become larger and larger, the members simply don't have enough in common to hold them together for as long. Individuals within larger groups don't have the same territorial memories, the same social bond strength, the same culture, and the same genetic lineage.

Fission-fusion patterns can appear at any scale, allowing members or subgroups from the same troop or pack to go their own ways intermittently. Why? Because again, they don't always have enough in common with their larger troop or pack to permanently remain together.

A more permanent fission in nature sometimes takes the form of an individual leaving its primary group. This might be by its own choice, or because it has been kicked out by the rest of the group. A casted out individual may or may not be able to join another group afterward.

It is important to recognize that group fission occurs when group costs outweigh benefits; this implies that when they decide to divide, individuals are sometimes choosing to forego significant group benefits because they are outweighed by even larger group costs.

If animals form groups, you might wonder who gets to *lead* the group. Who decides when and where the group will go? How do animal groups make decisions? Are some individuals compelled to go along with unfavorable decisions? Who should the group's decisions yield the most benefit for?

Biologists have investigated this topic among many animal species. However, as interesting as these questions are, this type of science is exceedingly difficult to conduct. We can't ask an animal, "How happy are you with this group? What do you *truly* feel like doing right now? Are there certain members of the group that you generally follow? Are there certain members of the group whose welfare is a higher concern for you, compared to other members? How well do you know what the other members of your group want?"

Partly because this research is so difficult to conduct, biologists also rely upon mathematical modeling to augment their understanding of group behaviors and decision making. This is done in reference to previously established biological concepts, such as genetic kinship, natural selection, physical dominance, etc. In spite of challenges, biologists have been able to make some headway in this interesting field, and although interpretations vary, there are some apparent trends.

The fission-fusion behavior, or the ability to gather or divide at will, is one of the primary ways in which animals manage collective decisions: when there are strong and divergent needs among the group, the group can simply split to allow those needs to be pursued separately. Also recall that among animals, the larger the group, the less cohesive the group tends to be. An implication of this is that larger and larger groups have less and less opportunity or need to make common decisions, and when they reach these junctures, they are more likely to split up than are smaller groups.

But in any case, whether large or small, groups do form and they do make decisions that help them to stay together. How then do they make decisions? Who figures out whether it's time to rest, time to patrol the territorial boundary, or time to drink? It appears that when groups are small, group decision-making is based on collective consensus or majority rule, or based on the discretion of one or more dominant or elder individuals. In other words, small group decision-making authority runs from collective to concentrated.

However, for highly taught situations, where resource competition within or between groups is tight, and when resource availability is sporadic, small-group decision making swings in the hierarchical direction (perhaps wolves, lions, and hyenas provide examples of this). But when resources are ample, the decision-making swings more toward collective consensus or majority rule.

In comparison, as groups get larger and larger, group decisions are based more and more on collective consensus and majority rule, and much less emphasis is based upon the preference of a few dominant or elder individuals. To understand why these decision-making patterns would emerge, we have to look more closely at the situation animals find themselves in.

A view of life through the laws of natural selection suggests that the traits that most strongly promote survival and reproduction are passed along to each new generation, and over time, these traits become predominant within a species. But just what does this mean?

We might make an *assumption* that through natural selection, individuals who are self-centered and self-serving will survive the longest and reproduce the most. The end result of this would be that each species will eventually be honed down to the most self-centered and self-serving behaviors and genetic predispositions.

However, this is probably an oversimplified view of a more intricate or evolved situation. Yes, a self-centered individual might reproduce more, sometimes, but will its genetic legacy carry forward if it is so self-centered that it allows its family to suffer or die? What if it competes too severely against its extended family? What if an individual undermines the health of the group that it is a member of? Will it suffer once its group dies out or disbands?

Clearly, self-promotion alone is not always a winning strategy. From a biological and evolutionary point of view, an individual certainly must hold its own needs very high, but it has to do this while considering its own family as well as the group (or groups) that it is depending on to help it thrive. Similarly, if a group is undermining some of its members, is that good for the group? Generally, no. What is bad for many members of the group is, in the long run, not good for the group, and we can expect such groups to be less successful and less coherent. And similarly again, what is bad for family and kin of an individual is also often bad for that individual's long-term outlook or family-level genetic legacy.

Given all of this, *decision making in animal groups that are remaining together has to account at some level for both individual needs and group needs.*

With this understanding, even decisions that seem to primarily benefit one individual can often be optimal for the group over the longer run. For example, in the most highly taught animal pack situations, decision making might skew in the hierarchical (dominant individual) direction. Although hierarchical decision making might appear to only benefit dominant individuals, it may actually benefit the whole group because dissent will lead to death of the pack (i.e., the death of all group members), or because a dominant individual must eat well and be exceptionally healthy in order to guard the pack's territory. So in some cases, emphasizing the value of one individual might allow the pack to have the overall highest probability of surviving.

So it would seem that in nature, a "good" group decision is one that is best for the members of that group, and a "good" decision for an individual is also, more often than not, good for the other group members.

If this general rule (i.e., what is good for individuals is also good for group members, as well as for the group) provides the structure of a good decision, *then the key components of a good group decision are firstly—knowing about the alternative choices, and secondly—knowing what other group members need*. Knowledge about alternative choices could pertain to things like migration, and which route holds fewer predators hiding in the brush. Knowledge about the needs of other group members could pertain to their condition, and whether some might starve if the longer migration route is taken.

With these issues in perspective, we can look back at decision-making patterns in different sized animal groups, and it starts to make sense. The first criteria of a good decision is knowing about the alternative choices that the group faces. Wisdom regarding the alternative choices often comes with experience—this experience is often held by dominant or elder individual(s).

The second criteria of a good decision is knowing the needs of the group members—this can come with experience held by dominant or elder individual(s); or it can come via high social awareness of a dominant or leader individual; or it can be determined without leaders, through subtle communication between the animals.

In smaller groups, the odds of having one or two much more experienced individuals is high—these unique individuals can understand the choices facing the group and the needs of the group members. This could explain why small-group decision making is often performed by one or just a few dominant individuals. Think of this like a human family with several children. The two parents are clearly more experienced, and they often know the needs and quirks of their children. Thus, we might get a family decision (a small group decision) made by one or two parents (experienced leaders).

In cases of small animal-groups in which there isn't a single individual that is more experienced or aware, the best group decisions can be made collectively, with no single leader. Think of this like a group of friends making a plan. When none of the friends are much more informed than the others, they might *all* need to give input to set the plan on its best course. In summary, small-group decision-making can be done in a variety of ways, so long as the selected method yields the best integration of information regarding the needs of the group members and the situation the group faces.

But in large groups, the dynamics can change. As group size grows larger and larger, there will be many elders and many leaders, and it becomes more and more difficult for elders and leaders to keep track of individual needs among the larger group. This helps explain why decision making in large groups becomes more and more of a collective or consensus-driven activity—no single individual or small minority has enough wisdom to take the decision on.

While large-group decision-making tends to be collective or consensus driven, a twist on this theme can occur with *representation*. It some cases, a small number of individuals are allowed to make decisions on behalf of a much larger group. This only occurs when first—those deciders are privy to exceptional information (they know something that the rest of the group doesn't), and second—those deciders have exceptional knowledge of group needs (they know the needs of the group members just as well as the group members themselves). This is apparently the case with honeybees, in which scouts choose a hive site for the whole colony. They can only do this because they have exclusively evaluated alternative hive sites and they are knowledgeable regarding the precise geometric needs of a good hive that will suitable for all the other bees.

> Pattern 21: Animal groups form when individual animals perceive that group benefits outweigh group costs. However, individuals or subgroups disband when grouping is not beneficial or when the group costs outweigh the benefits. In general, the larger the animal group, the less net benefit there is to maintain that group or to maintain it continually over time.

Essential Translation: Individuals within sustainable complex systems form groups when those individuals perceive that group benefits outweigh group costs. However, individuals or subgroups disband when grouping is not beneficial or when the group costs outweigh the benefits. In general, the larger the group, the less net benefit there is to maintain that group or to maintain it continually over time.

This pattern can now be translated into the context of human systems:

Community Translation: Sustainable community patterns allow people to form groups or engage within groups when they believe it is beneficial for them to do so. However, people or subgroups disband when benefits of the group dissipate or when the group costs outweigh the benefits. In general, the larger the group, the less net benefit there is to maintain that group or to maintain it continually over time.

For example, the McKays have two children. Just going through their daily meal, chore, and activity routines allows them to spend about 30 hours per week together as a family. The McKays are close friends with three local families who have children of similar ages. They see these friends about once a week at the park, for sports, trips, or other get-togethers. In addition, they have a larger circle of friends that includes another dozen families and couples. They aggregate with this larger group every few months at parties, celebrations, or other planned events.

In another example, Randy grew up in a small city, where he had good friends and extended family nearby. But in his early teens, he felt less and less like he really fit in. He had a love of open space and liked working with animals. In his late teens, he was able to spend a few months as a ranch hand hundreds of miles from home. His family thought this ranch experience might completely satisfy his yearning for the outdoors, but it only made his yearning grow. By his mid-twenties, he decided to move far from his home city and return to ranching. This choice was somewhat painful, but it was one he felt he had to make. He could no longer see his closest friends and family on a regular basis. Eventually in his new home, he joined a group of friends with similar backgrounds—single men who had moved away from their families.

Government Translation: Sustainable government bodies are only established in response to groupings and commonality, and their authority is only based upon commonality. In cases where they are in excess of commonality, they allow for opt-out alternatives whenever possible. The larger the grouping, the less commonality there tends to be.

For example, nine small nations have common concerns over military security and environmental quality. Over several months of discussions, these nations determine that they indeed have common problems, and they establish a federation to develop unified and efficient responses to these problems. The nations agree that the "Federation" only has authority pertaining to security and the environment.

Eventually, terms are negotiated that allow the Federation to amass a military force. This requires that each nation provide financial support and military personnel. Negotiations over environmental quality are more difficult, since the types of pollution being created by each nation and the types of cleanup each wants vary (for example, some are more affected by a neighboring nation, some are more concerned about air pollution, and others have more severe water pollution issues).

Eventually, the Federation settles upon pollution restrictions and cleanup activities that each nation finds somewhat burdensome. However, upon review, each nation separately determines that the benefits of the collective environmental improvements would far outweigh their own individual costs, and would also far outweigh any improvements they could gain through their own individual (separate) action.

The Federation operates coherently through several decades, although there are occasional points of renegotiation. One of these points occurs when a nation requests to opt out of cooperation over military security. Other Federation members believe that doing so is ill advised because that nation would no longer be protected. The Federation members deliberate on the request, and determine that there was no inherent reason why a member should be forced to remain part of the security coalition. That nation eventually opts out of the security coalition, but when another nation requests to opt out a few years later, it is denied. This nation is located in the interior of the Federation, and after deliberating, the Federation determines that an interior nation would go on receiving Federation protection by default, and so a practical opt-out opportunity does not exist.

There are other pivotal points of decision making as well. Two adjoining nations later request to join the Federation. Following negotiations, it is determined that one of these prospective member nations can join the Federation on standard terms, but the other potential member lacks the environmental issues in common with the other Federation members. That nation is then allowed to join the Federation as part of the security coalition only.

In another decisive moment, two member nations that have strict religious edicts request that the whole Federation adopt those edicts as a requirement for all members. Upon deliberation, the Federation agrees that no nation can effectively compel another nation to follow its customs, and that each nation will have to make its own religious choices.

In yet another decisive moment, four of the nations have similar monetary systems, and they agree to create a common currency. They request that this new currency become a Federation standard. The Federation finds that all of its members would probably benefit, in theory, from a common currency. However, they agree that joining the Federation on the basis of security and the environment is not enough reason to also require them to use a common currency. The Federation instead recommends that each nation adopt the standard currency. Eventually, seven of them do adopt the currency. While the remaining few nations do not adopt the standard currency outright, they make internal economic provisions so that its use will be unencumbered. They don't do this as a "favor" to the other nations; they do it because they believe it is in their own best interest to do so.

When a small group does form, the processes that guide its decision-making can vary widely, including numerous hierarchical or collective alternatives.

> Pattern 22: Smaller groupings of individuals, i.e., about 150 or fewer, make group-level decisions in a variety of ways. The methods include decision by an elder, dominant, or extremely knowledgeable individual; by several experienced or dominant individuals; by subgroups within the group; by majority rule; by unanimous rule; etc. When competition for resources is very strong, decision making by a physically or socially dominant individual tends to be a favored mode.

Essential Translation: Sustainable decision making by groups of about 150 individuals or less can occur by a variety of modes. These include decision by an elder, dominant, or extremely knowledgeable individual; by several experienced or dominant individuals; by subgroups within the group; by majority rule; or by unanimous rule. Under more stressful or potentially life-or-death situations, the decision-making might fall upon dominant individuals.

This pattern can now be translated into the context of human systems:

Government Translation: Sustainable decision-making processes by groups of about 150 people or less can occur through a variety of modes. These include decision by an elder, dominant, or extremely knowledgeable individual; by several experienced or dominant individuals; by subgroups within the group; by majority rule; or by unanimous rule. Under more stressful or potentially life-or-death situations, decision-making might fall on dominant individuals.

Several examples are provided:

Donna is a divorced mother raising her two children, and she has also taken on two younger foster children. She is very busy between work and caring for these four. She doesn't try to "micromanage" the details of each child's life, but she has determined that she does have to essentially run the family, making important basic decisions such as bedtime, what will be for dinner, what activities are allowed, and when there needs to be discipline. She tries to spend some extra time with each child, and she makes sure she stays in tune with how they are all doing; she takes this insight into account as she makes decisions each day. Although she experimented with running the family based more upon consensus, she found that it just couldn't be counted on to keep things on course, perhaps due to the age of the children involved.

The smallest military units in one nation are fifty-soldier platoons under the command of a field sergeant. Military rules limit the sergeant's authority to training periods. However, in combat, that authority becomes much more comprehensive, and soldiers are required to follow sergeant's orders on a 24-hour basis.

Nine high school sophomore girls have become friends. When there is a school function such as a dance or a football game, they all go together, or they meet and then hang out as a group. But the girls in the group have different personalities, and when there isn't a predetermined function, they don't have such an easy time deciding what to do. When this lack of consensus occurs, Melanie, who is central in this network of friends, usually guides the others to a decision such as a movie or going out to dinner. She is able to do this because she is aware of the values of the other eight girls, and is able to steer the decision in a direction that she believes will be the most fair and acceptable to all.

A small sporting goods chain has stores in six cities and about 130 full-time employees. This company holds quarterly meetings that are attended by the company's founder, who is also the principle interest owner, and by the head managers/co-owners of each store. These meetings are held to make decisions regarding merchandise, sales, stocking, employee benefits, etc. In many instances, the store managers agree, but if at any point in a meeting the five managers don't agree, the company founder makes the decision. The closest competitor to this chain is another sporting goods chain with five stores and about 110 full-time employees. In this company, decisions regarding merchandise, sales, and stocking are made by two to three senior managers from each store at semi-annual meetings. If the store managers disagree, discussions continue until a compromise can be reached. When it comes to employee benefits, these senior managers review alternatives. If they find that more than one reasonable benefit alternative exists, they present the alternatives to all employees for a vote.

> Pattern 23: Larger groupings of individual animals, i.e., about 150 or more, make sustainable group-level decisions through collective means. Collective group decisions tend to account for (or attempt to account for) the needs of all in the group; allow for the expressions that any in the group might make; and result in a choice that a given individual may or may not prefer. Collective decisions can take the form of majority rule, unanimity, various levels of consensus, etc.

The primary difference between decision making that is hierarchical or dominance based (Pattern 22 in part)—and decision making that is collectively based (Pattern 22 in part, and Pattern 23 entirely)—is that the hierarchical or dominance type has a single decider or small number of deciders. In contrast, collectively based decision making has no individual or subgroup making the choice (a few individuals might count votes or report the group choice, but they don't *make* the choice). In any given situation, decisions made through hierarchical/dominance systems and decisions made through collective systems may or may not produce the same final choice for the group.

Pattern 23 Essential Translation: Larger groupings of individuals, i.e., about 150 or more, make sustainable group-level decisions through collective means. Collective group decisions tend to account for (or attempt to account for) the needs of all in the group; allow for the expressions that any in the group might make; and result in a choice that a given individual may or may not prefer. Collective decisions can take the form of majority rule, unanimity, various levels of consensus, etc.

This pattern can now be translated into the context of human systems:

Government Translation: Larger groupings of people, i.e. about 150 or more, make sustainable group-level decisions through collective means. Collective group decisions tend to account for (or attempt to account for) the needs of all people in the group; allow for the expressions that any in the group might make; and result in a choice that a given person may or may not prefer. Collective decisions can take the form of majority rule, unanimity, various levels of consensus, etc.

For example, in a somewhat remote area, there is an indigenous village that has been occupied for several hundred years. The 3,000 or so villagers make claim to about 70,000 acres of surrounding land. By tradition, these people manage and hold their lands collectively, so there is no single owner of a given area. An energy company has developed an interest in petroleum exploration on part of the villager's land, and they make a financial offer to conduct their exploration.

The villagers have a four-step process for considering the company's offer. First they hold informational meetings with one to two senior individuals from each of 140 village families. The meetings include a description and discussion of what land would be sold, what would occur on the land they sold, and the amount of money they'd receive in return. In the second step, each of the senior villagers returns to their own family and passes along what they've learned. The family is given several days to discuss their opinions together, and with other friends and families if they choose. The third step is gathering up an opinion from each family member (which includes children who are old enough to understand the topic). The opinions are counted and the most frequent opinion within a given family is taken to stand as that family's preference. In the fourth step, each of the senior family members then return to the council meeting to provide their family's preference (either "yes," "no," or "no but would consider an alternative offer"). When all the family preferences are tallied, about 50 families are fully opposed, 40 are fully in favor, and 50 would consider an alternative offer.

Because more than half either agreed to the offer or would consider an alternative, the village allows the company to make a different offer. A spokesman from the village delivers the news to the petroleum company. He tells the company they are welcome to make a new offer, and that most objections were based on the lack of a plan for site cleanup should spills occur, and the fact that a traditional burial area sits on the property under consideration. The company reviews the situation and makes a revised offer to the spokesman. With this revised offer in hand, the villagers go back through their four-step decision-making process.

Among animals, there is a modified version of the large-group collective decision-making. If one or more individuals (who only constitute a small percentage of the group) are privy to exceptional or otherwise unknown information, and if they are extremely well informed as to the need patterns of the members of their group, then that individual or small group will sometimes make a choice on behalf of the greater group. A well-known example of this occurs with honeybees. Another example can occur with bird flocks or schools of fish when the whole flock or school willingly reacts to a danger that only few are directly aware of.

Pattern 24: Larger groupings of individual animals (i.e., about 150 or more) can make group-level decisions through a relatively small number of decision makers. This can only occur when the animals making the decision have exceptional access to information *and* they have a maximal understanding of the needs of their group.

Essential Translation: Larger groupings of individuals (i.e., about 150 or more) can make group-level decisions through a relatively small number of decision makers. This can only occur sustainably when the individuals making the decision have exceptional access to information *and* they have a maximal understanding of the needs of their group.

This pattern can now be translated into the context of human systems:

Government Translation: Larger groupings of people (i.e., about 150 or more) can make sustainable group-level decisions through a relatively small number of decision makers. This type of decision-making requires that the people making the decision have exceptional access to information *and* a maximal understanding of the needs of their group.

For example, a city with a population of five million wants to determine if and how a particular new technology can be incorporated into its business and community fabric, and on that basis it wants to proactively block or allow the technology. The city approaches the decision by employing people to gather information and then develop a representative consensus.

To accomplish this, they hire twelve professionals to develop a detailed report regarding the new technology, including its current and potential uses. This team includes one technical writer and eleven others who have already used the new technology. Meanwhile, a team of twenty-five surveyors spends focused time with selected citizens (spanning all ages, genders, backgrounds, etc.). Another five surveyors spend their time with businesses. The two survey teams ask questions specifically designed to help establish a portrait of the condition and need of the city's people and businesses.

Finally, a sociological team of twelve, including two historians, three economists, and two mental health specialists, among others, meet with the professional group and thirty representatives from the survey group. This yields a final group of fifty-four. This final group meets five times to discuss findings and exchange opinions. At a sixth meeting, proposals for implementation of the technology are discussed. At a seventh meeting, the fifty-four individuals each weigh in with their preferred course of action.

If two-thirds (thirty-six or more) of the group can agree upon a course of action, that course becomes the city's choice. If a two-thirds consensus cannot be reached after three more meetings, the issue is tabled for two years, at which time city staff review the topic to determine if it is still germane and worthy of a repeated city review. If so, the city will reinstitute its fact-finding and decision-making process.

Next, we look at how nature manages competition levels.

Chapter 13 Expression, Protections, and Group Formation

Earlier we defined *competition* as the case in which two or more individuals or systems are attempting to, or poised to utilize the same limited resources or the same single context. Then we saw how competition in sustainable natural systems declined in the latter stages of succession. How does this happen? Why don't individuals and groups continue to experience more and more competition for fewer and fewer resources over time? Our next task is to look at competition more directly and understand just how it is managed within natural systems. We will focus more on competition at the individual level because it makes the discussion simpler. However, principles pertaining to competition can also apply to groups such as species.

Individuals can compete for things such as water, space, food, nutrients, sunlight, warmth, a nesting area, a mate, social status, etc. Competition often occurs between similar individuals. Two huckleberry shrubs growing near each other might compete for soil nutrients. Two hickory trees growing near each other might compete for light. Two robins might compete for some of the same worms.

But, competition can also occur between dissimilar individuals. A pine tree might compete for water with the wildflowers beneath it. An elk and a buffalo might compete for some of the same forage plants. A bobcat and a golden eagle with overlapping territories might compete against each other for prey. A desert mule deer coming to take a drink might compete slightly with grasses trying to use the same water at the edge of the water hole.

While competition can occur between similar individuals, or between dissimilar individuals, competition is usually more intense when it involves *similar* individuals. This is because creatures with the same form (such as two hickory trees) tend to have nearly the same resource requirements. So, the most threatening competitors for any particular individual, at least on average, are the individuals most similar to itself.

Two sunflowers growing right near each other not only try to gather light, water, and nutrients, but they have just about the same strategy for acquiring them. They both require light at the same time (especially from early summer and onward). They both try to draw water from about the same depth in the soil. They will both have similar nutrient requirements. They will even try to attract the same types of pollinators. Not only will two sunflowers tend to have nearly the same needs, they will also find that the environment that can meet those needs is the same environment. In other words, if they have the same needs, they will tend to live and grow best in the same locations. If there is a limited amount of their preferred habitat available, they are positioned to covet the exact same space.

So our sunflowers not only want the very same things, but they want to gather the *very same things from the same location*. This is the conundrum of competition faced by each individual: each one is attracted to the same spaces that very similar individuals are attracted to, but if they inhabit the exact same space, intense competition will mean that they cannot readily obtain the necessary resources to meet their survival needs. For plants and animals, this is a potential setup for extreme and ever-heightening competition in which every resource scrap is the subject of a battle.

Despite this logical argument for extreme and ever-heightening competition in nature, our earlier look at the process of succession suggested that competition is actually *reduced* as sustainable systems mature (see Patterns 15 and 16, Chapter 10)—*not* endlessly intensified. In addition, we saw in Pattern 17 that the length of time spent in the mature, less competitive stages is actually much longer than the time spent in very high competition.

How does nature do this? How does nature take what should logically be a setup for extreme, stifling, and ever-heightening competition and transform it into something that has lots of cooperation, and is favorable for many individuals?

The main way that nature addresses this conundrum is by having similar individuals occupy the same general areas, but then having them separate, spread out, or occur at a limited density within those areas. This ability to have similar individuals separate, spread out, or become thinned out, is a critical mechanism for the *suppression of extreme competition* in nature. The suppression of extreme competition can be thought of as the way that life for individuals is made more manageable. This idea is simple enough, but we need to know how competition is suppressed before we can understand it well.

In nature, all individuals instinctively want to survive and prosper. To put it another way, each individual has an ability or willingness to engage in activities that promote their own *survival, growth*, and *reproduction*. For example, a sunfish will avoid bass or egrets (survival), eat insects or small fish (growth), and lay eggs (reproduction). For any single sunfish, avoiding predators and growing generally requires good hiding areas with enough food nearby. If the sunfish can avoid predators and find enough food, it can grow big enough to reproduce during nesting season.

This is a fine plan for a single sunfish, but as a single sunfish goes through these efforts, there will be many other similarly skilled sunfish in the pond trying to do the very same thing. The sunfish might be numerous and they might form schools, but if they all congregate in the very best area, they'll all quickly run out of food. And if there are too many sunfish in the pond, there won't be enough food for them to all survive and reproduce. Thus, the sunfish are likely to find themselves in some degree of competition with each other.

How much competition? Here's where our investigation of this topic will be facilitated by some new terminology. We can think of the extent to which a particular sunfish is free from competition as the degree to which it can exercise self-*expression*.

We can lump the sunfish into four theoretical (self) expression groups that reflect competition intensity. This illustration of expression applies not only to sunfish, but to all plants and animals. Here are the four levels of expression:

1. No Expression: When competition is extreme (i.e., when it has not been suppressed) a typical sunfish will lack sufficient resources to meet its daily bodily needs. Under these circumstances, it will die sooner rather than later, never having grown to anywhere near its potential, and never having reproduced.
2. Minimized Expression: When competition is very high, a typical sunfish may be able to meet its daily bodily needs, but no more. It may survive, but its growth and reproduction will be nil or almost nil.
3. Significant Expression: When competition is moderate, a typical sunfish can normally meet its daily bodily needs, and will generally have enough resources left over to grow and reproduce as well. Although depending upon the exact level of competition, it won't grow quite as rapidly, or reproduce quite as much as it could have if there had been no competition at all.
4. Maximum Expression: When competition is absent or minimal, a typical sunfish can meet its daily bodily needs, and grow, and reproduce. Not only can it do all these, but if it is in a favorable location, it can do these at a maximum theoretical rate, biologically unencumbered by resource limitations that competition would otherwise cause.

Remember that we first recognized a conundrum in nature wherein the most alike individuals would not only prefer to be in just about the same place, but they would have the highest potential to compete. In theory, this sets them up for unceasing and extreme competition. Then we noted that instead of falling into a state of extreme competition, nature simply suppresses the competition by having similar creatures separate, spread out, or limit their densities.

If we look at the list of expression levels above, relatively undisturbed natural systems reduce similar individuals down to the approximate density required to achieve *Significant Expression*. This third expression category includes moderate competition but it leaves significant room to prosper.

If this is all true, then nature is enabling many individuals to have an opportunity for Significant Expression by having similar creatures separate, spread out, or limit their densities. This is a valuable understanding, but we aren't quite done yet. We still need to know *how* creatures are able to separate, spread out, or have lowered densities.

There are four mechanisms in nature that enable the density of similar creatures to be reduced, thus enabling individuals to experience Significant Expression. We will call these mechanisms "Protections." These Protections can be applied singly or in combination. Remember that we are focusing on competition between individuals, but in most cases these same concepts apply to groups:

1. Resource Use: When an individual (or group) uses resources, it is depriving other competing individuals from obtaining those resources. This can discourage or diminish a competitor, and thus reduce the immediate level of competition. The resources being used can be water, food, light, nutrients, etc. For example, when a coyote catches and eats a rabbit, it has just deprived its regional competitors from being able to catch that same rabbit. Or when a tree spreads its roots out and pulls nutrients out of the soil, it is depriving its neighboring plant competitors of the chance to use the nutrients it already extracted.
2. Self-Defenses: When an individual (or group) has adaptations or behaviors that are designed to thwart a potential competitor, it can discourage or diminish a competitor, and thus reduce the immediate level of competition. Self-defenses can take the form of fending against or fighting off a competitor, threatening a competitor, establishing a guarded territory, emitting poisons, etc. For example, when a coyote chases intruders from its territory, it is discouraging them from hunting in its vicinity. Or when a tree emits chemicals from its roots that slow the growth of other plants, it is diminishing potential competitors through a form of self-defense.
3. Restraint: When an individual (or group) has adaptations or behaviors that are designed to limit its own growth or expansion, it can reduce the immediate level of competition for itself and other individuals. These restraints can take the form of individuals/groups restricting themselves to a niche, population control (population suppression), geographic control (migratory suppression), mating suppression, territory avoidance, or threat avoidance. For example, when a coyote refuses to venture into another coyote's territory, it is avoiding potential competition with its neighbor for food or mating. When the birth rate of deer decreases following diminished food availability, they are reducing their own level of competition. Or when a tree directs branch and root growth away from nearby trees, or away from the root zone used by wildflowers, it is using a restraint mechanism to avoid potential competition with its neighbors for light, water, and nutrients. In some cases, predators and diseases within natural systems can also be considered restraint mechanisms, in that they limit the density or spread of individuals or groups.
4. Barriers: When an individual (or group) is situated such that a potential competitor is unable to breach a physical, chemical, logistical, or environmental barrier, it prevents the development of further competition that would otherwise ensue. This type of competition suppression is more apparent at the group level than the individual level, but it does apply to both levels. For example, a span of semi-desert that is inhospitable to wolves could keep a population of coyotes safe from competition against their larger relatives. Or, a tree growing on a

small island in a lake—or on a rock outcrop—will have a much-reduced likelihood of competing trees arriving.

These Protections are the four types of methods that nature uses to suppress extreme resource competition and avoid the *No Expression* and *Minimized Expression* alternatives for individuals. But very interestingly, from the perspective of any single individual, none of these four Protections are singularly dependable or always dependable.

The first one, *Resource Use*, is based on the fact that if an individual uses its resources, those resources won't be available to its neighbors. True, but unless they can be taken and stored, neither are they available later to the individual itself! Having the minimal amount of a resource will get an individual through today, but if it wants to live and grow the next day, and reproduce at some point, the individual must actually have access to a surplus of the resource. In other words, any individual trying to enjoy *Significant Expression* has to have access to more than a bare minimum of resources.

For the coyote, it means there better be more than one rabbit in its territory—there needs to be one for tomorrow, and another for the next day, and another for the next day, etc. For it to secure enough rabbits to have a possibility of future Significant Expression, its territory has to expand until it contains enough rabbits or other prey. Resource use is an important part of competition suppression, but it would be self-defeating for an individual to suppress competition by depleting *all* of its own resources—it would leave the individual unable to meet its own future needs.

What then about *self-defenses*? Can an individual rely only on self-defenses to suppress competition? This is another important mechanism, but self-defenses require lots of energy to employ. It takes time and energy to patrol a territorial boundary, to fight with intruders, to make protective chemicals, etc. And sometimes the more extreme self-defenses, such as fighting off intruders, can result in loss or death for the defending individual. So, while self-defenses are critically important for many individuals, they are costly and often have to be applied judiciously.

The third Protection that suppresses competition is *restraint*. This mechanism often relies on individuals to limit their spread and forays into each other's areas. Let's consider the logic of this mechanism. Through natural selection (i.e., the survival and reproduction of better-adapted individuals), we would expect that individuals with a "drive" or a "will" to survive and procreate will tend to survive and reproduce more than individuals that lack such a drive.

It may be somewhat circular logic, but because of this natural selection, we shouldn't expect to find a world full of living things that don't want to live, grow, and reproduce. Rather, we'd expect to find a world of things that *will* try to live, grow, and reproduce. Inherent then, is that whenever individuals can grow and reproduce into an unused or uncolonized area, they will. So, there is a very natural inclination among living things to fill spaces—spaces that on their terms appear unfilled. This is how life could spread across the planet in the first place. While nature has produced many restraint mechanisms for creatures to limit their own spread and reproduction, it would be antithetical for them to completely eliminate their impulse or their own will or ability to spread into nearby areas.

If predators or diseases are providing restraints, then a given individual only benefits if those predators or diseases are activated upon its competitors, rather than upon itself. However, a given individual has limited influence over the application of predation and disease. Overall then, a given individual cannot rely completely upon restraint behaviors, predators, or diseases to suppress extreme competition.

Barriers are the last Protection. Barriers occur in all sorts of situations. Oceanic islands can be famously isolated from continents. And in an inverse situation, inland lakes are often isolated from other waters by large spans of land. These forms of isolation can make it very difficult for competing individuals to show up. In some situations, time can even act as a barrier, or other resource requirements can effectively create a contact limiter between species or groups.

Most barriers exist on their own accord. For example, islands and isolated lakes occur not because individuals want their competition suppressed, but because geological forces establish them. Thus, any individual or group that relies on barriers for competition suppression is generally relying on a condition that is outside of their control, and could change. If a barrier has been relied on heavily, and the barrier is broken, outcomes for the exposed individual or group can be negative. Barriers are very influential in the biological world, and we will revisit them again for different reasons.

To recap up to this point, nature addresses the conundrum of potentially extreme competition by suppressing competition. The tools to accomplish this are four individual and group Protections. But none of these four Protections alone—resource use, self-defense, restraint, and barriers—are always dependable or feasible. Therefore, individuals and groups tend to rely on combinations of these four Protections.

Because a given individual or group cannot experience Significant Expression while relying only upon Protections comprised of resource use, and because restraint and barrier Protections are generally not under their own control, they are reliant upon self-defenses to make up for any gaps in the other three Protections.

Pattern 25: Extreme competition for resources or other limited opportunities is ultimately avoided in nature by suppressing competition. This is done by reducing the density or numbers of competitors until a balance point of moderate competition is achieved. At this level, individuals have sequestered enough resources for significant, but not maximum, expression. Significant Expression requires that an individual sequester resources for both a present and a future. This applies to individuals singularly, and when they are in groups.

Essential Translation: In sustainable complex systems, extreme competition for resources is ultimately avoided by suppressing extreme competition. This is done by reducing the density or numbers of competing individuals or groups of individuals until a balance point of moderate competition is achieved. At this level, individuals and groups have sequestered enough resources for significant, but not maximum, expression. Significant Expression requires that individuals and groups sequester resources for both a present and a future.

This pattern can now be translated into the context of human systems:

Economic Translation: In sustainable economies, extreme competition for resources is ultimately avoided by suppressing competition. This is done by reducing the density or numbers of competing individuals or groups of individuals until a balance point of moderate competition is achieved. At this level, individuals and groups have sequestered enough resources for significant, but not maximum, expression. Significant Expression requires that individuals and groups sequester both present and future resources.

Environmental Translation: In sustainable environments, extreme competition between individuals or groups of individuals is ultimately avoided by suppressing extreme competition. This is done by reducing the density or numbers of competing individuals or groups according to the resources available in a given land area or land-use type. The density of individuals and groups is suppressed until those that remain have sequestered enough resources for significant present and future expression, but not Maximum Expression.

Physiological Translation: Sustainable individual and group livelihood includes access to enough resources to achieve Significant Expression.

Examples pertaining to Pattern 25 will be included within those of the next pattern.

> Pattern 26: In natural systems, extreme competition is suppressed via four Protection avenues: resource use, self-defenses, restraint, and barriers. This applies at the individual and group levels. Each of the four Protection types is fallible or practically limited, which often leads to their use in combination.

Essential Translation: In sustainable complex systems, extreme competition is suppressed via four Protection avenues: resource use, self-defenses, restraint, and barriers. This applies at the individual and group levels. Each of the four Protection types is fallible or practically limited, which often requires that a combination of the Protections be used.

Much earlier it was noted that boundaries between complex systems are often subjective rather than objective. The subjectivity can result from a lack of complete separation between the systems, or from an intergradation of the boundary between them. This means that perspectives regarding where one complex system ends and where another begins will vary. Herein, we are discussing complex systems such as species, natural communities, economies, communities, and governments in order to develop a better understanding of how things work, not to argue that these systems exist in very clearly demarcated units (which, they apparently don't).

Trying to separate the seven complex human modalities is a subjective exercise, and in cases where they are very entwined, separating them can become more confusing than enlightening. The purchase of a wedding dress might fulfill economic, community, occupational, cultural, and physiological needs and objectives. Trying to describe that purchase otherwise might be so subjective that it is confusing.

When it comes to the Protections that suppress extreme competition, this categorical difficulty arises. Somebody might lock their door to protect *numerous* assets, such as money (economic Protection), family (community and physiological Protection), and their house (environmental Protection). In other words, the act of locking their door could be an effort to simultaneously protect four system modalities. It would be unrealistic to express this only as an effort in economic Protection. Protections are very complicated, often arising from at least two complex systems, and designed or able to protect other complex systems. In fact, locking the door in some families might also be done in order to keep young children in as much as it is to keep others out.

The following translation and examples relating to Protections are an attempt to recognize and allow for complexity and interconnections. The main point of the examples is to show a span of Protection types, and what utility they could have, not to suggest a single way of categorizing Protections.

This pattern can now be translated into the context of human systems with the seven complex human system modalities combined together:

Economic, Community, Occupational, Government, Environment, Cultural, and Physiological translation: The interests of people and groups of people living in and with sustainable economies, communities, occupations, governments, environments, cultures, and physiognomies are protected from extreme competition via four Protections: resource use, self-defenses, restraint, and barriers. Each of the four Protection types is fallible or practically limited, which often requires that a combination of the Protections be used.

Numerous merged examples regarding Patterns 25 and 26 are given below:

1. When he has to wait overnight until his next bank deposit, the owner of a neighborhood five-and-dime store keeps the extra cash in a steel safe that is bolted to a cement floor (economic self-defense and barrier).

2. A family-run business preferentially hires family members whenever possible. This helps to prevent outsiders from working their way into the business and steering it to serve other interests (economic, community, and physiological self-defense).

3. A hunter posts NO HUNTING signs on his own 35-acre property to give himself a better chance of success during turkey hunting season (economic and occupational self-defense and restraint).

4. A retired man fishes out on a small lake several times a week. He's always the first to break the ice in spring, and he seems to know the best fishing spots and tactics. When others visit the lake to fish, they often have only moderate success, as the retired fisherman has already caught most of the best fish. Because of this, the lake remains a relatively unpopular fishing spot (economic resource use).

5. A nation forbids exports of three food staples so that other (foreign) nations will be unable to outbid its own residents for these foods (economic, cultural, and physiological self-defense). This nation also invests in technologies that make the production of these three foods relatively efficient and profitable (economic and occupational resource use).

6. A religious community encourages its members to share with those who have less (a form of economic restraint).

7. A study shows that the human population of a township consumes about as much food as can readily grow within its borders. In an effort to prevent further population expansion, it caps the number of homes that can be added within its confines (economic, environmental, and physiological restraint and self-defense).

8. A group of people inhabiting a valley is offered free construction of schools and infrastructure in exchange for 100-year lease rights to its only prime farmland. However, the valley dwellers refuse this and any other offer that limits their ownership or rights to these farmlands (economic, occupational, government, environmental, and physiological self-defense).

9. The Senemété Tribe claims a homeland defined by three mountain peaks and a river. This homeland is bordered by three other tribes. The other tribes are almost always respectful of the Senemété boundaries (economic and cultural restraint). However when an uninvited person strays into their territory, the Senemété are quick to catch, detain, and ransom the individual back to their home tribe. This vigilance helps discourage additional trespasses (economic and cultural self-defense).

10. A government prohibits foreign companies from establishing within its boundaries unless at least 80 percent of the employees that are hired therein are its own citizens (economic, occupational, and cultural self-defense).

11. A homeowner puts a fence around a small vegetable garden to keep out deer and rabbits (economic self-defense).

12. On a remote island, food raised locally is relatively inexpensive. Whereas food imported by boat tends to be very expensive (economic and occupational barrier).

13. A sheep-herding village continually moves sheep to better pastures, but never brings its sheep to graze upon three particular mountains, because these mountains are left for deer and wolves (economic and environmental restraint). At the same time, the villagers guard their sheep from wolves and other predators with herding

dogs (economic and physiological self-defense). When wolves come within visual range of the village, they are immediately shot (economic, community, and physiological self-defense).

14. A state issues only a hundred moose hunting licenses in a given year, and is prepared to lower that number if the moose population significantly declines (economic, environmental, and cultural restraint).

15. A village is essentially hemmed in by the ocean, steep cliffs, and deserts. In this village, it is taboo for a woman to have more than three children (economic, community, and physiological restraint).

16. A small nation discovers a wealth of potash within its borders. Its government quickly establishes a national company to extract the potash (economic and occupational resource use).

17. A scenic lake with excellent fishing is perched in a mountain valley. Access to the lake is almost impossible except through a town and then across privately owned property. The townspeople and landowners never discuss this lake with outsiders, nor do the landowners allow anyone other than townspeople to cross their land unless they have permission from someone in town (economic, community, environmental barrier and self-defense).

18. A tribe has an age-old cultural tradition in which each year, the very wealthiest tribe member gives away all of their possessions except for their home and clothing (economic and community restraint).

19. A nation studies its economy, food base, and environmental challenges, and then establishes an upper population ceiling (economic and environmental restraint).

20. A cement company is aggressive and extremely competitive within a 20-mile range, but does not solicit work beyond this distance (economic restraint).

Clearly, Protections that serve economic purposes are common, but they aren't the only purpose of Protections.

As noted, the previous patterns can be viewed as applying to both individuals and groups. There isn't much conceptual difference between a single individual sequestering enough resources for itself/himself/herself, and a group of individuals sequestering enough resources for the whole group. In this way, both individuals and groups can be protected from extreme competition, in order that they might experience Significant Expression.

However, there can be additional, unique implications to Protections when whole groups are involved. Many groups can self-replicate—they can carry themselves forward through time indefinitely, and can morph considerably as they do. As we will see, it turns out that Protections are a key requirement for the development and sustainability of self-replicating group differences, or *diversity*.

The diversity of lifeforms on Earth seems astounding. *Millions* of different species crawl, run, swim, float, drift, fly, root, dig, spread, and cling. These different species have a huge myriad of physical, chemical, and behavioral mechanisms that allow them to accomplish so many different, amazing things.

Just taking a look at one group of creatures, biologists think there are about *ten-thousand species* of birds on Earth. That makes for a lot of different kinds of birds. There are bird species that stay aloft for hours or days, and others that can fly thousands of miles without ever landing. There are bird species that weigh well over twenty pounds, and yet can still fly. Another can incubate eggs in temperatures *lower than -20° Fahrenheit*.

Of course, these incredible behaviors are not limited to birds. There are types of fish that swim in caves and are completely blind; others that live at the bottom of the ocean where pressures are hundreds of times higher than at the surface; and still others that incubate their eggs in their mouths. Some fish can survive drought by burying themselves in mud. Some swim over one thousand miles *up*stream just to lay eggs.

There are desert plants that can survive more than a year without any rainfall, and there are trees that can live longer than three thousand or four thousand years. At the other extreme, some plants can germinate from seed and then grow to produce their own seeds in about one month. There are insects living in colonies of millions of individuals. There are bacteria that live in boiling hot vents, and others that live deep in ice and rock. This astounding list goes on and on.

How did such amazing and different species ever arise? From a scientific perspective, the concept of biological evolution is that with each new generation, a species of plant, animal, or microbe, can change ever so slightly. If changes keep occurring with each new generation, a species that starts out looking like one thing can eventually change into something that looks quite different. For example, over time, a small brown sparrow could evolve into a colorful, larger bird species the size of a crow. Or a small tree with wide leaves could evolve into a thorny shrub with narrow leaves.

These sorts of changes might seem impossible, or unlikely to ever occur. However, biologists believe that over long periods of time, sometimes many millions of years, these changes can and do occur. As an analogy, you might stand at one coast of a continent and feel that at a practical level, you could never walk across the entire continent to reach the other coast. However, if you can walk, you would be wrong. Taking only one step at a time, you could indeed make this journey, just like other people have. Biologists believe that life on Earth has been evolving, step by step, for so long that it has been able to diversify from a handful of tiny creatures into millions of creatures—those amazing birds, fish, fungi, plants, animals, and microbes that now inhabit the planet.

All that is needed for evolution to occur is a few simple ingredients. First, there need to be differences between the individuals in each new plant/animal/microbe generation.

Second, there has to be better survival and/or more reproduction by some of those individuals in the new generation.

For example, if brown colored beetles give rise to a generation of brown beetles and some occasional green beetles, and the green beetles are less conspicuous to hungry birds, then the green beetles might survive longer and produce more young. If this happens over and over, this brown beetle could evolve into a green beetle. Similar changes could occur in regard to the beetles' size, sensory abilities, flying abilities, food habits, and environmental tolerance. So for example, what began as a large brown beetle with poor eyesight, poor flying capability, and a year-round diet of dead leaves might, over time, evolve into a small green beetle with good eyesight, good flying capability, and a seasonal diet of green leaves. Even though this change might take thousands of years, it is conceivable from an evolutionary perspective.

Biologists believe that evolution not only allows a species to evolve into a new, different species, but it allows one species to evolve into two, three, ten, or even hundreds of unique species. For example, our original large brown beetle species could give rise to the small green beetle species. If the original brown beetle species is still present when the green species has formed, then one species has become two. Afterward, perhaps the large brown beetle species gives rise to a large red species, then a small brown species, then a black and yellow species, etc., etc. All of this could happen without the large brown species going extinct. If so, what began as one beetle type would have become five species. These five species could each give rise to yet more species.

The development of these different species is called *speciation*. Some people like to picture speciation as a spreading out of types from a central point, like spokes from a wheel. Other people picture new species like twigs separating off a main branch of a tree. Biologists believe that speciation is essentially responsible for all forms of life on Earth—that over great periods of time, a smaller number of species have evolved into hundreds, then thousands, and then millions of species.

The gist of the theory of evolution is pretty simple. Single species evolve into one or more new species over time. The case in which an original species creates a new species without going extinct, or similarly splits into two new species, is perhaps the most simple avenue through which biodiversity expands. In this way, where stood one thing now stand two.

Following this line of thinking, you might assume that over time, the number of creature types (i.e., *biodiversity*) will expand and expand and expand, resulting in an endless number of species. However, this is not the case. Biodiversity does not go on indefinitely on planet Earth.

For example, North America is very large. Certainly it has enough *space* to house dozens of deer species, dozens of bear species, hundreds of types of hawks or jays, and hundreds of oak, maple, and pine tree species. But if we add the species up, we see it doesn't. It can depend on which taxonomist you listen to, but regarding native species north of the Mexican border, there are "only" about three bear species, five species of deer (including elk, moose, and caribou); twenty-five species of eagle, hawk, or falcon; a dozen species of jays; eighty-five species of oaks; a dozen species of maples; and thirty-eight species of pines.

Evolution has the ability to eventually create species for each new situation and variation that arises, so what is preventing more and more and more species from appearing? We could imagine that perhaps there hasn't been enough time for evolution to occur, and we are simply looking at North America too early in the evolutionary process. However, jays, deer, and pine trees have been in North America for millions of years, which is enough time for them to differentiate into additional species. We could imagine that perhaps more species did evolve, but they subsequently went extinct. This is true to a degree, but it is probably not enough to explain why the number of species doesn't just increase and increase indefinitely. It turns out that forming a new species isn't always that easy.

In nature, there are two main ways in which plants, animals, and microbes can reproduce. The simplest way is when one individual essentially clones itself—that is, to spread by budding off, growing laterally, etc. This cloning process generally produces new young that are identical to the parent. The other form of reproduction is when some sort of breeding or mating occurs between two individuals—this is the norm for most plants and animals. Mating allows for a new genetic combination to arise, and it creates offspring that might be slightly different from both parents. This is one of the major facilitators of evolution—by allowing a generation of offspring to be slightly different from the individuals in the parental generation, new forms are available to support evolutionary changes through time.

But in spite of this, reproduction through genetic combination and recombination (mating) is actually a double-edged sword with respect to evolution of new species. While mating reproduction allows for new forms and innovations to arise, it also has a tendency to squash differences after they arise. How can this be?

Let's go back to our example of the brown beetle species. The brown colored beetles crawl around the forest floor, looking for dead leaves to eat. Maybe every once in a while, just by genetic chance or by imperfect genetic copying, two of these brown beetles breed together, and a few beetles that prefer to eat green, living leaves are born. These new beetles walk the forest floor, looking for green leaves rather than brown leaves, and their bodies seem to grow well on the green leaves.

There's just one problem for these atypical beetles. They are brown colored, and with their color contrast against green leaves, they tend to be seen and eaten by hungry birds and lizards that forage on the forest floor. Overall, these periodic brown beetles that eat green leaves don't fare well. They are always eaten before their genetic character is passed along to the next generation.

However, chance genetic combinations occur for thousands of years, and on very rare occasions, a beetle or two are born that not only like green leaves, but also have a greenish coloration. These beetles not only thrive on green leaves, they also tend to survive to adulthood and reach their own opportunity to breed. Will these green beetles simply breed together to form a new green species? No, they won't, because they won't breed with each other. When it is time to breed, these healthy green beetles that eat green leaves crawl around on the forest floor and almost always find a mate that is brown and eats dead leaves. The result of these matings is a generation with a mix of brownish-greenish beetles that eat dead leaves and green leaves. These young don't fare all that well, because they aren't very well camouflaged from predators while they eat brown leaves, or while they eat green leaves. So, all else being equal, the lucky green beetles that eat green leaves tend to yield a smaller number of surviving offspring, and consequently, their traits will not tend to increase in the gene pool.

From our point of view, it seems like the brown beetle species should somehow just transform into two different but successful species—the brown beetles that eat brown leaves and the green beetles that eat green leaves. But, the genetic recombination that allows both beetle types to arise in the first place also acts like a unifying magnet that restricts the amount of variation that can develop within a single species.

Because the two beetle types will keep interbreeding, and any green-brown offspring are quickly eaten by predators, the green beetles are continually washed back out of the gene pool each time they arise. Maybe every once in a great while there is a green-beetle to green-beetle mating, resulting in many prosperous little green beetles. Is this enough to create a new species? No, it isn't.

This forest floor contains mostly dead (brown) leaves, and despite the occasional appearance of healthy green beetles that eat green leaves, and even a mating between them, their young tend to be outnumbered by brown beetles, and when it comes to breeding time they end up pairing with another brown beetle (again stymieing the advent of a larger green beetle group that could form a species).

Maybe your mind is now running with this, trying to figure out how two species can form from one. The problem is that the green beetles are anchored to the brown beetles. Another way of stating it is that the important process of genetic transmission also acts to synchronize the group. The group is genetically designed to reproduce itself authentically into the future, and under these circumstances, it is unable to schism and simultaneously produce both a unified future and two different futures. While evolution allows for changes, it also holds the group together in a shared collective destiny.

Then how can a second species ever evolve? It turns out, there is a way. For the beetles to become two different species, what they need is a means of developing two separate destinies. To do this requires the two beetle groups to congregate and operate independently of one another—in other words, for the groups to *separate*.

For example, maybe after thousands of years of mostly brown beetles and occasional green beetles, two green beetles are born that not only feed on green leaves, *but* they also ascend to the tree canopy when it comes time to search for a mate. These two green beetles climb a tree and they find each other, leaving the brown beetles far below. The green beetles breed and give rise to a generation that is green, crawls around on the forest floor, and eats green leaves. Then when the time comes, these offspring beetles also have an urge to ascend a tree to find a mate. The green beetles are no longer constrained by back-mating with the brown types. Finally, a second and unique species can begin to form. By thereafter interbreeding only with each other, the green beetles can develop more and more unique genetic adaptations that allow them to specialize just in eating green leaves in their particular forest situation. Eventually, the green beetles could become smaller, develop better eyesight, good flying capability, and have a shorter feeding season.

This diversification/differentiation would not be possible without separation from the brown beetles. In nature, of course, the story can become even more complex, but the presence of some form of separation or segregation is almost always necessary for the expansion of biodiversity. If you recall the four Protections discussed earlier in this chapter, it is the last one—*barriers*—that are generally required to initiate the expansion of biodiversity. In the example with the beetles, ascending into the tree canopy during the mating period essentially creates a contact barrier between the two beetle groups during mating.

The barriers that allow new species to form can have a geographic source (like separations due to an expanse of ocean), a genetic source (like genetic incompatibility that results in nonviable offspring), or a behavioral source (like different mating calls that aren't attractive to the other group). While the presence of a barrier can benefit an individual by allowing it to achieve Significant Expression, a whole group can speciate with the assistance of barriers, thereby increasing the diversity of living things. These barriers do not have to be absolute for some diversification to occur. However, when there is only a weak barrier, a single species might be able to develop diversity in appearance and habit, but its diversity across the barrier will usually be contained or restricted within the limits required to maintain only one species. In other words, the stronger the barrier, the greater the diversification.

Barriers of different sorts can lead to the development of many, many species over time. When those barriers are strong enough, or in place long enough, they even allow the development of unique natural communities and ecosystems. The enabling role that barriers play in diversification at the species level or higher is the basis for the next pattern.

> Pattern 27: A single species can evolve in a certain direction, and it can diversify to a degree, but it cannot split into two or more species without moderate or strong forms of barriers. The numbers of natural community types and ecosystem types are also expanded by barriers.

Essential Translation: In sustainable complex systems, the diversification, or expansion of the number of self-replicating groups requires barrier separations between the groups. In general, the more complete the barrier or separation, the more pronounced can be the diversification between the groups.

As with the prior two patterns, the impacts of barriers/separations upon economies, communities, occupations, governments, environments, cultures, and physiognomies are so intertwined that treating them individually would be unhelpful.

This pattern can now be translated into the context of human systems:

Economic, Community, Occupational, Government, Environment, Cultural, and Physiological translation: Human groups living and operating within sustainable complex systems require barrier separations in order to create unique economies, communities, occupational groupings, governments, environments, cultures, and physiologies. The more complete the barrier/separation, the more the groups can become different.

Several examples are provided:

1. A culture of sea explorers seeks uninhabited islands. Over the course of a few generations, they discover five widely scattered southern islands and establish permanent settlements. These island settlers retain seafaring ability, but their home islands are separated by such great distances that they have almost no contact with each other or with other groups of people. After about 5,000 years have passed, these groups still have some vague cultural similarities. However, they speak five different languages, have different types of government and land tenure systems, and even look somewhat physically different—varying to some degree in height, body size, skin tone, and facial appearance.

2. Following a disagreement, four families leave their larger tribe and settle on the opposite side of a deep and wide canyon, where they establish a new village. Over centuries, as their village grows, they have regular contact with their original source tribe and with other neighboring tribes. About 900 years after its first founding, the people of the newer village can still communicate with the older village, though this requires some effort because some words don't translate or don't carry the same meaning for the other group. The people of the two villages dress mostly alike, but they engage in somewhat different religious customs and have a different economic emphasis on certain crops and game animals.

3. A nation determines that it needs to make higher education more widely available, and establishes four universities. The universities are established with the same overall charter, such as the teaching of fundamental subjects. However, no single body actually governs the operations or internal decisions of all the universities in unison. A review fifty years after establishment finds that each university is unique. For example they have different levels of course emphasis upon science, culture, and business, and they make internal decisions differently. Though they are all different, each seems to be performing successfully overall. A review after still another fifty years finds that the four universities have diverged further, with different specialized courses available at each, and with different social and academic cultures.

4. A large village is set between the coast and steep mountains. The two primary income-producing occupations available to villagers are fishing, and packing hundreds of miles over the mountains to trade dried fish for other goods. Due to the harsh local environment, both of these occupations are difficult and require significant skill. Also, these occupations have a three-month off-season at opposite times of year. Thus people cannot perform work in both fishing and packing, and those in one occupation may go months without seeing friends in the other. Fishermen have their own codes of conduct and customs that help to keep them alive and in good spirits. The pack traders also have codes of conduct and customs that serve a similar purpose, but they are very different from those of the fishermen. There is effectively a kinship and common understanding among fishermen, and among pack traders, but their understandings are so different that sometimes they are unable to recognize each other's views and philosophies.

5. A window manufacturing company finds that its windows are often installed improperly, and begins to provide an installation service. For the first twenty years, the installation service is housed at the window factory, where it is run by the factory managers and has a steady crew of about fifteen. During a factory expansion, the installation unit is then moved across town and given its own management team. The next twenty years witness a rapid expansion of the installation wing of the company—it expands into door installation, energy auditing, and then insulation retrofitting, and then it opens up two more of its own outlets. Twenty years after having moved out of the factory, the installation unit has 120 employees, a different corporate culture, and a different pay scale from the window factory.

6. Like many her age, Angela finished high school and went away to college. After college, she came back to her hometown and moved back in with her parents and brothers. At this point, she noticed that she had been developing some different tastes in the last few years—different preferences in eating, organizing, etc. However, it was not until moving into her own home after five more years that she was able to fully establish her own patterns, such as having a cleaning schedule, having certain types of furniture, and having certain types of food and cookware in the house.

We've just taken a look at how the diversity of groups grows, and now we will look at whether that diversity can be maintained.

Many people are familiar with the phrase "survival of the fittest," which is sometimes used interchangeably with the term, "natural selection." Although these phrases are well known, the scientific principles they refer to are often not well understood. For many people, the term "survival of the fittest" conjures up strong images—perhaps of giant bears or lions fighting to the death, or of one noble gorilla or giant tree surviving harsh events while those around them perish. However, these images are not especially realistic.

From the scientific perspective, survival of the fittest and natural selection simply mean that the most advantageous traits will have a higher likelihood of being passed on to the next generation. This is a very important component of the theory of evolution, because it helps explain how, over time and many generations, a species could shift toward a more advantageous form. The idea of natural selection is that individuals with more appropriate (more "fit") adaptations are more likely to live longer and produce surviving offspring than those with less appropriate (less fit) adaptations. As this selection process happens for generation after generation, individuals with the more fit adaptations would tend to increase in number while those with the less fit adaptations would tend to decrease in number.

For example, imagine that a grassland is becoming progressively drier as the climate slowly changes. A certain species of wildflower growing in this grassland might have some individuals with slightly thicker, waxier leaves, and some with slightly thinner, fragile leaves. Presumably, leaves that are thicker and waxier will be more resistant to water loss, although they also require more energy for a plant to produce. With a drying climate and less water available, the plants with thicker leaves might be able to resist evaporation and be better suited than the plants with thinner leaves. Over decades and centuries as the climate becomes drier, we could then expect that the plants with thicker leaves will live longer, produce more seed, and eventually become the dominant form of this species in this grassland.

Similarly, natural selection can also work in favor of stasis. If the climate stays consistently dry and the thick-leaved plants already predominate, natural selection might prevent thin-leaved plants from becoming numerous, even if they occasionally germinate.

Considering this example, the idea that natural selection can produce group change or stasis is simple. But what does natural selection mean for a single individual? Here is where the misunderstanding often arises. What applies to a group, on average, over time, does not always apply to an individual.

First of all, being very well adapted to a situation doesn't guarantee prosperity, because there is always an element of probability. So the best-adapted plant might be eaten by a goat, just because the goat happened to be walking nearby. Similarly, imperfect adaptation to a situation doesn't guarantee that an individual will die. Another factor is that conditions are always in flux anyway, so an individual who is "perfectly" adapted won't be perfectly adapted as soon as conditions change.

What this means is that there is no such thing as being perfectly fit or perfectly unfit. All individuals are really a mix of fit and unfit traits, and the more perfectly they are fit to one situation tends to mean they will be less perfectly fit to other situations.

There is also an aspect of momentum at work (see Pattern 19, Chapter 11). All else being equal, the individual most likely to prosper in any given situation is the one already occupying that situation. The established individual tends to have Protections in place that prevent competitors from easily ousting it (see Pattern 26). For example, as the climate slowly dries, a plant with thinner, more fragile leaves might not be as well adapted as plants with thicker, waxier leaves. But if a thin-leaved plant was already established, its Protections might make it hard for new plants to grow right in its sphere of influence. So back to the image of two huge bears fighting to the death…the actual expression of natural selection usually turns out to be much more nuanced and gradual than this.

As explained above, natural selection does not provide a certainty regarding which *individuals* will thrive and which will languish. However, in contrast to this, looking at longer time scales, natural selection *does* provide more of a guaranteed outcome. Over longer time periods, probabilities become very influential, and the most fit groups, traits, or genes will prosper most (species, traits, genes, and similar reproducible entities can be considered self-replicating groups).

Over the longer timeframes that are experienced by self-replicating groups such as species, it is the fittest groups or traits that will tend to survive, while the least fit groups or traits will tend to diminish. In the prior example, even if some individual wildflowers survive partly by luck during a drying climate, the more truly fit individuals will, on average, survive and reproduce more effectively. If the story is right, natural selection tells us that as the climate dries, the thin-leaved plants will tend to wane or diminish within the overall population, while the thick-leaved plants will disproportionately represent the population as time goes forward.

What this means is that individuals are never perfectly adapted, and don't have to be perfectly adapted in order to thrive (sure, it helps, but it isn't required). However, groups that are less fit than—or not adapted as well as—competing groups, are essentially *guaranteed to die out*, all else being equal. This is worth repeating—a less fit self-replicating group or trait *will* eventually be replaced by a fitter group or trait, all else being equal.

The fact is, though, that all else is often not equal due to group Protections (see Patterns 26 and 27). That is, if a group has secured barriers or other Protections, it is not so highly prone to being eliminated by a competitor. *But, if two or more self-replicating groups of creatures are poised to occupy the exact same space, and no Protections remain in place for one or both of them, then only one group will survive.*

So what are Protections from the self-replicating group perspective? They are often a bit more abstract than individual-level Protections. From an individual perspective, barriers prevent creatures from physically interacting due to time and space separations. From the self-replicating group perspective, barriers can also prevent creatures from physically interacting, but barriers can also be an inability to interbreed, or an inability to produce viable offspring. In general, self-defenses are used to ward off competitors, but for the self-replicating groups, self-defenses can also include an unwillingness to interact or interbreed with another group. In general, restraints are a turning away from competition. But at the self-replicating group level, restraints can manifest as a shifting of resource requirements (niche adaptations), avoidance of entire geographic regions, or a disease that acts unequally upon different groups. An example of a restraint could be a group that avoids eating the foods that a competing group is eating. Finally, resource use for an individual pertains to the resources it has directly obtained, but for a self-replicating group resource use represents the entire array of resource *types* used by the group, rather than the resources being directly captured in the field.

Putting these all together, Protections that can allow self-replicating groups to survive include an array of conditions, behaviors, and adaptations, such as physical or temporal separations from others; inability or unwillingness to interact or to interbreed; inability to produce viable offspring; all sorts of behavioral, chemical, or physical defenses; avoidance of entire resource types or geographic regions; and a tendency to use particular resources.

This all might seem complicated, so it might help to give some plain examples in nature. Let's go back to the grassland plants, and the wildflower species with a thin-leaved strain and a thick-leaved strain:

Scenario 1: The climate has been drying. With no biological Protections in place, we can expect our thinner-leaved plant populations to gradually die off and be wholly replaced by the thick-leaved strain of the same species. This replacement would occur whether or not the two strains were interbreeding.

Scenario 2: The climate has been drying, and some of the thin-leaved plants have begun to adapt by growing temporarily in mud flats following infrequent heavy rains. This might effectively separate the reproductive period between the thick-leaved strain and the thin-leaved strain (i.e., it provides a temporal barrier), allowing them to coexist indefinitely without interbreeding and without direct competition that might destroy the other.

Scenario 3: The climate has been drying in one area, but on the margins of the grassland, the climate is still moist. The thick-leaved strain of wildflowers prospers in the drier areas, and the thin-leaved strain prospers in the wetter areas. With this partial resource-use differentiation in place, the two strains could survive indefinitely without being extinguished by the other (Protections here could be seen as resource use, restraints, and barriers).

Scenario 4: The climate has been drying in one area, but on the margins of the grassland, the climate is still moist. This allows both plant strains to survive for many centuries. But then, the whole climate begins to become wetter again, making all of the grassland more suitable for the thin-leaved strain. As the climate changes, if the thick-leaved strain enjoys no new Protections from the thin-leaved strain, the thick-leaved strain will be forced to interbreed with the thin-leaved strain, and over time will be outnumbered, outperformed, and wholly replaced by the thin-leaved strain.

Scenario 5: As in Scenario 4, the climate has been drying in one area, but on the margins of the grassland, the climate is still moist, allowing the separate survival of the thin- and thick-leaved plant strains. Then the climate becomes progressively wetter again, forcing these two strains back into the same space. However, in the meantime, the thick-leaved strain has changed its flower shape, which attracts different pollinators and normally keeps it from interbreeding with the thin-leaved strain. Depending on the climate and the different local habitats available, this could allow both strains (now perhaps technically different species) to survive in the same region without extinguishing or displacing the other through direct competition or through interbreeding and the genetic synchronization it requires (Protections here could be seen as a combination of resource use, restraints, self-defenses, and barriers).

The next pattern addresses the likelihood or ability of self-replicating groups to persist when faced with a change in their Protection status.

> Pattern 28: If a species is comprised of different subspecies (strains), or if there are closely related species that are capable of interbreeding, those different groups require ongoing Protections in order for all of them to persist. Should their Protections all be breached, the different groups will face full competition that will result in a single surviving group. Contingent upon numerous factors, including fitness, the single surviving group may be any one of the different strains or species, or it may be a mixture of two or more of them.

Essential Translation: In sustainable complex systems, maintaining the diversity of potentially competing self-replicating groups requires Protections. If only a degree of group Protections are lost between these groups, some diversity of groups can be maintained—the amount depending on the level of Protections that remain. If all Protections are lost between self-replicating groups, diversity of group types will become unsustainable, and the resulting final group will be based on numerous factors, including relative fitness.

Recognize here that "diversity of groups" and similar phrases refer to the number of different groups, and how different those groups are from each other (like the number of completely unique languages that occur on a continent). These phrases are not pertaining to the diversity of forms that may occur within a single group (like the slightly different ways of talking that any two relatives have).

Obviously, the implication of this pattern is that when two or more self-replicating groups lose all of their Protections from each other, at least one of the groups has become unsustainable. As with the prior three patterns, the survival of different group economies, communities, occupations, governments, environments, cultures, and physiognomies is so intertwined with group survival that treating them individually would be unhelpful.

This pattern can now be translated into the context of human systems:

Economic, Community, Occupational, Government, Environment, Cultural, and Physiological translation: Sustainability of distinctive human groups requires maintenance of Protections. If only a degree of Protections are lost between human groups, some diversity of human groups can be maintained—the amount depending upon the level of Protections that remain. If all Protections are lost between human groups, diversity of human groups will become unsustainable, and the resulting final human group will be based upon numerous factors, including relative fitness.

Even though the origin of Pattern 28 is relative to groups, it can be applied to conditions within individual human lifetimes, due to the self-replicating systems (such as cultures) that humans utilize.

Numerous examples are provided:

1. A large island contains two groups of people: an agrarian and fishing people near the sea, and a herding people who live on its mountain slopes and in its narrow valleys. These two distinctive groups have inhabited the same island for many centuries. But when a long-dormant volcano on the island becomes active, it slowly begins to smother the island's higher lands where many of the herders live.

   At first, this pastoral group ekes out a living at lower and lower elevations on its smaller and smaller land base. But this becomes more and more difficult. After about 70 years of living off of a continually diminishing land base, the herders make an attempt to survive by bringing a series of land claims and complaints to the coastal villagers. The herders assert that they have the right to coastal lands as part of their island heritage. However, at this point the herders have little political or military leverage and are far outnumbered by the coastal group.

   The herders are faced with: a livelihood that is almost untenable in such a small area; an economic system that does not compete well against agriculture near sea level; being outnumbered; and being forced to live in close proximity to the coastal villagers. Over the next two centuries, the remaining herders disappear through attrition and assimilation with the agricultural and fishing villages.

2. An ethnic group inhabits dozens of widely scattered villages in a jungle. This ethnic group has ancient ties to the jungle. During a wave of regional colonization, conquering soldiers from several thousand miles away establish a post near three of the jungle villages. Although they are not treated especially well by the conquerors, the native villagers do not attempt to fight or resist the conquerors. Soon the conquerors bring reinforcements and family members as they begin trying to populate the jungle region. However, the conquerors continually meet unfamiliar conditions, dangers, and maladies to which they have little genetic resistance. Within about fifty years, the outpost has failed to produce any value for the conquerors or their homeland, and it is abandoned, leaving the native local villagers to themselves.

3. Three hundred miles of coastline are punctuated with six villages. The native villagers have different cultures and little historical contact outside of their immediate neighbors. However, global traders learn of potential precious stones owned by these the villagers, and they quickly attempt to establish contact with these isolated people. Each village responds differently to the traders, which helps to determine the play of events for many years to come:

Village 1: The first village welcomes the traders onto their shores, but the villagers won't allow them to stay overnight or establish any sort of residence. With their new trading riches, villagers prosper, but they are eventually stricken with disease, reducing them to a fraction of their original population. They survive, but the loss of elders and their knowledge weakens their culture. Meanwhile, in their weakened state, the traders establish a permanent settlement near the village, which has a further degrading impact upon their culture because the newcomers outnumber the villagers and regularly disregard their traditions and laws.

Village 2: The second village welcomes traders but prevents their direct access, allowing trade only to take place from boat to boat at sea. This village benefits from trading and grows more prosperous. Eventually they are well off and continue to defend their territory against outside entry.

Village 3: The third village bars the traders from entering. However, while they communicate this restriction verbally, they make no indication that they will back it up with force. Soon, a group of traders land and take a group of villagers hostage. With this as leverage, they establish a fort and trading post. The village survives, but with a changed economy that makes the villagers more beholden to outside forces and leaves them with a diminished sense of self-identity and determination.

Village 4: The fourth village also bars contact and trade, and does this along with a bold show of weapons. This village remains largely unchanged.

Village 5: The fifth village allows trade and has almost no restrictions on the behavior of its own people or the traders. This village is colonized by traders. Soon there is intermarriage, and eventually this village becomes an unplanned intermediate of the two races and cultures.

Village 6: The people of the sixth village hide their valuables and insist they only have low-value items to trade, such as thatch and chickens. They simultaneously proclaim that they want no visitors and display weapons, and they show a willingness to use them. This village remains largely unchanged.

4. A religious group has ancient traditions that maintain its coherence. These include a unique language, myths, and rituals, as well as behavioral rules that prevent intermarriage or extensive contact with other groups. The religious group is a minority within their home nation, and when government leaders begin to persecute them, they flee their homeland for three nearby kingdoms, where they attempt to reestablish themselves within enclaves. However, they are essentially a powerless minority within each kingdom, and each of the kingdoms display different levels of tolerance toward the minority.

One of the kingdoms essentially practices cultural restraint, allowing the religious minority to establish and operate how they please, and without being forced to assimilate. The second kingdom also practices restraint toward the minority—allowing them to practice their customs, but it will not allow them to aggregate in any one area. Thus, they are spread across the kingdom as nuclear families or extended families. The third kingdom allows the minority to establish their own enclaves, but will not allow them to speak their own language or practice their own customs.

The minority group in each kingdom fares quite differently over the course of the next two hundred years. With their own language, foods, crafts, and activities, they have become a vibrant and stable subgroup in the first kingdom—they have even assimilated their abandoned homeland into their current sense of identity. In the second kingdom, religious minority descendants live as widely scattered families, and after two hundred years, their culture and identity is quite variable, ranging from those that speak their original native tongue to many who can barely speak a phrase of their original language. Similarly, there is a broad range of religious practice, and the group as a whole has become slightly less distinguished from the majority group with each passing decade. In the third kingdom, the religious minority has remained in its own neighborhoods. However almost none of its members are able to speak their native language, and knowledge of their original customs has virtually disappeared. Although most group members in the third kingdom have not intermarried into the majority population, they are culturally almost indistinguishable from it.

5. Two modest-sized neighboring nations have practiced internal population controls for many decades. This has allowed each to maintain and enjoy bountiful resources for their own use. With no history of resource coveting and the threat it would pose, the shared border between these nations is peaceful and unguarded. However, the western nation begins undergoing political and cultural shifts, wherein it gradually abandons its population management activities.

   With a growing population and a shrinking resource base, emigrants begin leaving the western for the eastern nation. Initially, the influx of immigrants is ignored as inconsequential, but when the culture and resource base of the eastern nation begin to show signs of strain, there is an attempt to close the border to additional immigration. However, by that time the eastern nation contains a large and vocal western nation diaspora, and along with growing economic and military threats from the western nation, the eastern nation fails to employ an effective border response. Eventually the eastern nation's original ethnic group is a struggling minority that is unable to maintain a comfortable resource base or express significant self-determination.

6. For centuries, a very large swamp forest has been unexplored and unexploited. Eventually though, teams of surveyors enter the swamp interior, where they are surprised to find a large tribe of people. This tribe speaks an unknown language and practices unusual customs—at least in the view of the society that sees itself as having discovered them. Before long, a swarm of scientists, curiosity seekers, and

people hoping for business development opportunities visit the tribe and surrounding swamp. Most of the tribal members attempt to accommodate or negotiate with the visitors, although a handful quietly abandon the main village and found a new village in more hidden part of the swamp.

When seventy-five years have elapsed since their "discovery," the tribal society has undergone upheaval in almost every way. Their tribal economy, technology, social cohesion, and systems of leadership have changed. The changes include far more social strife, a weakening of traditions, and a lower population due to diseases and emigration of their youth. Meanwhile, the new village established by the secretive group that left the main tribe resembles the pre-contact tribe in almost every way.

7. Coal seams were discovered in a rural community, and for three hundred years, coal mining became the standard occupation for local men. The mine work became so ingrained that the miners had many of their own words and their own subculture to promote good luck and survival. When the coal ran out, the subculture and rituals that the miners had developed faded away within about 100 years.

8. In hopes of becoming more prosperous, a family of six moves to a new continent. Although the family members hold their native culture in great esteem, they also feel uncomfortable with being so different from those around them. Although they do not intentionally go to great lengths to re-culture themselves, neither do they strongly attempt to maintain their native customs. Within four generations, the family (descendants) has become culturally indistinguishable from those around them.

9. The Bergmans and their four children move into a communal situation where many things, such as a large kitchen, are shared with other families. Despite all the sharing, the Bergmans still have a desire to keep some sense of separate identity. They take regular trips together, go out to dinner, and maintain their own holiday traditions. Sometimes they have to work harder to fit these activities in, but they seem to make a difference. Even after ten years of communal living, their coherence as a family has not diminished at all.

10. Briana and Tony married in their mid-thirties. But well before that they had each lived alone long enough to developed their favorite habits, such as when to go to bed, when to clean up, when to have quiet time, etc. Living together turned out to be more of a struggle than they imagined, and both chose to compromise many of their preferred habits in order to meet in the middle or make peace.

11. Two parishes were connected only by dirt roads and donkey trails. When a new connector highway is built, local business owners anticipate a jump in sales and profits. This does turn out to be true in some cases, but at the same time, many small companies are driven out of business. This seemed to especially occur whenever two very similar businesses were located in each parish.

12. In order to work more efficiently within its three national markets, an international automobile company splits off into three national companies. Following the split, each company operates successfully within its nation. However, two decades later, international treaties force the three companies back into direct and open competition with each other. Each company quickly adjusts its service strategy to develop its niche. In addition to their regular automobile line, one company begins specializing more in trucks, the second more in economy cars, and the third in luxury cars. While the treaty change is a jolt to each company, the specialization allows them all to remain profitable.

That wraps up four more patterns. These patterns originate from the competition that occurs between different individuals, or between different groups. Natural systems have found ways to suppress extreme competition. So rather than being engulfed in mutual competition, many established groups or individuals are able to experience some bounty of resources. Extreme competition is suppressed through resource use, self-defenses, restraint, and barriers. These Protections also play a unique role for self-replicating groups, and they are necessary for both the creation and maintenance of those groups. In other words, diversity of group types could not arise, nor could it persist, without these Protections.

Chapter 14 Government

Nature is often viewed as very "wild," or *unrestrained*. In contrast, human civilization is often viewed as a place where there is order and predictability. People often imagine that when humans escaped the wilds for civilization, they left behind a chaotic environment where there were no rules, restraints, or laws—a place where plants and animals live in a state of virtual anarchy, doing whatever they please. After all, plants and animals have no government, and no laws, right?

It is true—they lack written laws and court systems. But are there really no guidelines for or restraints on the behavior of plants and animals? If we look closer, we will see that the wilds are not so crazy at all. Nature is often quite organized, and behavior of various creatures falls within bounds that are consistent on the whole. How can this happen? What is forcing or causing creatures to behave predictably and with restraint? Here we will look into how behavioral organization arises in nature.

The original root word of "government" is not in reference to a group of people telling others what to do; it is simply in regard *to steering*. Within this framework, "government" is discussed relative to this root concept of steering or guidance, rather than in reference to an occupational sector of people who make rules, or tell others how they should behave. Apparently, somehow, in some way, there is a type of *steering* government at work among wild creatures.

Have you ever seen a hawk perched along the side of a road, waiting for a mouse, rabbit, grasshopper, or other prey to pounce upon? It is a common behavior. Just what is it that prompts a hawk to do this? The fact that something is causing this fairly predictable behavior means that something isn't random, and that some sort of government, some sort of behavioral steering is at work.

Not only is the *hawk's* behavior somewhat predictable, but so is the mouse's, the grasshopper's, and the rabbit's. The mouse will come out at certain hours to feed; it will eat certain seeds or plants; and it will try to hide from predators while it feeds. The rabbit will come out of hiding at certain hours, and like the mouse, will eat certain preferred plants and watch out for predators. The grasshopper will feed on grass, bask in the sun, seek out a mate, and jump when predators are near.

It's not just the animals, either. Something is guiding the tree, and the grasses and flowers below. The tree will emerge from dormancy each spring, grow new leaves, and produce seed if it is able. The grasses will compete for light and nutrients as they spread their leaves in spring and summer, and they will try to set seeds before they get mowed. Wherever we look in nature, something is guiding how creatures behave, whether singly or in groups. It isn't just a wild free-for-all, after all. The next four patterns address the nature of this behavioral guidance, or *government*.

One property of government in nature seems to be that it is carried out at the individual level, or at least very close to the individual level. In other words, to the extent that they are physically and physiologically able, each individual controls its own behavior, and is free to behave in any way it can, or is capable of choosing. Thus, the hawk along the side of the road is free to sit and hunt, or fly down into the grass and take a stroll, or hop from branch to branch all day.

Of course, it probably won't take a stroll or hop around all day, but, it could if it chose to. The behavioral decision rests with *itself*. A possible exception to this individual freewill quality is that other animals might try to coerce the hawk. A second hawk might chase it away. Or, crows or blackbirds might harass the hawk till it leaves. This might be semantics, because the hawk itself is still choosing to respond to attacks from other birds, but in nature, government is carried out by acts of individual freewill or individual freewill under very local coercion of others under their own freewill.

Above, we noted that the behavior of plants and animals is often somewhat predictable—the hawk will perch in a hunting position, the rabbit will seek out certain plants and hide from predators, etc. But how can animal behaviors be predictable if those animals are operating with freewill? It must mean that for some reason, they are *choosing* these predictable behaviors. Why would they be consistent with their behavioral choices? Because in choosing those behaviors, they are attempting to respond with accountability to four realms, or areas of their existence. These four areas of accountability remain through the life of each creature, and they engender consistent behavioral responses that reflect this. Before we look at the nature of this accountability, let's define it:

> *Accountability* is the increased likelihood of receiving positive consequences (supporting or positive feedback) for certain behaviors, and the increased likelihood of receiving negative consequences (unsupporting or negative feedback) for other behaviors. Under accountability, the type of feedback cannot waver randomly for given behaviors—it must be associated non-randomly with specific behaviors, although specific feedback events may include an element of probability.

With this in mind, we can look at accountability in nature. First, each individual is *accountable to its own physiological demands*. The more its behavior supports its physiological requirements, the more likely an individual is to obtain resources, prosper, produce offspring, etc. However, the less its behavior supports its physiological requirements, the less likely that it will obtain resources, prosper, produce offspring, etc. In other words, the individual must align with its own physiological reality, or experience demise. For example, if the hawk is going to hop from branch to branch like a warbler, its body will grow hungry and tired, and its likelihood of survival and success will gradually drop.

Second, each individual is *accountable to its environment*. The more its behavior is aligned with its environment, the more likely an individual is to obtain resources, prosper, etc. However, the less its behavior is aligned with its environment, the less likely that it will obtain resources, prosper, and produce offspring. In other words, it must align with its environmental reality, or experience demise. For example, if the hawk is going to hunt during a rainstorm when its prey is inactive, it will likely be cold and unsuccessful, and will be in worse condition going into the next day.

Third, each individual is *accountable to its peers and associates*. Hawks are a lot less social than some other animals. Still, even hawks have to synchronize with their mates and offspring if they want to successfully reproduce. This means partaking in courtship rituals, nest building, and coordination according to the behavior of another. These activities require each to watch the other, monitor the status and location of the other, listen to their calls, respond to needs of fledglings and adolescent offspring, etc. Raising young hawks requires *two* hawk parents, even if that means overcoming an inborn independent and territorial streak. The failure of a male or female to communicate and coordinate with their mate and their young will lead to unsuccessful courtship, nest building, mating, egg care, and survival of young.

Finally, each individual is *accountable to its competitors and predators*. In nature, individuals with inferior skills or aptitude are more likely to be killed by predators, or to be replaced by more skillful members of their own species. A hawk that can successfully feed itself while avoiding or thwarting the territorial attacks of others has a higher likelihood of surviving its first few years and finding a mate. Similar scenarios could be painted for a variety of plant and animal species—sparrows, egrets, turtles, rabbits, bears, trees, and wildflowers. Yes, even plants have to coordinate with their peers and associates. For example, a plant that flowers at a different time of year than the rest of its own species might not be able to cross-pollinate and reproduce.

Continuing with the theme of government in nature, it further seems that the application of government and the response to government is nuanced by situation. In other words, rather than having a single hardened "law," accountability is applied non-uniformly upon individuals, and individual response to accountability is further nuanced according to the individual situation. For example, a larger hawk might be able to handle cold weather better than a small one, so accountability to the environment occurs for both, but is unequal. Or, a hawk alighting onto a competing hawk's favorite hunting perch might be attacked right away, but trespassing high over the same territory might go unnoticed. Or, a hawk parent might share more food with a vocalizing adolescent when it detects more urgency in its calls.

This nuanced characteristic is true for all sorts of plants and animals. A tree is accountable to its environment in that it must be able to withstand wind forces. But the narrower and shorter the tree's profile, the less wind force it will be subjected to; and the force will be much higher on trees near a ridge top than on trees growing in a hollow. Even on the same tree, the wind forces will be much higher under full leaf-out or with snow cover, and much lower when the canopy is bare. We could find countless of other examples in nature, wherein the intensity of accountability applied to each individual, and thus the necessary response by each individual, varies by situation.

Finally, it seems that in nature there is an alignment between government and system function. As natural systems become refined and operate sustainably, the natural response of each individual to accountability tends to be aligned with the function of the systems within which they live (see Pattern 16 in Chapter 10 for Stage 4 of succession). For example, the hawk will feed on many small animals—birds, mammals, lizards, etc. These animals breed at rates that are far beyond replacement rates, and they can quickly consume their entire food supply—which would cause their own starvation and community disruption.

Thus, in order to suppress extreme competition and experience Significant Expression (see Pattern 25, Chapter 13), many of these animals must be killed. The hawk is able to help accomplish this population reduction for these other species, and on behalf of the system they all live within. In fact, a favorite food item of the hawks are small mammals, and just when rodent numbers are exploding in late spring, the hawks are spending the most time hunting in order to feed their hungry nestlings. In other words, accountability of the individual (hawk) to its peers and associates (offspring) align it to behave more intently in a manner that helps maintain the balance (rodent control) of the system.

We could find other examples of this throughout nature. For example, excess nutrients are often flushed into wetlands in the spring. In early summer, it happens that cattails produce large amounts of windborne pollen. The pollen is high in nitrogen, and when it blows back out of the wetland, it reduces the nutrient load inside the wetland. Thus the activity of the cattails in responding to accountability (toward their physiological selves and competitors), is timed just right to help control excess nutrient levels in wetlands, which can help stabilize the water quality of aquatic ecosystems.

In nature, even though governmental authority generally lies with the individual, the processes of government can of course also apply to groups. Just as with an individual, a group of animals living out of synch with their environment, for example, will experience accountability (in this case negative feedback).

> Pattern 29: In nature, governmental authority, or choice making authority, is local. Generally, the governmental authority for each individual is either the individual itself, or a closely related or adjacent individual(s).

Essential Translation: In sustainable complex systems, governmental authority, or choice making authority, is local. Generally, the governmental authority for each individual is itself, or closely related or adjacent individuals.

This pattern can now be translated into the context of human systems:

Government Translation: Sustainable human governments have a fundamental component of individual- or very locally derived choice.

For example, in a sizeable city, many people do not feel like they receive enough exercise during their daily work routines. To address this, many of them have joined *City Wind*—a large health and exercise club with a gym, a track, an indoor skating rink, tennis courts, workout rooms, exercise equipment, two pools, saunas, steam rooms, and sports teams. Being a large facility with several thousand members, the club management believes that good patron etiquette is needed to help maintain a quality experience. Management asks all members to wear clean shoes into the locker room, refrain from monopolizing equipment, wipe off sweaty equipment after they finish, shower before entering the swimming pool, throw their own used towels into a bin, keep their own children from screaming, etc. Rather than have staff continually watching over each area, they ask each member to be conscious of maintaining good etiquette, and signs are posted reminding patrons of various house rules.

Recall that accountability was defined as the increased likelihood of receiving positive consequences (supporting or positive feedback) for certain behaviors, and the increased likelihood of receiving negative consequences (un-supporting or negative feedback) for other behaviors.

> Pattern 30: In nature, government—or behavioral steering—is affected by accountability to four areas. The four areas are the physiological self; the environment; peers and associates; and competitors and predators. Behaviors that are *in synch* with the true quality and function of these four areas tend to result in an increased likelihood of receiving positive consequences. Conversely, behaviors that are out of synch with the true quality and function of these four areas tend to result in an increased likelihood of receiving negative consequences.

Essential Translation: In sustainable complex systems, government—or behavioral steering—is affected by accountability to four areas. The four areas are the physiological self; the environment; peers and associates; and competitors and predators. Behaviors that are *in synch* with the true quality and function of these four areas tend to result in an increased likelihood of receiving positive consequences. Conversely, behaviors that are out of synch with the true quality and function of these four areas tend to result in an increased likelihood of receiving negative consequences.

This pattern can now be translated into the context of human systems:

Government Translation: Sustainable human governments are based on behavioral choice as modified by accountability to four areas. The four areas are as follows:

1. The physiological self
2. The environment
3. Peers and associates
4. Competitors and predators

Behaviors that are *in synch* with the true quality and function of these four areas tend to result in an increased likelihood of positive consequences. Conversely, behaviors that are *out of synch* with the true quality and function of these four areas tend to result in an increased likelihood of negative consequences.

The previously described City Wind health and exercise club is the setting for the next several examples:

1. Josh found that his lifestyle was becoming somewhat sedentary, and he joined City Wind in his thirties. By his forties, he had joined a tennis team and a hockey team, and tried to attend the club at least twice per week. Even though sometimes it was hard to make the time and muster the energy, he stuck with his efforts to remain active through the decades. Even though his health was not perfect, he remained quite nimble and alert, and felt a sense of well-being up to the time of his death in his late eighties.

    Josh's younger brother, Nathan, also joined the club in his thirties, but his ambition to remain active waned before he turned fifty. Although he also lived to be almost ninety years old, he spent much of his last twenty-five years struggling with health problems such as high blood pressure, irritability, digestive difficulties, weak immunity, and forgetfulness.

2. There are about fifty runners and ten swimmers who exercise outdoors most of the year, and only attend City Wind during the winter season. This is because the local winters are cold, and these individuals prefer not to challenge the environment during that time—risking injury when evenings are dark, paths are icy, winter winds chill the muscles, and the nearby lake is frigid.

3. In most years, City Wind hosts sixteen men's basketball teams. One particular team is comprised of seven long-time friends. This team, having named themselves the "Swifts," is actually older, slower, and slightly less athletic than most of the other teams. However, because they keep returning to the league year in and year out, they are very experienced as a group. Younger teams with more spring and more showmen find—to their frustration—that the Swifts are hard to beat because of their superior teamwork and coordination.

4. While City Wind's management expects members to be courteous and use good judgment in their use of the facilities, they are prepared for exceptions. Members are encouraged to remind other members of rules if they see someone behaving inappropriately, such as leaving wads of their hair beside a sink. Members are also allowed to register a complaint against others for such behaviors. If there are three independent complaints against a member in one year, management will discuss the issues with that person. If complaints continue beyond this, the individual will have their membership revoked.

5. There are three other health clubs in the city. One of them is clear across town from City Wind, but the other two are within fifteen miles. City Wind takes these nearby competitors seriously. They monitor what the other clubs charge for membership and what they provide. They don't necessarily try to outdo these competitors, but they make sure that in most ways, their own facility provides at least a comparable experience for the cost. Beyond that, they keep up with what's new in the health and exercise scene, and they do make a point of providing some specialized services and equipment that the other clubs don't.

> Pattern 31: In nature, the unfolding of government is often nuanced according to each individual situation. This includes nuance in the conditions that each group or individual plant or animal faces, the accountability and consequences they are faced with, and how they respond behaviorally to that accountability.

Essential Translation: In sustainable complex systems, the unfolding of government is often nuanced according to each individual situation. This includes nuance in the conditions that each group or individual faces, the accountability and consequences they are faced with, and how they respond behaviorally to that accountability.

This pattern can now be translated into the context of human systems:

Government Translation: The unfolding of sustainable government is often nuanced according to each individual situation. This includes nuance in the conditions that each group or individual faces, the accountability and consequences they are faced with, and how they respond behaviorally to that accountability.

Several examples are provided:

1. Dan and Jennifer are married and they often come to City Wind together on the weekend, but once inside, they don't spend much time together. Dan wants to work hard, and he likes to work alone. His favorite routine is the stair machine, followed by lap swimming and the hot tub. Jennifer likes group exercise. She often signs up for yoga or various aerobic classes. If Dan misses his exercise, he doesn't gain weight, but he's grumpy and lethargic going into the start of his work week. If Jennifer misses her exercise, she feels alright, but she tends to gain weight, and that leads to her feeling self-conscious.

2. City Wind has recreational and team basketball, tennis, and ice hockey, but it has to juggle their location and timing. The local climate allows an outdoor hockey rink for about four months per year, so the hockey teams play then. In the spring, after ice hockey has finished, the rink is transformed for recreational roller-skating and roller blading. Meanwhile, winter basketball and tennis teams effectively share an indoor arena by alternating usage each day. But in spring, summer, and fall, the outdoor tennis courts open up. This frees up more space for indoor basketball, so expanded recreational and team basketball is available for the year's seven warmer months each year.

3. Jeff is six feet and four inches tall—the tallest man on the Swifts basketball team. He does about half the Swift's rebounding, and his teammates expect this as a matter of course. If he gets under ten rebounds in a game, his teammates are disappointed, and those are the times the Swifts are more apt to lose. Ned is five feet and six inches, making him the shortest Swifts player. He is not expected to rebound, but on the occasion when he does pull one down, he usually gets a pat on the back.

4. One of City Wind's main competitors is The Players Club, which is located in an upscale area near downtown. The Players Club offers a gym, weight rooms, racquetball courts, a cardio room, a small track, saunas, and steam rooms. However, The Players Club has no room for expansion. Its managers realized decades ago that their location meant membership would be expensive, and that they would not be able to offer the variety of facilities offered by other clubs (such as swimming pools, hockey rinks, tennis courts, etc.). In order to remain competitive, The Players Club instead offers the highest quality service. For example, membership includes free drinks, locker room attendants, and free training coaches. In contrast, the managers at City Wind recognize that while they cannot provide the best location and service, they can provide a wider variety of facilities. They keep a close eye on The Players Club, and always make sure they charge a lower price and provide comparatively more facilities.

> Pattern 32: In sustainable natural systems, there is an alignment between government and system function. That is, the pressure of accountability applied upon each individual and group of individuals tends to be aligned with the function of the systems within which they live.

Essential Translation: In sustainable complex systems, there is an alignment between government and system function. That is, the pressure of accountability applied upon each individual and group of individuals tends to be aligned with the function of the systems within which they live.

This pattern can now be translated into the context of human systems:

Government Translation: In sustainable human governments, there is an alignment between forces of accountability and system function. That is, the pressure of accountability applied upon each individual and group of individuals tends to be aligned with the function of the systems within which they live.

For example, along with the other health clubs in town, City Wind helps to absorb a significant amount of regional sports and fitness demand. Without these clubs, many citizens would eventually insist that public parks have a bigger emphasis on sports. This could result in some conflict, since these parks already serve many other important social needs such as local greenways, areas of quiet reflection, playgrounds, and neighborhood special events and meeting areas.

There are other ways in which the community is supported by the presence of City Wind. Many good acquaintances, friendships, and even marriages have resulted from chance meetings over the years at the club. City Wind also employs 140 people, which improves local employment and tax figures. There are even outside businesses that probably wouldn't exist were it not for City Wind. Within one block of the club, a juice bar, a pizza parlor, and a take-out Chinese restaurant all owe about half of their business to patrons on their way to or from the club. In effect, while the city itself would certainly not collapse without City Wind and the other health clubs, the actions undertaken by patrons responding to accountability toward their physiological selves has become part of maintaining the city's environmental, social, and economic systems.

Chapter 15 Resilience

If we are told that something is "resilient," we might imagine that it would be able to withstand some sort of abuse or attack and still come out in one piece. Other images for resilience are to be able to "bounce back," to "shake it off," or to "bend without breaking." The idea is that even when things change or become challenging, a resilient system or structure can survive.

Resilience may have an intuitive simplicity—the image of something bending but not breaking, or someone being knocked down but getting back on their feet. While images of resilience may be simple, here we will be looking more deeply in hope that we may ultimately understand resilience in situations that are more complicated and unfamiliar. Our goal will be to gain some understanding of the source of resilience, to predict whether a system will be resilient, and to understand how resilience can be improved or supported. Our study of resilience will require some new terms and care in the usage of certain terms.

Before we dive right into resilience, let's review where we are so that its connection to our theme is clear. Do you recall what led us on this journey? It was an attempt to figure out what could be sustainable, or what sustainability of human systems, structures, and institutions might look like. Much earlier, a definition of sustainability was given:

> *Sustainability* (noun) is the degree to which a particular system, structure, or institution is able to begin and continue to accomplish a particular objective under a given set of external conditions.

We have been analyzing sustainable complex natural systems for some time, looking at the characteristics they seem to have in common. We have touched upon many patterns (32 already), but in doing so, it usually hasn't been clear whether these patterns are *causing* sustainability, or are *caused by* sustainability, or, are simply *associated* with sustainability by design or chance. Attempting to determine these causal relationships (or lack thereof) is mostly beyond the scope of this book, but *resilience* is different. Why? Because in many situations, *resilience is clearly required* for sustainability.

The following is our working definition of resilience:

> *Resilience* (noun) is the degree to which a particular system, structure, or institution is able to continue to accomplish a particular objective, without undergoing a massive change in constitution or internal relationships, following a change in internal or external conditions.

If you look closely now at the definitions of sustainability and resilience, you will notice some things in common. Both require that a system, structure, or institution be able to continue to accomplish a particular objective. But resilience goes a step further. For the system, structure, or institution to be resilient, it must not only continue to accomplish a particular objective, it must be able to do so *after conditions have changed, and it must be able to do so without being greatly transformed or replaced by some new system.* Another way to state this is that a system may be sustainable under a given set of external conditions, but if the system is not resilient to a change in those external conditions, it will not persist and continue to perform after those conditions have changed.

Looking at resilience is sort of like looking at sustainability from a different angle. Until this point, we have been concerning ourselves with trying to conceptualize what complex sustainable systems might look like. But with resilience, we are assuming that we already have a complex sustainable system on hand, and we are wondering whether a *change* in internal or external conditions will throw the system off course, forcing it either to fail in its stated objective(s), or forcing it to massively reorganize into a different system in order to keep accomplishing its objective.

Actually, resilience ties in very well with our early efforts to understand the physics of sustainability. Do you remember what is required for *anything* to function for, or toward, a given objective? A system can function for, or toward a given objective if and only if it has internal designs, conditions, and forces that support the objective; and/or external designs, conditions, and forces that support the objective. As we consider resilience, we are merely wondering if a weakening in the level of internal or external support will render the system either nonfunctional or functional only after massive alteration.

Here are some examples of the application of this topic: Will a car dealership be resilient to an auto production strike? In other words, will it survive the strike as the same coherent company, while still earning money? Will it be resilient to a doubling of prices by the car manufacturer? Will a particular relative be physically resilient in the face of a new viral disease? Will a friend be emotionally resilient to the sudden loss of their spouse? Will a region be environmentally resilient to a 25 percent increase in precipitation levels? Will a nation be economically resilient to a trade embargo?

Let's move on to the definition of *disturbance*. After that, we will go through examples that portray resilience, and lack thereof, in natural systems.

> We will define *disturbance* as a change in conditions. A disturbance can originate internally or externally relative to a system's boundaries.

For the purposes of discussion, we can divide disturbances into two general categories—*inconsequential* and *consequential*.

> *Inconsequential disturbances* are weak, small, or peripheral events relative to a system and the processes it uses to achieve an objective. Because inconsequential disturbances lack the ability to alter the trajectory of a sustainable system, systems can be considered resilient to inconsequential disturbances.

A 4-mph wind blows across a forest. Will the forest be resilient to this disturbance? Of course it will; this gentle breeze is too weak to destabilize the forest. A successful businessman loses a quarter while trying to plug a parking meter. Will this greatly affect his economic situation? Of course not; the scale of this event is too small relative to his available resources to have a significant impact. In a county where farming is the dominant land use, will an increase in illegal drug use by teenagers alter the land-use environment? Probably not—drug use among teens is tangential to the ecology and economics of how land is used.

> *Consequential disturbances* are a strength, scale, and type capable of interfering with a system's ability to remain and achieve its objectives. A system met with a consequential disturbance may or may not respond with resilience.

What will the forest look like after a storm with 115-mph winds blows through the area? If the forest is still standing, many of the individual trees will have fallen. What will happen to the businessman's personal economic situation after his company is successfully sued for patent infringement? Even if his company stays in business, the man's standard of living might drop. What will happen to land use in a farming area if trade policies lead to a drop in food prices and a spike in the price of illegal drugs? Land that once grew grains and vegetables might begin shifting toward crops that supply those illegal drugs.

Herein, we will learn how to predict (at least in some cases) whether a disturbance will be consequential or inconsequential. Systems are resilient to inconsequential disturbances, so if we can determine that a disturbance is inconsequential, it allows us to cross it off of our list of concerns, and focus on more important issues. On the other hand, a system will not necessarily be resilient to a given consequential disturbance. The amount of system resilience to consequential disturbances involves many factors. Ultimately, we will learn some of those factors, and we will learn how a system can develop more resilience to a given consequential disturbance.

Hmmm…predict whether a disturbance is consequential? Suggest how something can become more resilient? You might wonder, "Why not go further?" Why not just analyze the system and determine exactly how it will react when it is faced with a particular disturbance? That will give us the highest level of certainty, right?

The problem with trying to do that is the same one we encountered when trying to understand sustainability of a complex system through tedious logical analysis. Do you recall the village school issue? We took a reductionist approach to a school, analyzing the parts of the school system from all sorts of angles. We included independent forces and feedback loops in the mix. While this helped us greatly in understanding some aspects of the potential school, we were ultimately unable to determine whether any particular design was sustainable because the situation was too complex.

Complex systems are indeterminate. They cannot be controlled completely from any given leverage point, and they have no predetermined future. This, in turn, means that resilience of complex system is rarely a precise property that can be predetermined with perfect accuracy. So, we will look to patterns and predictors of resilience, but here we will not attempt to determine exactly how a complex system and all of its parts will respond to a particular disturbance.

With that understanding, we can now begin to look at resilience in nature. We already know that when a system is met with an inconsequential disturbance (such as the gentle breeze through the forest), it will respond with resilience. But what about when the system is met with a disturbance that is more substantial, a consequential disturbance that is capable of bending, shifting, toppling, or breaking the system? Will the system be resilient? Or will the system be destroyed? Will it fail to accomplish its objective? Will it be massively transformed? In some cases, natural systems survive consequential disturbances and ultimately respond resiliently to them by *immunizing* themselves. We will look at two hypothetical examples of this immunization process.

Imagine there is a wild landscape named *Newcastor*. Perhaps it is a very large island or a peninsula on a continent. Along the seashores of Newcastor, the land is level. Inland, there are gentle hills that rise up, and the deep interior of Newcastor is somewhat rugged and mountainous. The land cover of Newcastor is more than 99 percent forested, with open areas confined to the seashores and a few rocky mountaintops. In this land, untamed rivers flow down off the mountains, through valleys, and between hills before reaching the sea. Several thousand species of plants and animals inhabit Newcastor.

For the sake of discussion, we can assume that the objective of Newcastor's ecosystem is its ability to house thousands of plant and animal species, and its ability to maintain healthy soils and produce clean water. We will look at several hypothetical events and alternative scenarios on Newcastor to help demonstrate resilience in natural systems.

In our first scenario, under a stable climate, Newcastor has remained rather unchanged for about 25,000 years. However, through natural planetary fluctuation, the local climate begins to shift toward slightly warmer, drier conditions. This begins with an occasional drier-than-normal summer. But over some decades, those drier summers become the standard, and winters also become slightly warmer (meaning a bit more evaporation occurs), and soils are slightly drier, on average, as the warm growing season begins.

If the weather were simply showing variability, this would be an inconsequential disturbance, and there wouldn't be much impact upon the ecosystem. But with a consistent directional change in climate, Newcastor's 25,000-year-old ecosystem is now a bit out of synch with external conditions (see Patterns 19 in Chapter 9, Pattern 30 Chapter 14, and Pattern 2 in part, Chapter 7). Because temperature and water availability impact many species of plants and animals, and because this change occurs across all of Newcastor, it is both of a type and scale that is capable of altering the ecosystem, and it can be considered potentially consequential. Here's how the ecosystem responds:

Many plant species growing in a particular location are no longer the best-adapted species for that location. Competition begins to cause many species that favor wetter, cooler weather to begin shifting their distributions incrementally toward shaded valleys, toward higher elevations, and toward cooler forests on slopes facing away from the equator. Other plant species that occupy sunny, windy outcrops with thin soil begin to expand their ranges, and they become dominant wherever droughty conditions are frequent, such as on steeper slopes that face the equator. Headwater streams that were perennial become ephemeral in places—only running with water during wetter and cooler times of the year. Most fish species in the rivers shift their distributions slightly downstream to maintain their preferred water temperature and depth in rivers that have lowered average water levels.

While many plant and animal species shift directly in response to the climate's change, others shift as an indirect response. Otters and other fish predators abandon portions of their range that have become fishless ephemeral streams. Specialized pollinating and herbivorous insects shift as their host food sources shift. With some vegetation changes, three new species of bird that used to migrate through Newcastor begin to stay and nest. Over a few thousand years, two new plant species arrive by wind and take hold on dry slopes.

However, under these drier conditions, a few species are no longer able to find a survivable amount of habitat on Newcastor. Over these few thousand years, one species of frog, two insect species, one species of tree, and one species of vine go extinct. Overall, a few colonizations and a few extinctions have occurred. This number is fairly small, given that several thousand species inhabit Newcastor. The overall character of Newcastor is similar to what it had been before. The main difference is that habitats have shifted.

The system (Newcastor's ecosystem) has responded to the change in climate (a consequential disturbance) without massive reconstitution or reconfiguration. Furthermore, if we consider the ecosystem's objective to be its ability to house thousands of plant and animal species, and its ability to maintain healthy soils and produce clean water, then it has also continued to accomplish its objective. So although it took some adjustment time, our system was able to respond with resilience to this particular consequential disturbance.

However, more events lay ahead for Newcastor. We will watch those events and see where they lead in relation to resilience.

Over a few thousand years, Newcastor has shown itself to be resilient to a gradual shift in climate towards warmer and drier conditions. Ten thousand years later, after this climate situation is stabilized, the stage is set for another significant disturbance. Following a large storm, two beavers arrive on the shore of Newcastor. These beavers make their way up a forested valley and begin doing exactly what they are good at—cutting down trees and damming the river. Even though both are female, one of them was already pregnant and gives birth a few weeks later. The beavers find essentially perfect river habitat and an unlimited food supply. Under these ideal conditions, the beavers breed and spread rapidly across Newcastor.

When only a hundred years have passed since the beavers' first arrival, Newcastor is looking like a much different place. Beavers chop trees down in order to feed on branches and to build dams with logs and sticks. Each dam creates a pond averaging several acres in size (about the size of a city block). In addition to the trees the beavers cut down, many more trees are killed when the pond water rises around them. When a beaver's local food supply runs out, it will abandon its lodge and dam site. The unmaintained dams then can sometimes fail (often during periods of high flow, when water pressure is highest). The dam burst can release a torrent of logs, mud, rocks, and floodwater. The debris released in this torrent might be deposited on nearby banks, and meanwhile, the rushing water deluge will carve away at other banks and loosen rocks from the riverbed. The drained pond behind each broken dam becomes a warm, muddy basin filled with scattered dead tree trunks. This scene is repeated over hundreds of rivers and streams and over many thousands of acres.

This activity is detrimental to many other species. In addition to all the trees and other vegetation killed along the rivers, fish that need cold running waters have less habitat, and they decline. Forest browsing animals such as deer, and tree-climbing animals such as squirrels have less useable habitat as well, and their populations drop. The areas of newly disturbed muddy ground invite a flush of weeds, scavengers, and opportunistic flies. The beavers have essentially created thousands of acres in Stage 1 of succession (see Pattern 13, Chapter 10).

However, even when the destruction is at a peak, the ecosystem begins a gradual immunizing response to this situation (see Pattern 2, Chapter 7). It begins working with and around the new situation. Marsh and meadow plants that were originally confined to slivers of river sandbar and deltas begin to colonize the large muddy meadows appearing across the landscape. Pollinating and herbivorous insects follow closely behind these marsh and meadow plants, and the proliferation of insects invites predatory amphibians and songbirds. Two species of fish that were previously found only in the larger, warmer river deltas begin showing up in the beaver ponds. The presence of these fish eventually attracts egrets, osprey, and fishing eagles to the edges of these ponds. A type of marsh duck that had never previously bred in Newcastor seems to have found these ponds to be ideal nesting grounds. And a type of woodpecker that would previously only migrate through Newcastor begins nesting near the groves of dead trees behind the beaver dams. Several wetland tree and shrub species that were originally confined to the edges of rivers redistribute over several decades and become far more common along pond edges and in other areas subject to periodic inundation.

Other changes occur gradually over several centuries. A type of orchid with white and purple flowers is displaced from some of its prime habitats along moist river corridors, but it is able to survive on some of the tiny drainages that are too small for beavers to dam, and it reestablishes along some of the more stable pond/forest edges. A type of secretive forest deer finds that some of the best browsing can be found in and along these ponds, and begins feeding there at night. With all of the beavers about the landscape, the small local wildcats and fishing eagles also gradually begin learning how to successfully hunt these beavers. In doing so, they are actually able to reduce the number of beavers slightly.

Ecosystem changes and adjustments continue, but eventually the pace slows. Through this disturbance process, only five species end up going extinct due to the beavers (two small plants, a snail, a mussel, and one fish). What began looking like a catastrophe appears more like a mild disturbance after about 2,000 years have elapsed, and after about 5,000 years have passed, the ecosystem is again relatively stable.

Whether looking at the warmer climate, or the arrival of beavers that occurred after that, the ecosystem went through a similar process. In order to reach a new stability point after experiencing a change in conditions (disturbance), the ecosystem had to reconfigure, adjust, and compromise in order to *incorporate* the disturbance. The ecosystem essentially "took actions" that would have been completely unwarranted were it not for the change in climate, or the beavers. In other words, adjusting for these disturbances meant that the ecosystem had to shift away from what would have otherwise been its ideal configuration. And, once the disturbance had been incorporated, the ecosystem was essentially basing its configuration upon that disturbance.

In fact, the latest ecosystem configuration *expects* the beavers to kill trees, make dams, and then leave dams in disrepair. It is now immunized to the presence of beavers, and it has become resilient to beaver activities. What would happen if the beavers were all eliminated after the ecosystem had incorporated their presence? Given that the ecosystem has now reconfigured (warped itself) to include the beavers, removal of the beavers would actually present a new disturbance for the ecosystem! If it helps to think of it this way, the ecosystem becomes functionally dependent upon the very disturbance processes it manages to immunize to.

The response of Newcastor's ecosystem to climate change and the arrival of beavers provide examples of immunity development through *incorporation*. The two other ways of developing immunity are through *resistance* and through *avoidance*. Staying in Newcastor for a bit longer, maybe the beavers find that one particular species of tree is sometimes slightly bitter tasting. With the arrival of beavers, these trees quickly undergo natural selection (see discussions prior to Pattern 28), becoming even more bitter. As this particular tree species is eaten less and less by beavers, it is able to increase its population. The increasingly bitter trees are an example of immunization through resistance.

Then what about immunization through avoidance? Maybe there is a tree snake species that lives in dense forest canopies. Perhaps these snakes avoid open areas because when they are forced to descend and travel along the ground, they become prey for hawks or weasels. When these tree snakes find that their forest is becoming thinned out due to flooding and beaver felling of trees, they move upslope into denser, slightly drier forests. Because the snakes have relocated outside the zone of disturbance, it is an example of immunization through avoidance.

The method used by a natural system or group of creatures to develop immunity—incorporation, resistance, or avoidance—can depend upon the tools that are available. Can a tree do anything to avoid or resist the coming of cold weather? No, but it might be able to incorporate the weather into its biology. Can a small songbird incorporate subzero temperatures into its biology? Maybe not, but it might be able to migrate to a predictably warmer place, and thereby immunize through avoidance. There isn't always a clear line between incorporation, resistance, and avoidance. What one person considers a case of incorporation, another may consider avoidance. Regardless, the development of immunity, no matter what the form, does require the expenditure of energy or the effective loss of energy by requiring behaviors that would otherwise not be ideal.

The next pattern addresses how a complex system can respond resiliently to a consequential disturbance through the process of immunization.

> Pattern 33: Individuals, species, species classes, natural communities, ecosystems, and the biosphere are capable of responding resiliently to a given consequential disturbance by *immunizing* themselves. This occurs through repeated adjustments following prolonged or repeated exposure to the disturbance. The adjustments can take the form of resistance, avoidance, or incorporation. The development of immunity requires the expenditure of energy or the effective loss of energy by requiring behaviors that would otherwise be undesirable. Once immunity has occurred, depending upon the situation, the natural system will no longer experience the disturbance as a stress event, but as a normal or even necessary event.

Essential Translation: Sustainable complex systems are capable of responding resiliently to a given consequential disturbance by *immunizing* themselves. This occurs through repeated adjustments following prolonged or repeated exposure to the disturbance. The adjustments can take the form of resistance, avoidance, or incorporation. The development of immunity requires the expenditure of energy or the effective loss of energy by requiring behaviors that would otherwise not be preferable. Once immunity has occurred, depending upon the situation, the system will no longer experience the disturbance as a stress event, but as a normal or even necessary event.

This pattern can now be translated into the context of human systems:

Economic Translation: Sustainable economies are capable of responding resiliently to a given consequential disturbance by *immunizing* themselves. This occurs through repeated adjustments following prolonged or repeated exposure to the disturbance. The adjustments can take the form of resistance, avoidance, or incorporation. The development of immunity requires the expenditure of energy or resources, or the effective loss of energy or resources, by requiring economic patterns that would otherwise not be preferable. Once immunity has occurred, depending upon the situation, the economy will no longer experience the disturbance as a stress event, but as a normal or even necessary event.

For example, a nation adopts a new fuel for its heating, transportation, and power generation needs. This fuel is derived from a specialized crop. A few years after the new fuel has become the dominant energy source, a crop failure causes an unanticipated supply shortage, which sends serious shocks through the economy and triggers a year of recession. The fuel shortage is viewed by most as a fluke event, but nine years later, during an economic boom, there is a fuel refinery explosion followed by a worker strike. This causes another fuel shortage and another temporary recession. Ten years later, the nation is involved in a war, which increases its use of the fuel. During the war, several pipelines are bombed, causing yet another economic shock due to a fuel supply shortage. However, this third shock is not as intense as the prior two, because activities that immunize the economy had already begun—this includes protected fuel storage reserves, creative fuel conservation capacities, a new refinery, access to different sources of energy, and overseas crop production. By the time another unanticipated crop failure occurs four years later, there is localized economic stress and slight impact upon the national economy, but it does not result in anything that would be considered a national economic shock.

Community Translation: Sustainable communities are capable of responding resiliently to a given consequential disturbance by *immunizing* themselves. This occurs through repeated adjustments by the community following prolonged or repeated exposure to the disturbance. The adjustments can take the form of resistance, avoidance, or incorporation. The development of immunity requires the expenditure of energy or the effective loss of energy by requiring community behaviors that would otherwise not be preferable. Once immunity has occurred, depending upon the situation, the community will no longer experience the disturbance as a stress event, but as a normal or even necessary event.

For example, in one ethnic region, the number of manufacturing jobs has declined. Most of the families respond by sending their men to work abroad for stints of about a year. The ethnic group and its communities have been established and fairly stable for hundreds of years, and sending the working age men to faraway lands is a very stressful change for families and friends, and for important community religious activities. In the first few years that these men are absent, overwhelmed mothers spend less time with their own children, gangs of rowdy young men form, many of the elderly are neglected, and the working age men who remain are spread so thin that they are unable to have much quality time within their social circles. However, the community begins to respond by steering its members toward new social groupings. Wives and mothers form groups to watch the younger children collectively. This allows them to take turns visiting husbands who are abroad. Under the mentorship of elderly and adult men, teen clubs and sports teams are established to help channel the energy of this age group.

These teens are also paid for taking on extraordinary responsibilities and performing critical labor. This helps to relieve pressure upon the low number of working age men in the community, allowing them to focus upon more critical projects. With these changes, the community structure is quite different from what it had been, although after some decades pass, most of the people are quite pleased with the social structure of their lives, and would never expect it to be different.

Occupational Translation: Sustainable occupational systems are capable of responding resiliently to a given consequential disturbance by *immunizing* themselves. This occurs through prolonged or repeated adjustments of the occupational system following repeated exposure to the disturbance. The adjustments can take the form of resistance, avoidance, or incorporation. The development of immunity requires reconfiguration or the effective loss of energy by requiring occupational configurations that would otherwise not be preferable. Once immunity has occurred, depending upon the situation, the occupational sector or system will no longer experience the disturbance as a stress event, but as a normal or even necessary event.

For example, several extended families run lucrative fish markets near a medium-sized coastal city. The markets are open twice per week, and they sell the latest catch from the boats that they own. All is well for these people until one year, a trading ship arrives from a nearby country. The ship carries tons of salted, smoked, and frozen fish. It docks for five weeks. The traders sell their goods at a low price until their stores are emptied. The families that operated the original fresh fish markets are greatly damaged by this event—they have tons of spoiled fish and significant financial losses. The families are relieved when the ship leaves, but are dismayed when it returns again in one year, and they go through the same loss event. When the ship returns for a third year, selling its preserved fish at a low cost, the families that operated the fresh fish markets are a bit more prepared. While the competing ship is in the harbor, they close market and stop fishing. Although they go for several weeks with no income, at least they don't lose money. The ship that sells preserved fish keeps arriving every year. By the sixth year, the fishing families have begun amassing clothing and jewelry to sell to the traders on the ship while it is in dock. This clothing and jewelry business expands annually as the ship's crew learns to take these items back to their homeland and resell them there. Twenty years after the foreign ship first arrived, some individuals in the fishing families actually specialize in the clothing and jewelry trade, and the total incomes of the families are similar to what they were decades earlier, before the competing boat arrived.

Government Translation: Sustainable governments are capable of responding resiliently to a given consequential disturbance by *immunizing* themselves. This occurs through repeated government adjustment following prolonged or repeated exposure to the disturbance. The adjustments can take the form of resistance, avoidance, or incorporation. The development of immunity requires the expenditure of energy or the effective loss of energy by requiring behaviors that would otherwise not be preferable. Once immunity has occurred, depending upon the situation, the governmental system will no longer experience the disturbance as a stress event, but as a normal or even necessary event.

For example, a previously unknown disease sweeps through a region, sickening about two-thirds of the populace and killing many who are young, old, or already weakened. The specific disease mechanisms and chemistry are not understood. However, when the disease sweeps through the region again six years later, some people discover that two potent herbs seem to help ameliorate the disease, and they observe that people living in highland areas seemed to avoid the sickness. People begin raising and storing their own medicinal herbs. When the disease hits again, in addition to the early formation of some internal physiological immunity, people consume their herbs and flee lowlands for cabins in the hills until the disease passes. At this point and beyond, far fewer people catch the disease, and among those who do, there are far fewer deaths or serious illnesses.

Environmental Translation: Sustainable environments are capable of responding resiliently to a given consequential disturbance by *immunizing* themselves. This occurs through repeated adjustments following prolonged or repeated exposure to the disturbance. The adjustments can take the form of resistance, avoidance, or incorporation. The development of immunity requires the expenditure of energy or the effective loss of energy by requiring behaviors or configurations that would otherwise not be preferable. Once immunity has occurred, depending upon the situation, the environment will no longer experience the disturbance as a stress event, but as a normal or even necessary event.

For example, dozens of prosperous towns line the base of a mountain range. The townspeople rely upon selective logging on the mountainsides for building material and fuel, hunting in mountain forests for meat, and upon croplands on the nearby plains for the remainder of their food. Then, against the objections of the townspeople, the national government institutes an intensive logging program in the mountains. The clearcutting of mountain faces alters hydrology and destabilizes soil. Just a few years after the logging begins, exceptionally large floods begin to occur in the local rivers, and not long after that, devastating mudslides begin occurring. Many homes are destroyed and many people are killed as mudslides hit towns. The local land-use and survival systems are thrown into chaos: logging and hunting opportunities on the mountainsides are gone, and towns are forced to relocate. The next two decades are difficult, but people begin adjusting their land uses to the new conditions. To avoid mudslides, they move villages further onto the plains, sometimes on top of where their crop fields were. They begin raising goats on pasture grasses that are taking hold on the mountainsides. They plant trees along river edges and around the villages. Wildlife are attracted to the planted trees, and some hunting returns. Townspeople discover that the mudflows are especially fertile, and begin planting crops there. Seventy years after the intensive mountainside logging began, homes are safely located, people have goats and wildlife as sources of meat, crop yields are good, and there are enough trees to serve for fuel and building. There are still periodic mudflows, but this just leads to a partial and temporary crop loss.

Cultural Translation: Sustainable cultures are capable of responding resiliently to a given consequential disturbance by *immunizing* themselves. This occurs through repeated adjustments following prolonged or repeated exposure to the disturbance. The adjustments can take the form of resistance, avoidance, or incorporation. The development of immunity requires the expenditure of energy or the effective loss of energy by requiring behaviors that would otherwise not be preferable. Once immunity has occurred, depending upon the situation, the culture will no longer experience the disturbance as a stress event, but as a normal or even necessary event.

For example, an authoritarian family from an ethnic minority takes control of a country. These new leaders ban many important cultural activities enjoyed by the majority population—including traditional religious gatherings, sports, and wedding events. Soon after these bans take effect, there are signs of significant depression in society. Alcohol use skyrockets. Family strife and violence increases, and suicide rates double. However, local communities begin quietly improving their lot. Although teaching religion directly is effectively banned, religious devotees learn to translate their principle values into songs. In this way, they are able to continue passing the religion forward to children. Wedding ceremonies continue through small, secret ceremonies in private homes followed by large community feasts with no mention of the wedding event. Young men learn that they can practice and compete in their traditional horseback sports by gathering unannounced on remote ranges and farmlands, far from the eyes of the authoritarian government. Even though it is nearly a century before the authoritarian regime is ousted, the culture of the majority remains quite intact, and some of the cultural practices that were implemented only because of the authoritarian regime are valued and retained.

Physiological Translation: Human groups and individuals are capable of developing sustainable resilience, or *immunizing* themselves, in relation to a given consequential disturbance. This occurs through repeated adjustments following prolonged or repeated exposure to the disturbance. The adjustments can take the form of resistance, avoidance, or incorporation. The development of immunity requires the expenditure of energy or the effective loss of energy by requiring behaviors that would otherwise not be preferable. Once immunity has occurred, depending upon the situation, the people will no longer experience the disturbance as a stress event, but as a normal or even necessary event.

For example, on one continent, small nation-states engage in continual battle over resources. The continual fighting is especially stressful for the states located between two or three other hostile groups. One of these states expends significant effort in being war-ready. Beginning at age four, boys are very gradually introduced to physical stresses that are based on war-like conditions. This includes games that emphasize physical fitness and short periods of deprivation from food, water, adequate shelter, etc. These stress events are continued through childhood and are gradually increased in intensity. By the age of about twelve, boys who do not perform these stress events satisfactorily are removed from potential military service. Training for the remaining boys continues with increased intensity of workouts and stress situations. By their late teens, the boys are stronger, faster, and more physically and psychologically capable than most adult men in the surrounding states. While many citizens of this nation do not enjoy preparing their children for such difficulty, they recognize that they owe their survival to these preparations.

In this chapter, we are looking at resilience. First we defined *resilience*—a system, structure, or institution is resilient when it is met with a change in internal or external conditions, and is still able to accomplish a particular objective without undergoing a massive change in constitution or internal relationships. Clearly, resilience of a given system is only meaningful in the face of a given change in internal or external conditions, which we then termed a "disturbance."

Next we defined *consequential disturbances* to be those of a strength, scale, and type capable of interfering with a system's ability to remain, and to achieve its objectives. We defined *inconsequential disturbances* as weak, small, or peripheral events relative to a system and the processes it uses to achieve an objective. Tiny or un-impactful disturbances are happening around us all the time—the gentle breeze through the forest, the businessman's lost quarter at a parking meter, an increase in drug use set against land uses in an entire region. Clearly, we aren't concerned about the tiny or unrelated, inconsequential disturbances, and we'd rather focus our energy on the consequential disturbances. But how do we know the difference between them?

Sometimes, we have enough everyday knowledge to know the difference. In extreme or simplified situations, it is often obvious whether a disturbance is consequential or inconsequential. For example, we'd naturally assume that an energy bill that is $5.00 higher than normal will be an inconsequential disturbance for a middle class family. At the other extreme, we can be pretty certain that a permanent loss of work income will be a consequential disturbance event for the same middle class family. We intuitively know consequential from inconsequential disturbances in very familiar or extreme situations.

But can we tell the difference in unfamiliar or intermediate situations? Will a 10 percent income reduction be a consequential or inconsequential disturbance for the middle class family? With more information, we might be able to do some computations and make a solid prediction regarding this family's financial situation after the income drop. But as situations become even more complex and unfamiliar, the distinction between consequential and inconsequential and disturbances becomes even less clear. How will the state economy respond to a 15 percent drop in tax revenue? The financial analysis that is possible upon the middle class family becomes far more uncertain and difficult to conduct as the system or situation grows more complex.

However, in many cases there is a way of determining beforehand whether a potential disturbance should be considered consequential or inconsequential. This predictive ability is based on a useful concept that has been developed by ecologists. This concept—*historic range of variability*—can often help predict whether a disturbance falls into the consequential or inconsequential realm, and thus whether we even need to have any concern over a system's resiliency to that disturbance.

If we determine that a particular disturbance is consequential, *historic range of variability* will even help us further in estimating whether a system will ultimately respond with resilience or not. Historic range of variability (sometimes abbreviated as

"HRV") refers to the range of conditions that a group of plants and animals have previously been subjected to.

For example, if a natural prairie occurs where the annual high and low temperatures are about 100 degrees and 0 degrees respectively, we would know that the historic range of temperature variability for that prairie is 100 degrees to 0 degrees. Then if climate change shifts those annual temperature extremes upward to 120 degrees and 20 degrees, we could say that some of the time the prairie is operating *outside of* its historic range of variability.

It turns out that natural systems generally experience events that are well within their historic range of variability as *inconsequential*—not threatening to the system form or objective. In contrast, events that are near the limits of their historic range of variability, or beyond their historic range of variability, tend to be experienced as *consequential*—threatening to system form or objective.

In general, the further a disturbance pulls a system outside of its historic range of variability, the less likely the system will respond with resilience. Furthermore, natural systems tend to be more resilient to new disturbances if they occur while that system is operating well within its historic range of variability. In contrast, natural systems tend to have less resilience to new or additional disturbances if they are already operating on the periphery or outside of their historic range of variability.

Let's look back to Newcastor and see an example of this. We will look at scenarios involving ecosystem resilience while that ecosystem is operating from different states relative to its historic range of variability.

The change in climate on Newcastor to somewhat warmer and drier conditions was a disturbance that pulled Newcastor outside the limits of its historic range of variability. This was a consequential, somewhat disruptive event, but on the whole, and with the passage of time, it happened that the ecosystem was able to resiliently immunize to the shifted climate, and only five species on Newcastor were lost as a result. The arrival of beavers on Newcastor was also a disturbance that essentially pulled portions of the ecosystem outside of its historic range of variability. This, too, was a consequential, disruptive event, but with the passage of time, the ecosystem was able to resiliently immunize to the presence of beavers, and as with a shifted climate, only five native species were lost as a result. Furthermore, the beavers hardly had any impact on habitats outside of Newcastor's river corridors.

Let's look at Newcastor again well after it has immunized to the warmer climate and the presence of the beavers—perhaps 8,000 years after the beavers first arrived. Weather during this era follows a typical pattern. Winters are cool with rain and periodic snow. Summers are warm and often fairly dry. Weather is fairly consistent, though like snowflakes, each year varies to some degree, and there are occasional years or groups of years that are especially warm, cold, dry, or wet. About five or ten times per century, there will be a year or string of years where the weather steps outside of its typical pattern. We will look at just one of these instances:

Following twenty years of relatively typical weather, there are two years in a row when the weather is especially favorable for plant growth. In those two years, the dry season never really occurs. Instead, rains are ample during the normally dry months. Then, when it is time for the cold and wet season, the normal rains fall, but the temperatures don't become as cold as usual. This means that for two years, plants have a virtual growth party. This, in turn, means that it is also a great time to be a beaver. Every beaver is well fed, and a few marginal areas of stream expand and become inhabitable, which effectively creates even more beaver habitat in Newcastor. When late spring comes, beaver litters are numerous and a bit larger than normal. Looked at from the perspective of beaver populations, these two years are essentially a beaver baby boom. As a result, there are about 30 percent more beavers running around the landscape than there were before the two years with exceptional plant growth.

Our interest here is to try to predict whether this increase in beavers will constitute a significant or an insignificant disturbance for Newcastor. If you wanted to figure this out in the direct sense, you'd have to study beaver population dynamics and then predict how these beavers and the ecosystem will interact. But here, the historic range of variability gives us another way of addressing the question. Because Newcastor's weather does periodically break from its typical patterns, we can assume that many temporary aberrations in weather lasting just a few years have already occurred. Therefore, with these two years of good growing conditions, the ecosystem is operating well within its historic range of variability. We can imagine that over this 8,000-year period, the ecosystem has already had many opportunities to respond to spikes (or drops) in the beaver population due to temporary weather changes. Therefore, we would predict that the beaver baby boom in Newcastor will be an *inconsequential disturbance* at the ecosystem level. That means, if we are monitoring the ecosystem and concerned about its welfare, we can essentially remove the beaver baby boom from our list of concerns.

In general, knowing that a disturbance is well within the historic range of variability of a complex system means the system has dealt with that disturbance before, and it has had many opportunities to immunize to that disturbance. Thus we can assume that it will be a relatively inconsequential disturbance to the system as a whole.

But now let's add more to this story. Our very first scenario saw a climate shift that the ecosystem was able to immunize to through incorporation. Our second scenario saw the arrival of beavers—which the ecosystem was also able to immunize to through incorporation. However, in this new scenario, instead of now carrying on happily ever after, there is another unexpected event.

After the ecosystem is immunized to the beavers, *a second species of beaver* arrives. This is a *giant* form of beaver weighing around 200 pounds and requiring significantly more food. Just like their smaller relatives, the giant beavers also build dams, but they rely more upon aquatic and marsh plants for food. The giant beaver species finds suitable conditions in lower river zones, often not far from where rivers meet the sea. As they establish, they often take over areas that their smaller beaver competitors had already inhabited. As they take over these areas of lower elevation, they build dams and cut trees—landscape impacts that are not very different from those caused by the smaller beavers. The ecosystem responds somewhat smoothly to the presence of these large rodents. Where the giant beavers operate, the marshy plants move in, warm water fish move in, and the other plants and animals that associate with them follow.

In this case, the giant beavers were impacting the landscape in much the same way that the smaller beavers had been impacting the landscape for several thousand years. The ecosystem merely had to apply its previously developed coping strategies to any new areas affected by giant beavers, and life continues pretty much as it had before. No additional plant or animal species go extinct as a result of the arrival of the giant beavers, although numerous native species on Newcastor change their ranges or populations somewhat as a result of the giant beavers.

Because the giant beavers were new to the island, they were a disturbance. But their similarity to the pre-existing smaller beavers meant that the ecosystem was not pulled all that far out of its historic range of variability (we might say they pulled the system to the approximate edge of that historic range). If a system is pulled to the edge of its historic range of variability, and not too much further, then there are no certainties, but the system has a reasonable likelihood of responding with resilience. In the case of the giant beavers, that's just what happened.

When an event pulls a system's to the edge of—or outside of—its historic range of variability, it should be considered a *consequential disturbance* (or at least a potentially consequential disturbance). Once a disturbance has been identified as consequential, we know that the system is at some risk of failure. The risk of failure may be high or low depending on many complicated factors, but the further that disturbance is from the system's historic range of variability, the higher that risk of failure will usually be.

Now we will add to this story yet again with an even larger disturbance. Under a stable climate and no significant disturbances, Newcastor has remained rather unchanged for about 25,000 years. But then, the planetary climate shifts. Increasing cold at the poles is causing a thickening of ice caps, and this is causing a lowering of sea levels. As sea levels drop, shallow sea beds that can be forded by land animals are exposed. During the episode of exposed sea beds, Newcastor becomes just a 3-mile swim away from the edge of another land mass. Both small beavers and giant beavers arrive on Newcastor at the same time. In addition, within the next two hundred years, a species of elephant, a wild pig, and a type of wolf also arrive on a land that has never known their presence. All of these animals find suitable habitat and quickly begin to populate Newcastor.

Before the arrival of these new animals, lightning strikes would periodically occur on Newcastor. Sometimes a lightning strike would start a small fire, but such fires would never burn more than one or two trees and a few square feet before dying on the shady forest floor due to lack of fuel. After the elephants become common, however, fires behave differently. As the elephants feed and knock over trees, grassy openings form. When fires reach one of these grassy patches, they burn hotter, and trees are killed over a few acres. Gradually, the combination of elephant browsing and bigger fires opens up large portions of Newcastor's dense forest canopy, changing it into a more open woodland.

This dramatic change in habitat is stressful for most of the forest plant and animal species, although a small percentage of the native species are highly favored by the new conditions, and they spread at the expense of the majority.

Wild pigs find the land quite to their liking as well. Newcastor abounds with tasty, succulent herbs, roots, seeds, fungi, and small birds and animals. As the pigs forage, they churn up thousands of acres of fertile soil in search of food. The pigs especially prefer certain plants, and they easily find bird and turtle nests, and these creatures grow scarce. In the place of these dwindling species, less palatable, weedier plants and omnivorous rodents tend to establish.

The wolves find hospitable conditions too. They feed on the recently arrived pigs, beavers, and even young elephants. In addition, they find easy pickings among the native fauna. They readily hunt down the local deer and flightless birds, which are not adapted to evading wolves. As they do this, the ecological governance that these native animals used to place upon the nearby ecosystem, in terms of limiting some plant species and helping others to spread, ceases. And on top of this, we know the story of the beaver impacts—cutting trees, building dams, drowning forest patches, etc., etc.

About a thousand years after the beavers, elephants, wild pigs, and wolves have arrived, Newcastor looks very different. It is a grassier landscape with ponds, sediment deposits below dam bursts, patches of churned up soil, and far fewer species. The plant and animal species directly consumed by these five new mammals have undergone huge population declines. A smaller number of species that were unpalatable, or which were able to quickly take advantage of these new conditions, have taken their place.

Where there were a moderate number of many species, there are now large numbers of just a few species. Unable to find ground safe from the fires, the foraging pigs, and the flooding beavers, our orchid with white and purple flowers has gone extinct. Unable to evade hungry wolves and pigs, and unable to find as much of its favored foods, the secretive forest deer is also extinct. Many other plants and animals, each with their own unique ecological history and needs, have also disappeared. Along with them, the trophic dependencies that they helped maintain no longer operate. Many other plants, insects, fungi, and animals no longer have the food sources or conditions they require, and they follow with similar declines. After 2,000 years, not only have hundreds of plants and animals gone extinct, but the relationships of those that remain are often reconfigured. For example, many insect species that did not go extinct are found in smaller numbers, or associate with different plants or habitats than they had previously.

Newcastor's ecosystem experienced the arrival of these five mammal species as a consequential disturbance. Not only was each new species well outside Newcastor's historic range of variability and therefore consequential, each invading species pulled the ecosystem outside of its historic range of variability *at the same time*. The ecosystem was unable to gracefully adapt to (immunize to) the arrival of these new mammals all at the same time, and instead, it had to undergo a *massive* change in constitution and internal relationships.

Overall it was unable to accommodate these five species with resilience. The prior ecosystem organization is mainly gone, and a new ecosystem that is resilient and immunized to the presence of these five mammals is slowly forming in its place.

If Newcastor's ecosystem was resilient to the arrival of just beavers (a resilient, immunizing response to a consequential disturbance), why couldn't it also be resilient to the arrival of giant beavers, elephants, wild pigs, and wolves? Indeed, it possibly could have been resilient to all of these new animals, but *not all at the same time*. For the creatures and the environment (soils, waters, landforms, etc.) to make balancing adjustments to any one of these new animals would have taken more time. Perhaps in a few thousand years, the adjustment to the beavers could occur. Then if the elephants arrived after that, the ecosystem might have been able to adjust in another several thousand years.

If these disturbances could all be taken one at a time in this manner, then the ecosystem might have been able to ultimately respond to all the disturbances with resilience. However, in this case, the ecosystem was faced with five simultaneous consequential disturbances. As such, it was forced to collapse and reorganize. After collapsing, it will slowly come to a new harmony or balance. Perhaps 4,000 years after these animals have arrived, the ecosystem is restabilizing in its new configuration with a smaller set of species. Gradually, new species will form or colonize to help refine and perfect the balance within Newcastor's new ecosystem (see Patterns 15 and 16, Chapter 10). Perhaps after another 70,000 years have passed, Newcastor's number of species (often termed *biodiversity*) will be about the same as it was prior to the influx of the common beavers, giant beavers, elephants, wild pigs, and wolves.

This story shows how a system can essentially collapse when pushed too far outside of its historic range of variability.

Even if you do not understand all the workings of a system, or understand exactly how the system will respond to a disturbance, you can determine whether particular disturbances are inside or outside of the system's historic range of variability. If the disturbances are near the edge of the historic range of variability, or outside the historic range of variability, you can suspect that they will be consequential disturbances.

The further that any particular disturbance is outside a system's historic range of variability, the less likely the system will respond with resilience. The higher the number of consequential disturbances that are heaped onto a system all at once, the less likely the system will respond with resilience. This is because as the system is pulled further and further outside of its historic range of variability, its internal processes are further and further out of synch with its external conditions (and/or its ability to achieve an objective). Eventually, the internal processes, external conditions, and system objectives are so mismatched that practical adjustments cannot be made to create rebalance. Instead of then maintaining functionality through adjustments, the intricate system essentially breaks down, and very gradually reconvenes with a new strategy that addresses the new conditions.

One of the key factors here is that a resilient system response becomes less likely not only as the magnitude of a particular disturbance grows, but as the *number of* separate disturbances grows. In other words, the capacity for a resilient response to any single disturbance is lowered by the presence of other disturbances, even if those other disturbances are unrelated or of independent origin.

The secretive forest deer could have adjusted to the beavers by abandoning flooded habitats and then learning to feed in those areas at night. They could have adjusted to the opening of the forest canopy by gradually adjusting their preferred foods and retreating to darker, protected valleys and nooks of Newcastor. They could have adjusted to forage competition with wild pigs if their favored plants were given more time to redistribute after the pigs had arrived. They could have adjusted to predation from wolves and wild pigs by learning to herd together or evolving mechanisms to go unseen and un-smelled.

However, the deer were hit simultaneously by all these disturbances, and they were unable to make all of these adjustments in time to stave off extinction. In fact, not only were the deer unable to make all of the necessary adjustments, each challenge that the deer faced weakened their response to the other challenges. When some deer had to retreat from flooded areas, more competitive pressure was placed on the deer whose home range they invaded. Being hungrier, the deer then had to over-consume their preferred foods, which were already beginning to decline due to the activities of elephants and wild pigs. As their prime food supply declined, they produced fewer fawns—this just when their fawns were being increasingly eaten by wild pigs. Meanwhile, the adults were being predated by wolves, and their total numbers crashed. We could imagine that a handful of the deer survived in a hidden nook for another two centuries, but this small group became inbred and lost its genetic vigor. After a fire swept through this hidden nook, hungry wild pigs and wolves eventually did away with the remaining few deer.

Once pulled outside of its historic range of variability, the ecosystem (and many of its individuals and species) loses a lot of its capacity for resilience. If the intensity of the disturbance grows further, or if the number of different disturbances is multiplied, the ecosystem is pulled even further outside of its historic range of variability, and it loses even more capacity for resilience. As the capacity for resilience becomes smaller and smaller, the likelihood for catastrophic collapse and reorganization grows. What applies to ecosystems applies similarly at other scales.

Can the economy of a middle class family respond with resilience to a 25 percent decrease in income? Maybe it can. Can it respond with resilience to a doubling in the price of its energy and tuition bills? Maybe it can. Can it respond with resilience to the income cut and the higher bills at the same time? These are unrelated events, but their combination has a higher likelihood of causing economic catastrophe for the family than either event would alone, or if they happened in different decades.

> Pattern 34: The more an individual, species, species class, natural community, ecosystem, or biosphere is operating *within* its internal and external historic range of variability, the *more* resilience it will have in responding to continued or additional disturbances. The more an individual, species, species class, natural community, ecosystem, or biosphere is operating *near the periphery* or *outside* of its internal and external historic range of variability, the *less* resilience it will have in responding to continued or additional disturbances.

Essential Translation: The more a sustainable complex system is operating *within* its internal and external historic range of variability, the *more* resilience it will have in responding to continued or additional disturbances. The more a complex system is operating *near the periphery* or *outside* of its internal and external historic range of variability, the *less* resilience it will have in responding to continued or additional disturbances.

The implication here is that a sustainable system becomes unsustainable once it is unable to respond to a disturbance with resilience.

This pattern can now be translated into the context of human systems:

Economic Translation: The more a sustainable economy is operating *within* its internal and external historic range of variability, the *more* resilience it will have in responding to continued or additional disturbances. The more it is operating *near the periphery* or *outside* of its internal and external historic range of variability, the *less* resilience it will have in responding to continued or additional disturbances.

Examples provided for Pattern 34 are based on those given under Pattern 33.

For example, a nation adopts a new fuel for its heating, transportation, and power generation needs. The fuel is derived from a specialized crop. A few years after the fuel has become the dominant energy source, a crop failure causes an unanticipated supply shortage, sending serious shocks through the economy and triggering a year of recession. Within two years, crop production has returned to normal levels, and the recession has ended.

Then, eleven years later, during an economic downturn, there is a fuel refinery explosion followed by a worker strike. This occurs just when the government is slashing spending due to a debt crisis. The combination of the economic downturn, fuel shortage, and government spending decline causes a steep dive in commerce, which initiates a massive rise in unemployment and a drop in household incomes. The nation is mired in debt and depression for over a decade. When it emerges from the depression, the economy is smaller than it had been, and many companies have disappeared.

Community Translation: The more a sustainable community is operating *within* its internal and external historic range of variability, the *more* resilience it will have in responding to continued or additional disturbances. The more it is operating *near the periphery* or *outside* of its internal and external historic range of variability, the *less* resilience the community will have in responding to continued or additional disturbances.

For example, in one ethnic region the number of manufacturing jobs has declined. Many of the families respond by sending their men to work abroad for stints of about a year. The ethnic group and its communities have been established and fairly stable for hundreds of years, and sending the working age men to faraway lands is a very stressful change for families and friends, and for important community religious activities. In the first few years that these men are absent, overwhelmed mothers spend less time with their own children, gangs of rowdy young men form, many of the elderly are neglected, and the working age men who remain are spread so thin that they are unable to have much quality time within their social circles. The community begins to respond to these difficulties by resiliently reorganizing some of its social groupings.

But before the community can re-stabilize, armed attacks from a neighboring group gradually drive these people out of their traditional lands. Without ideal locations to migrate to, members of the ethnic group end up scattered among eight other nations—some of which are thousands of miles away. With its families separated and its people scattered, the ethnic group is unable to maintain strong cohesion. Twenty years later, a peace agreement allows the ethnic group to resettle in its homeland. Only about half of them return to the homeland, but even after they return, they are forced to live on a smaller land base. The ethnic group still clearly identifies itself as such, and its major ceremonial traditions remain in practice. However, the community clan system has broken down, along with its associated customs. In the place of a clan system is a nuclear family system similar to the surrounding ethnic group.

Occupational Translation: The more a sustainable occupational system is operating *within* its internal and external historic range of variability, the *more* resilience it will have in responding to continued or additional disturbances. The more an occupational system is operating *near the periphery* or *outside* of its internal and external historic range of variability, the *less* resilience it will have in responding to continued or additional disturbances.

For example, several extended families run lucrative fish markets near a medium-sized coastal city. The markets are open twice per week, and they sell the latest catch from the boats that these families own. All is well for these people until one year, a trading ship arrives from a nearby country. The ship carries tons of salted, smoked, and frozen fish. It docks for five weeks. The traders sell their goods at a low price until their stores are emptied. The families that operated the fresh fish markets are greatly damaged by this event—they have tons of spoiled fish and significant financial losses. The families are relieved when the ship leaves, but are dismayed when it returns again in a year, and they go through the same loss event.

The ship returns again in the third year. But two weeks after it leaves, there is a massive toxic chemical spill nearby, and for health reasons, a fishing ban is enacted for nine months. When fishing does resume, continued safety concerns require that the fish be caught at least 100 miles from shore. Shortly after fishing has resumed, one of the extended families is accused of running contraband, and their main vessel is seized for two years. Another family has its main vessel damaged in a boating accident just before their accountant loses nearly all the business savings in a gambling binge. This tumultuous period eventually results in the total bankruptcy of two fishing families, and a significant downsizing of a third. In their wake, the foreign fishermen and two of the original fishing families have expanded to fill the void left by these collapsed businesses.

Government Translation: The more a sustainable system of government is operating *within* its internal and external historic range of variability, the *more* resilience it will have in responding to continued or additional disturbances. The more a system of government is operating *near the periphery* or *outside* of its internal and external historic range of variability, the *less* resilience it will have in responding to continued or additional disturbances.

For example, a previously unknown disease sweeps through a region, sickening about two-thirds of the populace and killing many who are young, old, or already weakened. The specific disease mechanisms and chemistry are not understood. However, when the disease sweeps through the region again six years later, some people discover that two potent herbs seem to help ameliorate the disease, and they observe that people living in highland areas seemed to avoid the sickness. People begin raising and storing their own medicinal herbs. When the disease hits for a third time, in addition to the early formation of some internal physiological immunity, people consume these herbs, and flee lowlands for cabins in the hills until the disease passes.

The fourth time that the disease sweeps through the region is during a period of food shortage caused by an unusual and prolonged drought. Many of the people are unable to move into the highlands because they would have to abandon their meager food supplies and risk starvation. During the drought, people have also been less committed to growing medicinal herbs, focusing instead on raising whatever food they could. As a result, the death toll is higher than in the prior two disease outbreaks, which both occurred during periods of ample food supplies.

Environmental Translation: The more a sustainable land-use system is operating *within* its internal and external historic range of variability, the *more* resilience it will have in responding to continued or additional disturbances. The more a land use system is operating *near the periphery* or *outside* of its internal and external historic range of variability, the *less* resilience it will have in responding to continued or additional disturbances.

For example, dozens of prosperous towns line the base of a mountain range. The townspeople rely on selective logging on the mountainsides for building material and fuel, hunting in mountain forests for meat, and on croplands on the nearby plains for the remainder of their food. Then, against the objections of the townspeople, the national government institutes an intensive logging program in the mountains. The clearcutting of mountain faces alters hydrology and destabilizes soil. Just a few years after the logging begins, exceptionally large floods begin to occur in the local rivers, and not long after that, devastating mudslides begin occurring. Many homes are destroyed and many people are killed as mudslides hit towns. The local land-use and survival systems are thrown into chaos: logging and hunting opportunities on the mountainsides are gone, and towns are forced to relocate.

During this chaotic relocation period, mountain villagers—who are also fleeing a degraded environment—are relocated by the government into makeshift camps. These camps are also near the base of the mountains, and in many cases, they're on land that townspeople need for survival. The combination of environmental degradation and resettlement means there is not enough food and water for everyone. While townspeople spar with refugee groups over land rights, it becomes almost impossible for them to make the rapid and coherent land-use changes that would be necessary to continue serving their own needs.

For most of the towns, the next two decades are marked by infighting, crimes, group-against-group hostilities, and unplanned land uses. Rather than reorganizing into a useable landscape system, individual families are scattered in a hodgepodge fashion, and are essentially forced to eat away at their own resource capital to survive year to year. They do this at the same time that they are dealing with continued floods, landslides, and a deforested landscape. Over time, the situation stabilizes, but not before the environment is significantly damaged and somewhat depopulated.

Cultural Translation: The more a sustainable culture is operating *within* its internal and external historic range of variability, the *more* resilience it will have in responding to continued or additional disturbances. The more a culture is operating *near the periphery* or *outside* of its internal and external historic range of variability, the *less* resilience it will have in responding to continued or additional disturbances.

For example, in one region, the cultural and religious traditions of the majority ethnic population are rooted in thousands of years of life in rural settings, but due to changes in trade and technology, these people are experiencing a relatively rapid population shift from farms and villages to cities. Because of this shift, the culture is undergoing some loss and reconfiguration as people move away from their family and landscape contexts.

While this demographic shift is underway, an authoritarian family from an ethnic minority takes control of a country. These new leaders ban many important cultural activities enjoyed by the majority population—including traditional religious gatherings, sports, and wedding events. Soon after these bans take effect, there are signs of significant depression in society. Alcohol use skyrockets. Family strife and violence increases, and suicide rates double. Communities try to counteract the bans in a variety of ways. They change their mode of transferring many of their religious stories. They meet secretly for certain special occasions, etc. Despite these efforts to save the culture, in many cases the younger generation has moved away, and there simply aren't enough opportunities to pass along the culture while government bans are also in force. Eventually, the authoritarian regime is ousted, but by that point the cultural practices of the majority ethnic groups are not as widely recalled, understood, or practiced as they had been just seventy years prior.

Physiological Translation: The more that individual people's bodies are operating *within* their internal and external historic range of variability, the *more* resilience their bodies will have in responding to continued or additional disturbances. The more that their bodies are operating *near the periphery* or *outside* of their internal and external historic range of variability, the *less* resilience they will have in responding to continued or additional disturbances. Sustainability in the face of a disturbance requires resilience. What applies here to individual bodies also applies to ethnic groups, racial groups, and to all humankind as a group.

For example, on one continent small nation-states engage in continual battle over resources. The continual fighting is especially stressful for the states located between two or three other hostile groups. One of these states expends significant effort in being war-ready. Beginning at age four, boys are very gradually introduced to physical stresses that are based on war-like conditions. This includes games that emphasize physical fitness and short periods of deprivation from food, water, adequate shelter, etc. These stress events are continued through childhood and are gradually increased in intensity. By the age of about twelve, boys who do not perform these stress events satisfactorily are removed from potential military service. Training for the remaining boys continues with increased intensity of workouts and stress situations.

Those in charge of training are selective regarding stress timing. They are careful to avoid heaping multiple stresses on the boys simultaneously, and they are careful to allow the boys recovery time between intensive training or stress events. They believe that as long as the stress and training situations are applied singly, and with recovery time in between, that the boys will be challenged but strengthened, rather than worn down. By their late teens, the boys are stronger, faster, and more physically and psychologically capable than most adult men in the surrounding states. While many citizens of this nation do not enjoy preparing their children for such difficulty, they recognize that they owe their survival to these preparations.

Before continuing, it is important to point out that resilience is not a black-and-white digital attribute or occurrence. People tend to use simple words such as "resilience" in a way that ignores the complexity and nuance that they entail. The truth is, pretty much any and every disturbance causes a system to lose some of its ability to accomplish its objective, or causes the system to reconfigure to some degree.

So rather than a world full of completely resilient systems on one side and completely non-resilient systems on the other, the world is full of systems that are partially resilient to many different potential disturbances. If a system is almost never 100 percent resilient or 0 percent resilient, it becomes semantic to call it resilient, or to call it non-resilient. Instead, our more important quest becomes trying to understand what factors increase resilience and what factors decrease resilience. Bearing this in mind, we can continue forward, using simplified language, but recognizing that the real story is often one of gradation or relativity.

We are using the word "resilience" to mean the degree to which *a system* is able to continue functioning without massive reformation following a change in internal or external conditions. But soon after this, the quality of *disturbances* was discussed. The windstorm that blows down one forest patch might not topple another forest nearby. Similarly, one particular windstorm might be too weak to topple a forest, but a second, stronger windstorm might blow that forest down. Certainly, it isn't only the system, or only the disturbance, that determines resilience. Rather, what determines resilience is the system properties *relative to* the disturbance characteristics.

Our next discussions pertain to invasive species, a topic that highlights the importance of the relative matchup between the particular system and the particular disturbance.

One of the disturbances endured by natural areas is the introduction of a new species. This can happen through an act of nature, like when chance weather events wash some plant seeds up onto the shore of a land where they have never grown before. Although a species can be introduced simply by natural processes, on the contemporary landscape, far more species are introduced by humans. People bring in shrubs from another continent because they like the way they look. They bring in birds from another land so they can raise them as pets. They bring in foreign livestock. They bring weed seeds in the mud that sticks to the bottoms of their boots. They bring insects stowed on ships. They accidentally bring viruses and diseases that could not have crossed the oceans, mountains, or deserts without human assistance. When these plants, animals, or microorganisms land in a new place, they encounter species, species groups, natural communities, and ecosystems that have never before known their presence.

Any particular species is obviously considered native in its original location and habitat, and is only considered to be an introduced species outside of those original natural limits. Depending upon where you look on the planet, the introduced species include all sorts of things, such as various sparrows, swans, rabbits, house cats, mice, horses, snakes, turtles, beetles, worms, ants, carp, eels, and hundreds of species of trees, shrubs, grasses, herbs, and aquatic plants.

Many such species have been—and are being—introduced. Some of these introduced species have little impact upon the existing natural systems that they meet—they are a disturbance to those natural systems, but they are more of an *inconsequential* disturbance. Other introduced species have a huge impact upon the existing natural systems that they meet—effectively becoming *consequential* disturbances upon those systems.

We can refer to these consequential introductions as "invasive species." We can learn some important things by taking a look at the characteristics of invasive species, and comparing them to species that are introduced but never become invasive and consequential.

Let's look at how this works from the start. First, a species is introduced, either naturally or by humans. If the species is introduced but then immediately dies out without reproducing, it will be an inconsequential disturbance and a noninvasive species. But if the introduced species survives and reproduces, then it has established. There are probably many factors that help determine whether a species will actually establish after introduction, including the number and gender of introduced individuals, the health of the introduced individuals, where and when they are introduced, the availability of suitable habitat, and probably a number of other chance or luck factors.

For those species that do establish, they will either remain inconsequential, or will eventually move on to become invasive and consequential. What is the difference between those that stay inconsequential and those that become invasive? The answer to this probably involves several factors, but here we focus on answering this question through the lens of *Protections*—specifically, lost Protections.

Do you recall Protections? In Patterns 25 through 28 we saw how individuals, groups, and species assure themselves adequate resources through Protections. The four Protection avenues that species are able to employ include **resource use, self-defenses, restraint, and barriers**. None of these four protects are singly reliable all of the time, so most species are protected by combinations of two, three, or all four of them.

In the very act of introducing a new species, the last of these Protections—barriers—are breeched as the spatial separations between the species are eliminated. This leaves three Protection avenues, but one of these—restraint—is also often lost during a species introduction. Here is why: Many of the mechanism by which species effectively behave with restraint are actually pests, diseases, or parasites upon them.

For example, an herb species might be able to grow prolifically, except that a certain beetle feeds on the herb's roots wherever the herb becomes very densely populated. The beetle provides a negative feedback loop for the herb—the more common it becomes, the less likely it is to continue spreading. This helps keep the herb at moderate numbers, which in turn leaves space for other species. But if the herb is introduced to a new continent, it will probably be introduced without the beetle that feeds upon it. So as our herb breeches barriers Protections, it does so without its natural restraint mechanisms.

This often leaves the native species with only the remaining two Protections relative to the introduced species—self-defenses and resource use. But the story of lost Protections continues. The native species sometimes have rather weak or useless defenses to employ against the introduced species. This is because the self-defenses of the native species are often honed to the other native species with which they have historically competed; their self-defenses are *not* as geared toward species that have not evolved alongside themselves. For example, maybe numerous native ant species on an island have developed population balances based on the ability of each species to fend off the other. But when a more aggressive ant species is introduced, the native species are unable to put up an effective defense, and they disappear.

While our discussion about relations between resilience, Protections, and invasive species is slanted toward competition dynamics, many of Earth's native species have famously gone extinct when they were faced with invasive predators to which they lacked self-defense mechanisms (recall the secretive forest deer on Newcastor that were highly vulnerable to introduced wolves and wild pigs). You might think that the native species will be able to evolve new self-defenses toward an introduced species, and indeed they can, but this could take anywhere from decades to hundreds of thousands of years for this evolution to occur.

At least that leaves the native species with one Protection, which is resource use, right? Actually no, not necessarily. Recall that when a species employs resource use as a Protection, it is depriving other species of access to that resource. However, a species employing resource use must refrain from using all of its resources, or it will have no future. Thus, there is always some amount of unused resources, even when resource use is an active Protection. In addition to all of this, the incoming, introduced species might have very different ways of behaving and using resources. It might be able to utilize things that none of the native species view as direct resources. As an analogy, imagine an entrepreneur eyeing a scenic and popular sandy beach and envisioning glass factories (producing sand out of glass) instead of seeing all the tourism. That entrepreneur might make plans that will short-circuit the existing tourism economy.

If the introduced species is able to utilize a resource in a whole new way, then it can short-circuit the existing food and energy networks, and lay claim to resources that native species had previously relied upon, even if indirectly. In this way, the Protection of resource use (which is already limited in its utility) can be rather ineffective when a new species is introduced.

So that's it; if the introduced species is accompanied by restraints, is affected by the locally employed self-defenses, and has conventional modes of resource use, it will tend to be inconsequential. In cases where restraints of an introduced species are weak, self-defenses employed against it are ineffective, and the introduced species has unconventional modes of resource use, the introduced species will be consequential, and many of the native species will probably find themselves on the losing end of an extreme competition scenario.

For at least many situations, we can now understand why some introduced species are inconsequential, and why some are consequential. But what good is this if we can't predict the invasive potential of a species ahead of time? Actually, if we look further, we will see that we do have some predictive power.

- When a species is introduced, the loss of barriers is a given.
- Restraints will be absent to the degree that the new species lacks any intrinsic or extrinsic mechanisms for self-limitation, or loses any such mechanism on the journey.
- Self-defenses employed against the introduced species will tend to be ineffective to the degree that the introduced species is *different from* the native species. This is because the self-defenses employed by native species are designed to work against other native species, and are not as effective when the introduced species is very unlike those native species.
- Similarly, resource use employed against the introduced species will tend to be ineffective to the degree that the introduced species uses resources very

differently from the native species. In this case, "different" can translate to being genetically unrelated, being of a different size, shape, behavior, biochemistry, or having different timing of activities, means of reproducing, etc.

We can flip the idea and use the term "similarity" to indicate how much an introduced species differs from the native species it encounters. If an introduced species is a high-similarity species, it is more likely to remain an inconsequential, benign introduction. Alternatively, if an introduced species is a low-similarity species, it has many differences from the native species it is being introduced to, and it has a stronger chance of becoming a consequential, invasive introduction.

There are many common examples of introduced species that remain benign and inconsequential. For example, the common dandelion is an introduced species in North America. It can be found growing out in lawns, gardens, fields, and along edges of woods and wetlands. Although widespread, dandelions are at rather low densities in natural areas, and they do not disrupt native habitats to a significant degree. Another example in North America is ring-necked pheasants, which were introduced from Asia and are found across the countryside. Even though they are common in places and utilize natural habitats much of the time, they do not seem to displace native species or greatly change the ecosystems they inhabit.

Why were dandelions and pheasants relatively inconsequential introductions? Perhaps in part because they both happened to have a high **similarity** to the native species they encountered.

When dandelions entered North America, they encountered hundreds of other short, flowering herbs, including many in their own family (the sunflower family). Dandelions grew in similar places and ways compared to native species, and native animals such as deer and rabbits found that they could eat them. Because they were close cousins to many native asters, hawkweeds, fleabanes, etc. they could easily share pollinators and native predatory insects and caterpillars. In other words, their high degree of similarity allowed them to fit into the fabric without tearing a hole or changing the texture of the whole cloth.

A similar situation probably occurred with pheasants. They entered a land that already had many seed and insect eaters (such as sparrows, quails, turkeys, and mice). They nest on the ground like several other native songbirds and quails do, and native predators such as foxes found them to be good eating. Again, they were able to weave into the fabric of their new land without greatly altering its texture.

Then there are many examples of introduced species that became invasive and consequential. One example of an invasive species is the Australian paperbark tree in Florida, which forms dense thickets and smothers out sunlight from native wetland plant communities. Another example is the zebra mussel, which is native to Eurasia and has established itself in the Great Lakes. Zebra mussels have altered the Great Lake aquatic ecosystems. Due to the zebra mussels, numerous species of fish and native mussels are much less abundant than they once were.

Why have these introduced species become so invasive and consequential? Maybe because they had low similarity to the native species they encountered. Australian paperbark trees were introduced to South Florida—a land that had less than a dozen of species in its particular plant family. They were brought without the associated insects that feed on them. Furthermore, they had colonial behaviors and combinations of adaptations to fires and flooding that were unlike the adaptations that most native plants had. When they entered the fabric of South Florida, they had low similarity to the native species, and they began changing the weave wherever they colonized.

Zebra mussels have adaptations that allow them to form dense colonies where native mussels could not, and they, too, were introduced without all of the creatures that eat or parasitize them. The zebra mussels, rather than add to the weave, changed the whole fabric of the systems they invaded.

Do we *know for certain* which species will become invasive? Actually, we don't. Once introduced, we often don't even know if a species will even survive at all in the new location. If it does survive and reproduce, a variety of factors determine whether it will become invasive. Similarity is probably just one of the main factors underlying invasive potential.

In Pattern 34, we saw how historic range of variability was a useful tool in predicting whether a disturbance was consequential and whether a system would respond with resilience. In fact, species introductions and similarity are just a special case of historic range of variability—when the introduced species has high similarity, it is really just another way of saying that its form and behavior fall well within the system's historic range of variability. When the introduced species has low similarity, it is really just another way of saying that its form and behavior fall near the periphery or outside of the system's historic range of variability.

Pattern 35: When a "new" species is introduced to other species, species groups, natural communities, and ecosystems, the tendency for the new species to behave disruptively is often associated with its tendency to behave with restraint, and with its similarity to the native species that are already present. The more restraint mechanisms that the introduced species has, the less invasively and disruptively it will tend to behave. The more similar the new species is to existing species, the more likely it is that its introduction will be inconsequential and non-disruptive. The less similar the new species is to existing species, the more likely it is that its introduction will be consequential and disruptive. Similarity between introduced and native species is just a special category of a natural system's historic range of variability.

Essential Translation: When a "new" entity or group is introduced into a system, the tendency for the new entity or group to behave disruptively within the system is often associated with its tendency to behave with restraint, and with its similarity to the other occupants that are already present. The more restraint mechanisms that the introduced entity or group has, the less disruptively it will tend to behave. The more similar the new entity or group is to existing system occupants, the more likely it is that its introduction will be inconsequential and non-disruptive. The less similar the new entity or group is to existing system occupants, the more likely it is that its introduction will be consequential and disruptive. Consequential and disruptive events lower the likelihood that a receiving system will remain sustainable. Similarity between introduced and indigenous entities and groups in a system is just a special category of a system's historic range of variability.

Obviously, the more consequential and disruptive an introduction is, the less likely it is that the receiving system will remain sustainable. This pattern can now be translated into the context of human systems:

Economic Translation: When a "new" economic entity or process is introduced into an existing economy, the tendency for the new entity or process to behave disruptively within that economy is often associated with its level of restraint, and with its similarity to the economic entities or processes that are already present. The more restraint mechanisms that the introduced entity or process has, the less disruption it will tend to cause in the existing economy. The more similar the new entity or process is to existing economic entities or processes, the more likely it is that its introduction will be inconsequential and non-disruptive. The less similar the new entity or process is to economic entities or processes that are already present, the more likely it is that its introduction will be consequential and disruptive. Consequential and disruptive events lower the likelihood that a receiving economy will remain sustainable. Similarity between introduced and indigenous economic entities or processes is just a special category of an indigenous economy's historic range of variability.

For example, the nation of Verdenia uses a fuel derived from a specialized crop for most of its heating, transportation, and power generation needs. Over time, scientists and plant breeders try introducing other crop varieties for fuel use. These other varieties are essentially interchangeable crops, and some are used locally, but none of them have a widespread effect upon Verdenia's energy economy and infrastructure, which has been relatively stable for over a century.

Meanwhile, in a foreign nation, a team of researchers discovers a way to produce energy from isotopic minerals that can be mined from certain rock layers. Once mined, it is a cheap fuel, but it cannot be used on a small scale for local power generation. Foreign corporations begin producing this very cheap and abundant energy source. With financial support from their own national government, these corporations enter Verdenia's energy market.

The combination of foreign government support and an inexhaustible energy supply means that the new energy source is cheaper than Verdenia's crop-derived fuels. This sets off relatively rapid changes in Verdenia's energy economy and infrastructure. Verdenia's native energy producers (farmers, distillers, and pipeline operators) are put under great economic stress, and end up for the most part out of business. At the same time, Verdenia's vehicle design and manufacturing have to be retooled with respect to this new energy source.

Community Translation: When a "new" person/group is introduced into a community, the tendency for the new person/group to behave disruptively within the existing community is often associated with their tendency to behave with restraint, and with their similarity to the other community members that are already present. The more restraint mechanisms the introduced person/group has, the less disruptively they will tend to behave within that community. The more similar the new person/group is to the existing community, the more likely it is that their presence will be inconsequential and non-disruptive. The less similar the new person/group is to the existing community, the more likely it is that their presence will be consequential and disruptive. Consequential and disruptive events lower the likelihood that a receiving community will remain sustainable. Similarity between introduced and indigenous people and groups is just a special category of an indigenous human community's historic range of variability.

For example, an ethnic group believes in gender segregation from the age of six until marriage. Within this ethnic group's homeland, there is a worker shortage at nearby mines and ore processing facilities. In response, the national government locates hundreds of families from a second ethnic group into the towns occupied by the indigenous ethnic group.

This second group does not practice gender segregation, and furthermore, they are religiously opposed to it. This second group has a proselytizing tendency, and rather than try to conform to local beliefs, they behave assertively, making sure that boys and girls have public contact. This is very disruptive for the indigenous group, which is unable to practice its customs while girls are mixing with their boys, and while boys are mixing with their girls. The indigenous ethnic group considers taking actions against the second group, but declines to do so because if conflict arises, the second ethnic group is likely to receive the backing of the national government.

Occupational Translation: When an occupational group is introduced into a preexisting occupational establishment, the tendency for the new group to behave disruptively within that occupational establishment is often associated with the group's tendency to behave with restraint, and with its similarity to the other occupational groups that are already present. The more restraint mechanisms the introduced occupational group has, the less disruptively it will tend to behave. The more similar the new occupational group is to existing occupational groups, the more likely it is that its introduction will be inconsequential and non-disruptive. The less similar the new occupational group is to existing occupational groups, the more likely it is that its introduction will be consequential and disruptive. Consequential and disruptive events lower the likelihood that a receiving occupational establishment will remain sustainable. Similarity between introduced and indigenous occupational groups is just a special category of an indigenous occupational system's historic range of variability.

For example, a civilization has many small- to medium-sized businesses, and within its business culture there is no history of commercial advertising. When immigrants from distant civilizations begin to arrive, they bring their own trades and professions with them—including advertising. Without a prior tradition of commercial advertising, the civilization has no rules or regulations pertaining to truth and slander. In very little time, competing businesses that had reached a market equilibrium are thrown, as some businesses benefit from advertising at the expense of others. Within four decades, many businesses have gone bankrupt, and the businesses that remain are typically larger and more aggressive than they were previously. And as far as occupations go, advertising is the third most commonly selected profession among young people.

Governmental Translation: When a second, new human group is introduced into an indigenous, preexisting group, the tendency for the new group to behave disruptively among the indigenous group is often associated with its tendency to behave with restraint, and with its behavioral similarity to indigenous group. The more restraint mechanisms the new group has, the less disruptive it will tend to be for the indigenous group. The more similar the new group is to the indigenous group, the more likely it is that its introduction will be inconsequential and non-disruptive. The less similar the new group is to the indigenous group, the more likely it is that its introduction will be consequential and disruptive. Consequential and disruptive events lower the likelihood that an indigenous governance system will remain sustainable. Behavioral similarity between newly introduced and indigenous groups is just a special category of an indigenous society's historic range of variability.

For example, a small nation is attacked and gradually colonized by an empire. The citizens of the empire who move to the invaded country are given numerous special privileges. They are immune from prosecution from many behaviors that the locals see as crimes, and are afforded extra legal rights when they are in conflict with the local people. Because of this, they are almost completely unaccountable in their behavior toward the locals.

Eventually, local protesters begin a subversive rebellion. At the same time, antagonism begins to erupt between locals who protest the colonizers, and those who cooperate with the colonizers. The tension and resulting lack of safety begins to affect commerce, and along with despair and low morale among many people, a large percentage of the younger native population begins to emigrate to nearby countries. The original local group is left struggling for their legal rights, for a sense of control, and for their sense of identity.

Environmental Translation: When a "new" human group or land-use practice is introduced into a region, the tendency for the new group or land use to cause disruption within the region is often associated with its tendency for restraint, and with the similarity between the new land uses and (pre)existing land uses. The more restraint mechanisms the introduced group/land use has, the less disruptively they/it will tend to be relative to existing land uses. The more similar the new group/land use is to existing land uses, the more likely it is that their introduction will be inconsequential and non-disruptive. The less similar the new group/land use is to existing land uses, the more likely it is that their introduction will be consequential and disruptive. Consequential and disruptive events lower the likelihood that a receiving land-use system will remain sustainable. Similarity between introduced and indigenous land uses is just a special category of the indigenous land-use system's historic range of variability.

For example, dozens of prosperous towns line the base of a mountain range. The townspeople rely on selective logging on the mountainsides for building material and fuel, hunting in mountain forests for meat, and upon croplands on the nearby plains for the remainder of their food. Then, the national government institutes a logging program in the mountains.

The logging eventually leaves unrelated mountain villagers homeless. These displaced people migrate to the base of the mountains, where they are unable to practice their traditional hunting and gathering lifestyle. The displaced villagers lack national government protections and are weakened from being uprooted. In order to survive, they quickly conform to the local land-use activities of the townspeople. This includes growing crops, cutting trees for buildings and fuel, and hunting periodically in the mountains for meat. With the burden of higher numbers of people, there is some conflict and stress over land uses and land rights during the next decade. However, relocation of the mountain villagers goes relatively smoothly overall, considering the difficulty it could have caused for themselves and the local townspeople.

Pattern 33 showed how over time, a system can develop resilience to a consequential (i.e., significant) disturbance. Patterns 34 and 35 showed how, in some instances, one can predict whether a given system will experience a given disturbance as consequential or inconsequential. The next pattern shows how a system's resilience to potential future disturbances can be strengthened.

Do you remember the story of what happened when beavers first showed up on Newcastor? The behavior of the beavers was outside the ecosystem's historic range of variability, and thus the beavers were a consequential disturbance. Forest patches were killed. Numerous animals and cold-water fish declined in numbers. River channels were scoured out after beaver dam bursts. What had been a stable and sustainable ecosystem was thrown off track. But after getting "knocked down," so to speak, the ecosystem responded in a variety of unique ways. Marsh and meadow plants that were originally confined to slivers of river sandbar and deltas began colonizing ponds and mudflats. Native insects that lived off of these plants also increased. Amphibians and songbirds in turn could prey upon those insects, and they increased. Species of fish that could tolerate warmer water began appearing in the beaver ponds. Egrets, osprey, and fishing eagles could then hunt for these fish and amphibians in the ponds. New species of duck and woodpecker began to breed in Newcastor. Species of trees and shrubs changed their distributions in order to colonize wetland edges. One tree species was able to develop chemical resistance to the beavers, and could grow more successfully where beavers had cleared other trees. Deer changed their foraging patterns. Wildcats and fishing eagles learned how to hunt the abundant beavers.

What would have happened to Newcastor without all of these native plant, insect, fish, bird, and mammal species, and their unique responses to the landscape changes? We don't exactly know, because complex systems are not deterministic, and in the same manner that we can't predict the exact outcome to current events, we can't predict the exact outcome to alternative scenarios either. However, without the response of these species—their changes in behavior, distribution, culture, physiology, etc., it is likely that Newcastor would have remained in a state of imbalance for even longer.

Without the wetland plants moving into beaver ponds and mudflats, unstable, nutrient-heavy mud would have continued sloshing around river basins. Without the new duck species, there would have been less balance in competition between plant species, and these plants would have more difficulty achieving Significant Expression (see Pattern 25, Chapter 13). Without the new fish coming to live in the beaver ponds, there would be few feeding egrets, osprey, and eagles, and thus the system would have more difficulty recapturing extra nutrients and returning them to the nearby forest (in the form of dead fish). Without the response of chemical resistance to the tree, there would be even less tendency for stabilizing roots along streams and rivers, and riverbanks would erode more than they did. Without all of these species making adjustments within the ecosystem, the ecosystem is less and less able to stabilize while recycling (rather than bleeding) its nutrients and other raw materials (see Pattern 12, Chapter 9).

When a significant disturbance occurred, the ecosystem relied partially upon its *diversity* of species to gather itself back together. With these different species and their great variety of interconnections, the ecosystem was able to mend the tear in its fabric. Without all of these species, it would have been able to mend its tear, but it would have taken a lot longer, and would have been less graceful.

Then, can you recall what happened when giant beavers arrived on Newcastor, as described prior to Pattern 34? These giant beavers arrived on an ecosystem that had never known their presence before. However, it was *after* the much earlier arrival of smaller beavers, and the ecosystem already had a chance to immunize to (become resilient to) the smaller beavers. By the time the larger beavers arrived, they were merely a larger but similar version of the existing, smaller beavers. Again, via similarity, the presence of the small beavers was a supporting factor in being able to cope with the larger beavers. In this situation, the ecosystem experienced only a small fabric tear—one that it could quickly mend. This was made possible by the presence of one extra species—the smaller beavers.

Both of these cases demonstrate something. Whether it is widespread physical disruption or an introduced species, the ecosystem often carries immunity through the species it already has, or it ends up immunizing itself through their assistance. When it comes to matching similar native species to an introduced species, an ecosystem with more species has a better chance of providing a similar match than one with fewer species. And if there is major disruption, an ecosystem with more overall species can generally respond and re-stabilize more rapidly and with more grace, compared to an ecosystem with fewer species.

The number of species that an area contains is sometimes referred to as its *biodiversity*. Biodiversity to an ecosystem is like a tool chest to a handyman. The more tools he has in his box, the more rapidly he will tend to perform his work, and the more likely it is that he will be able to tackle the odd project that requires special equipment.

Biodiversity, in many cases, improves resilience in ecosystems. Although biodiversity is often understood as the number of *species* a place has, it can also refer to the amount of biological variation at other levels—such as the number of natural communities or ecosystems in an area, the range of species groups present in an ecosystem, the genetic diversity present within a given species, etc. All levels of biodiversity are likely to contribute, in general, to resilience. The primary exception to biodiversity's support for resilience, however, is when that diversity is in the form of introduced species, rather than native, indigenous components of an ecosystem.

> Pattern 36: Individuals, species, species groups, natural communities, ecosystems, and the biosphere are generally made more resilient to disturbance by their inherent forms of biodiversity, and are made less resilient to disturbance if operating from a state of reduced biodiversity.

Essential Translation: Individuals and other complex systems are generally made more resilient to disturbance, and thus more sustainable, by an inherent diversity of members or active components and/or ranges of potential behavior, and are made less resilient to disturbance, and thus less sustainable, if operating from a state of reduced membership or numbers of active components and/or reduced ranges of potential behavior.

As with several other patterns, separate translations of this pattern into economic, community, occupational, governmental, environmental, cultural, and physiological modalities are somewhat artificial. In nature and among humans, these modalities often occur together inseparably (a new species carries with it a new economy, a new community disposition, a new occupation, etc.). However, herein each modality will be translated individually in order to help make the point of diversity's resilience role relative to that modality.

This pattern can now be translated into the context of human systems:

Economic Translation: Economies are generally made more resilient to disturbance, and thus more sustainable, by an inherent diversity of sub-economic activities and/or ranges of potential economic activity, and are made less resilient to disturbance, and thus less sustainable, if operating from a state of reduced diversity of sub-economic activities and/or reduced ranges of potential economic activity.

For example, two mid-sized nations are side by side. One of these nations derives over 90 percent of its energy from a specialized crop. The second nation uses this same crop for about a quarter of its energy production; the other three-quarters of its energy is sourced from waste material, wind power, nuclear power, and animal power (i.e., beasts of burden). When a new pest infestation leads to a near total loss of the specialized energy crop, the national economy of the first nation is devastated. Commerce almost shuts down. Many people become jobless and go hungry, and it takes a decade for the national economy to get back in order. The second nation is also impacted by the crop loss—energy prices rise quickly, and for several years everyone feels the pinch. The second nation experiences a significant recession, but as the other power sources (waste, wind, nuclear, and animal) are ramped up, energy prices nearly come back to prior levels. Within three years, the second nation is back to normal levels of economic activity.

Community Translation: Individuals, groups, and other complex systems are generally made more resilient to disturbance, and thus more sustainable, by an inherent diversity of community member types, and are made less resilient to disturbance, and thus less sustainable, if operating from a state of fewer community member types.

For example, in one ethnic region, the number of manufacturing jobs has declined. Many of the families respond by sending their men to work abroad for stints of about a year. One of thousands of families to participate in this behavior lives in an apartment-like setting with about ten other closely related nuclear families (aunts, uncles, cousins, grandparents, parents, etc.). There are eight related teenage boys in this complex. The complex generally buzzes with activity and conversations between every combination of relations. All eight boys have strong relationships with various brothers, sisters, cousins, uncles, aunts, and grandparents. The extended loss of their fathers is saddening for each of the eight boys, although against a backdrop of the other relatives in their lives, none of the boys are significantly destabilized emotionally or derailed from their life ambitions.

Just on the other side of the river is another family that sent the father abroad for work. The father's two boys are left with only their mother in the home. Both boys were generally well behaved, yet they have great difficulty with the loss of their father. One of the boys becomes very quiet and detached, unwilling to engage much with his mother or his school. The other boy works very hard in school and to help maintain his small family, but he has outbursts of anger, and plans to leave home as soon as he can afford to.

Occupational Translation: Individuals, groups, and other complex systems are generally made more resilient to disturbance, and thus more sustainable, by an inherent diversity of occupations or potential occupations, and are made less resilient to disturbance, and thus less sustainable, if operating from a state of reduced diversity of occupations or potential occupations.

For example, several extended families run lucrative fish markets near a medium-sized coastal city. One year, a massive typhoon runs along the coast, smashing the harbor and then flooding the city. Nearly everyone is able to reach safety away from the coast, but almost every ship and fishing boat is sunk or destroyed beyond salvation. In addition, there is significant financial damage to the city's infrastructure. This city has a relatively diverse workforce that includes farmers, loggers, carpenters, engineers, construction workers, accountants, communications workers, planners, vendors of all sorts, and medical workers. These people are able to conduct emergency operations, followed by repairs and rebuilding. The city essentially repairs its infrastructure within six months, although shipping and fishing don't return to normalcy for about three years, as boats are recovered and built from scratch.

Further down the coast is a smaller city that is even more reliant upon fishing as a major food source. This city and its harbor are also badly damaged during the typhoon. The city's moderately diverse workforce, which includes many fishermen and a few farmers, loggers, carpenters, construction workers, vendors, and medical workers, are able to meet their emergency infrastructure and medical needs, but they struggle to repair many of the complex portions of their infrastructure. To speed up repairs, many people need to receive rapid technical training, but there are still significant delays. The fishermen—who represent a relatively high proportion of the workforce, have to work in construction or various other professions while boats are gradually repaired and rebuilt. Within about three years, normalcy has returned to the town, but full-scale fishing does not resume for another two years.

Government Translation: Individuals and other complex systems are generally made more resilient to disturbance, and thus more sustainable, by an existing or potential diversity for (self) governance, and are made less resilient to disturbance, and thus less sustainable, if operating from a state of reduced diversity in existing or potential governance.

For example, a previously unknown disease sweeps through a region, sickening about two-thirds of the populace and killing many who are young, old, or already weakened. The specific disease mechanisms and chemistry are not understood. However, when the disease sweeps through the region again six years later, some people discover that two potent herbs seem to help ameliorate the disease, and they observe that people living in highland areas seemed to avoid the sickness. Some people make the effort of raising these herbs afterward, while others feel it isn't worth the trouble.

When the disease hits again, people who had grown the medicinal herbs consume them. Also, many people flee lowlands for the hills. They do this despite the fact that it is often inconvenient—because they must abandon home, and often have to camp or live in old shacks for several months. The rate of serious illness and death is much lower among those who raise the medicinal herbs and those who put up with poor living conditions in the hills.

Environmental Translation: Individuals, groups, and other complex systems are generally made more resilient to disturbance, and thus more sustainable, by an inherent diversity of land uses or potential land uses, and are made less resilient to disturbance, and thus less sustainable, if operating from a state of few land uses or few potential land uses.

For example, eight prosperous towns line the base of a mountain range. People in the five most northern towns obtain their building material and fuel via selective logging of mountainsides, and their food from penned livestock and crops on the grasslands below. People in the three southerly towns also obtain their building material and fuel via mountainside logging, but their other needs are met with a more diverse set of activities.

They obtain food from hunting in mountain forests, fishing in nearby rivers, and livestock grazing and crop production on the plains. In addition, they produce their own electricity with windmills staged near their croplands.

Then, when an intensive national government logging program is instituted, the clearcutting of mountain faces leads to altered hydrology and soil destabilization. Exceptionally large floods begin to occur in the local rivers, and not long after that, devastating mudslides begin occurring. Many homes are destroyed, and many people are killed as mudslides hit towns. Each town's land use and survival systems are disrupted.

Wood can no longer be gathered in the desired quantities off the mountainsides. The five northern towns that had used this wood as their only source of fuel are forced to burn crop residue to heat and light their homes. This reduces the food available for their livestock, and because their livestock were penned, they aren't well suited to an open grazing system. Furthermore, the people in these towns have rusty hunting and fishing skills. Overall, people in these five towns have a difficult time rebalancing their fuel, crop, and meat sources.

People in the three southern towns are quick to shift extra livestock onto the deforested mountain slopes to help make up for diminished hunting opportunities. They are able to burn some of their crop residue for fuel (residue that their free-range animals were not reliant upon). And although they lost their wood fuels and building material, their ongoing access to power from windmills makes it much easier to rebuild damaged infrastructure.

Cultural Translation: Individuals and groups are generally made more resilient to disturbance, and thus more sustainable, by internal diversity in culture or potential diversity in culture, and are made less resilient to disturbance, and thus less sustainable, if operating from a state of reduced internal cultural diversity or a reduced range of potential culture.

For example, an authoritarian family from an ethnic minority takes control of a country. These new leaders ban many important cultural activities enjoyed by the majority population—including traditional religious gatherings, sports, and wedding events. Soon after these bans take effect, there are signs of significant depression in society. Alcohol use skyrockets. Family strife and violence increases, and suicide rates double. However, there is diversity in cultural practice among this majority group, and certain clans experience much less strife. These clans have cultural practice variations that are much harder for an authoritarian government to find and suppress. Their practices include unwritten religious songs; small, in-house weddings; impromptu sporting competitions; and community feasts unaccompanied by religious activity. These alternative practices become "contagious" and spread quickly through the majority culture, helping the majority ethnic group to maintain its self-identity and self-determination during a century of repression.

Physiological Translation: Individuals and groups are generally made more resilient to disturbance, and thus more sustainable, by an inherent diversity of physical capabilities and characteristics or ranges of potential capabilities and characteristics, and are made less resilient to disturbance, and thus less sustainable, if operating from a state of reduced physical capabilities and characteristics or ranges of potential capabilities and characteristics.

For example, on one continent, small nation-states engage in continual battle over resources. The continual fighting is especially stressful for the states located between two or three other hostile groups. One of these states expends significant effort in being war-ready. Beginning at age four, boys are very gradually introduced to physical stresses that are based on war-like conditions. This includes games that emphasize physical fitness and short periods of deprivation from food, water, adequate shelter, etc. These stress events are continued through childhood and are gradually increased in intensity.

By the age of about twelve, boys who do not perform these stress events satisfactorily are removed from potential military service. Although they lack the ideal physical disposition for battle, boys that are removed from military service are often intelligent, and with training, many become skilled military strategists. Boys that do remain in physical military training are sure to develop well-rounded skills such as marksmanship, wielding weaponry, weight bearing, climbing, swimming, sprinting, distance running, and wrestling. With the uncertainty of future conflicts, no boy knows exactly which skills will become life-saving in the future. This broad training has allowed this nation to consistently protect its territory and bear lower casualty levels than its enemies and neighbors.

Next, we will continue our discussion with a broader look at diversity.

Chapter 16 Diversity

In biological science, the diversity of types of living things is referred to as *biodiversity*. Although biodiversity is often viewed as the number of *species* in a particular area, it can also refer to the amount of biological variation at other levels—such as the number of natural communities or ecosystems in an area, the range of species groups present in an ecosystem, the genetic diversity present within a given species, etc.

Biodiversity plays an influential role in the behavior of natural systems. Biodiversity effects resilience and stability. It also relates to competition, expression and Protections, group decision-making, succession, lag recovery time, etc.

If you were designing your own complex system, and you wanted the system to be sustainable, you'd want it to be diverse enough to be resilient, but you would not want it to be unsustainably or inappropriately diverse. How do you know when the amount of diversity present is appropriate for the system at hand?

Perhaps this question can be answered using some very complicated math formula. But rather than take that route, we will look to nature to give us a sense of how much relative diversity is fitting for different situations. Looking toward nature won't tell us anything extremely specific, such as exactly how many law firms should be in a local city, but it could provide a relative sense of which city can sustain the most law firms.

Wherever we look in nature, we see lots of different species. There are millions of them across the planet—birds, mammals, reptiles, amphibians, insects, plants, algae, fungi, etc. We will consider diversity by looking at just how many different *species there are in any one place*. If we were to take a look at nature in different locations, we would see that, in fact, some places have more species than others.

If we go to the tropical Amazon Rainforest of South America, we can find somewhere between 10,000 and 20,000 tree species (note that this isn't the number of individual trees, which is much higher; this is the number of *types* of trees). In contrast, if we head up to the boreal forests of North America, we will find thirty or forty species of trees, or perhaps twice this figure if we are including tree species along the fringe of the boreal forest. Or, we could go to the tropical island of Puerto Rico, where there are somewhere around 500 species of trees.

Even places that are right next to each other can have very different levels of species diversity. If we compare the somewhat central states of Colorado and Kansas, we see that Colorado has roughly 2,500 native plant species, whereas Kansas has closer to 1,500 native plant species—just about two-thirds the number that Colorado has. The above comparisons are made regarding plant diversity, but similar comparisons can be made with birds, fungi, fish, bats, snails, etc. Why is there such a contrast in species diversity in these different situations?

In some cases, biodiversity differences are simply being caused by *scale*, or the amount of available area. A very small area cannot host a species that requires vast amounts of space.

Consider a simple scenario—perhaps you would like to host some rhinoceros on your property (this is probably *not* a good idea, but we can imagine it for the sake of argument). Of course, you'd recognize that you'll need more than one for reproduction purposes. Maybe you own 10 acres of grassy and brushy field. The grass and brush in the field might keep one rhinoceros happy, but the field won't be big enough to support a whole family of rhinos. It should also be noted that rhinos have a territorial streak, and might not tolerate each other in close proximity. So realizing this, you could buy some more land—let's say 120 acres of grass and brush. For the sake of discussion let's assume that 120 acres is just enough space to feed and contain the territorial ambitions of five rhinos.

After your land purchase, you obtain two male rhinos and three female rhinos. This might work out for a while, but the problem you will eventually have is that five rhinos can't sustain a breeding population over the long term. Very small populations tend to develop genetic defects due to inbreeding (which is the accumulation of unwanted genetic mutations). And anyway, as successive births and deaths occur, even if you start with two males and three females, it's just a matter of time before your population of five or six animals breeds itself into a dead end—with either all males or all females.

The rhinos make for a very simplified example, but they help to illustrate some of what biologists have found in the fields of "population genetics" and "island biogeography." Essentially, a species needs enough room to survive, not just temporarily as a few individuals, but over long periods as a species with many individuals. Then when you factor in the effects of resource availability, competition, and expression (see Patterns 25 and 26, Chapter 13), it turns out that a species usually needs much more space than would be dictated by its physical size alone relative to the land or habitat area.

This principle of scale helps to explain why even with similar habitat types, the number of tree species in the tropical Amazon rainforest is so much higher than on the island of Puerto Rico. It isn't that there's something particularly missing from Puerto Rico, but being several hundred times smaller, it simply can't sustain the number of species that the Amazon Rainforest can.

That explains a lot, but it can't explain why the species diversity levels can be so different in areas of comparable size, such as Colorado and Kansas. Why are there many more plant species in Colorado? The primary reason for this is probably that Colorado has so many more *habitat types*. In Colorado, you can go from plains to prairie to mountain forest and alpine tundra. Plus, there's also a bunch of desert and canyon country. Kansas has a more uniform terrain. It does have hills, streamside woodlands, moist prairies, and dry prairies. But Kansas lacks the snowy mountains and very dry rocky lands that are in Colorado. Due to Colorado's diversity in geography and the plants that associate with those particular climates and soils, Colorado has more habitat niches to fill than Kansas does. Again, Colorado is not overburdened with species, and Kansas is not inherently species deficient. The flora of both states are fitting. But being somewhat more topographically and geologically varied, Colorado holds more species than Kansas.

That explains even more of the difference in the Earth's biodiversity as we go from place to place. But how can we account for the fact that the Amazon rainforest contains so many more tree species than the boreal forest of Canada? And again, it isn't just trees. We would find higher diversity in the Amazon Rainforest if we looked at other plants, insects, fish, etc. Both of these forests are huge, so we can't really explain it through size difference. Both forests span across different temperatures, moisture levels, and terrains. So, our second explanation—habitat variability—can't explain this great difference in species diversity either.

One factor that could explain this contrast is *favorability* or *suitability* of habitats. The boreal forest is huge, but life there is accompanied by some significant stresses too. Winters are cold and dark. In addition to the direct stress of low temperature in winter, the trees cannot make much new energy (from dim sunlight) or absorb much water (which has turned to ice). Also, fires periodically move through the boreal forest and burn large areas. In contrast, the Amazon rainforest is almost always moist and warm—excellent conditions for tree growth.

It seems that there are many cases of this across the planet. Where base environmental conditions are more stressful for a given form of life, such as for fish, birds, cacti, or grasses, those forms of life often still occur, but they are represented by a lower diversity of species.

Interesting maybe, but what causes this? The cause is not fully understood. Though one explanation could be that when base environmental conditions are stressful, there are a more limited number of workable physical adaptations available to the biological world. It is perhaps a big hurdle for a species to develop a new adaptation to an environmental challenge, and maybe when there are a limited number of potentially viable adaptations, there are a more limited number of species that can develop those adaptations in the face of simultaneous competition. On the other hand, we know that a species that is already adapted to a situation can radiate into many other species, but we saw that this isn't so easy unless Protections are available (see Patterns 27 and 28, Chapter 13). Whatever the cause though, biological diversity does seem to be correlated with favorability of habitat.

We now have three explanations (or at least associative patterns) for the difference in species diversity from place to place. But there's at least one more common reason why diversity can be much higher in some places compared to others.

Remember back to the land of Newcastor? Due to the arrival of two species of beavers, and also of elephants, wolves, and wild pigs, there was a consequential disturbance to the ecosystem. Hundreds of plant and animal species went extinct from Newcastor. In that story, Newcastor regained its original level of biodiversity after 70,000 more years elapsed. How could it have regained biodiversity? Gradually—very gradually—new plants, mammals, and insects will find Newcastor through migration. If they arrive and establish without causing other species to go extinct (see discussions prior to Patterns 33 and 34 for consequential disturbances), then they increase the total level of biodiversity. In addition, new species can evolve from the species that are already living in Newcastor, resulting in two species where there was once one (see discussions prior to Pattern 27).

In these ways, over time, the number of species can increase. The longer that time passes after a severe disruption or consequential disturbance (species invasion, flood, volcano, glaciation, etc.) that resets succession (see Patterns 13 through 16, Chapter 10), the larger the biodiversity can generally become. This process takes a long time, of course (see Pattern 19 in Chapter 11 for lag time), but it does result in more biodiversity.

If you go to a recently formed or exposed island in the ocean (island formation is a rare event in a human lifetime, but frequent enough over the planetary lifetime), you would find it to be nearly bereft of lifeforms. If you could return to the island hundreds or thousands of years later, you would probably find that a much greater number of species have found that island, and have begun forming their own complex natural community.

Does this maturation requirement for an increase in biodiversity help to explain why the Amazon rainforest has so many more trees than the boreal forest? You might think not, since neither region is an island, and both have ancient stands of trees. However, in the geological scheme of things, scientists believe that the boreal forest has not existed for very long, at least in its present location. Much of the existing land inhabited by the boreal forest was under a mountain of glacial ice just 10,000 or 15,000 years ago. So, the existing boreal forest is effectively a colonizing ecosystem on this newly exposed land.

In comparison, there is evidence that the Amazon Rainforest has been in place for far longer—perhaps for millions of years. This age difference is probably a contributing factor to the difference in biodiversity between the boreal forest and the Amazon Rainforest.

Herein, "biodiversity" and "species diversity" have often been used interchangeably. However, it was also noted that biodiversity pertains to other hierarchical levels such as the diversity of genetic types, the diversity of natural communities, of ecosystems, etc.

As you consider the patterns pertaining to biodiversity, keep in mind that while species diversity might be easier to visualize, these patterns would also usually apply to these other hierarchical levels. For example, factors that are associated with higher diversity of species are usually also associated with higher genetic diversity within a given species.

> Pattern 37: Larger areas of habitat can sustain more biodiversity (diversity of individuals, species, species groups, natural communities, and ecosystems) than smaller areas of otherwise similar habitat.

Essential Translation: A larger system context can sustain more types of systems or subsystems than a smaller but otherwise similar context.

This pattern can now be translated into the context of human systems:

Economic Translation: A larger population and/or region can sustain a higher diversity of economic activities and conditions than a smaller but otherwise similar population and/or region.

For example, the island of Canellia is about 40 miles long and has an area of about 900 square miles. Most of Canellia's 120,000 residents live in one of five small cities, although the majority of its land is used for agriculture. Just 100 miles to the east lies Dryer Island. Dryer Island is about 5,500 square miles, which is about six times larger than Canellia. Dryer Island has a population of about 750,000 people, and like Canellia, has most of its residents in cities but most of its land is agriculture.

Three-quarters of Canellia's economic activity occurs within six sectors: agriculture, fishing, construction, professional services, retail sales, and hospitality. Dryer Island's economy is certainly larger, but it is also a bit more diverse. Nine sectors make up 75 percent of its economic activity. These include the six dominant sectors on Canellia, plus manufacturing, transportation, and mining.

The next six examples pertain to the previously described island of Canellia and Dryer Island.

Community Translation: A larger human population can sustain more community groups than a smaller but otherwise similar human population.

For example, it is common for residents on Canellia to join one or two interest groups or teams. Some of the larger groups are religious study clubs, soccer teams, and crafting hobby groups. The situation is very similar on Dryer Island, where there are hundreds of teams and interest groups. However, people on the island of Canellia that have the most uncommon interests, such as rock painting, wooden utensil carving, or sand ball, sometimes take special trips to Dryer Island just to comingle with groups that don't exist on their own island.

Occupational Translation: A larger population and/or region can sustain a higher diversity of occupational groups than a smaller but otherwise similar population and/or region.

For example, a large majority of young adults on the island of Canellia stay near their homes and eventually become gainfully employed. However, some natives of Canellia want to pursue niche interests that are not taught in local academic or training schools. This can include topics such as astronomy, entomology, ancient history, or microbiology. People with these niche interests often leave home to pursue a higher education on Dryer Island, or at universities even further away. People that remain in their niche fields for their career work are sometimes able to obtain employment at the main hospital or university on Dryer Island, but it is rare for them to find employment that is central to their interest back home on Canellia.

Government Translation: A larger population and/or region can sustain a higher diversity of behaviors than a smaller but otherwise similar population and/or region.

For example, all vehicle speed limits on Canellia are between 20 mph and 55 mph. Most speed limits on Dryer Island also fall within that range, but there are two particular roads through old town centers with speed limits of only 10 mph, and there is a 25-mile highway straightaway where speeds of 65 mph are allowed.

Environmental Translation: A larger region can sustain a higher diversity of land uses than a smaller but otherwise similar land region.

For example, residents of both Canellia and Dryer Island enjoy access to recreational parks with accommodations for many uses, including picnicking, nature viewing, hiking, camping, trail riding, field sports, and water sports. However, Dryer Island has three additional public areas that are specially designed for other uses. One of these is a large park dedicated to game hunting, another is for aerial uses (hang gliding, ballooning, model craft flight, etc.), and the third is a solitude nature park, with a wild interior and no machinery noises.

Cultural Translation: A larger population and/or region can sustain a higher diversity of cultural values and practices than a smaller but otherwise similar population and/or region.

For example, the culture on the island of Canellia can be described as generally hard working, and with a stable commitment to family time and traditions that span many centuries. Every family and every town on Canellia are unique, but other than some differences between rural and town folks, there aren't any strong cultural differences from region to region across Canellia.

The culture on Dryer Island is slightly more variable from place to place. People on the west side of Dryer Island are known more for their focus on economic progress, whereas the people on the east side of the island have a reputation for focusing more on play and pleasure. In addition, a central portion of Dryer Island contains four villages that speak a completely different language than the rest of the island. These people, who call themselves the Stehek'n, claim to be the original inhabitants of Dryer Island. The Stehek'n practice their own unique ceremonies and have their own cultural tales explaining their place in the universe.

Physiological Translation: A larger population and/or region can sustain a higher diversity of human physiological variations than a smaller but otherwise similar population and/or region.

For example, a team of researchers attempts to learn more about the people of Dryer Island and Canellia. Through a combination of anthropological, physiological, genetic, and linguistic evidence, they come to believe that the Stehek'n are most closely related to a continental group about 2,000 miles away. From the perspective of ancestry, Dryer Island's majority (non-Stehek'n) population and the residents on the island of Canellia are determined to be closely related, although the residents on Dryer Island seem to have a slightly more diverse physiology and genome than the population on Canellia does.

In addition to area size, biodiversity in nature also tends to increase in tandem with differences in the landscape or the supporting environment. An area of a given size will tend to be more biologically diverse if it has more differences in climate, terrain, geology, etc.

Pattern 38: Areas with more habitat types can sustain more biodiversity (diversity of individuals, species, species groups, natural communities, and ecosystems) than similar sized areas with more habitat uniformity.

Essential Translation: A system context that is more diverse can sustain more types of systems or subsystems than a similarly sized but otherwise more uniform system context.

This pattern can now be translated into the context of human systems:

Economic Translation: A resource or asset distribution that is more diverse can sustain more types of economic activity than a similarly sized but otherwise more uniform resource or asset distribution.

For example, Hall County and Ravinia County are adjacent and of similar size. Hall County occurs on a very gently sloped plain formation. It contains fertile farmland interspersed with forested patches, wetlands, and towns. About two-thirds of Ravinia County also occurs on this plain formation and is also dominated by agriculture. However, part of Ravinia County slopes down to meet a large river that drains into the ocean, where Ravinia County contains 15 miles of sea coast. More than two-thirds of the annual income received by citizens of Hall County is generated through agricultural activity. Incomes in Ravinia County are primarily supported through three sectors: agriculture (40%), recreation and tourism (20%), and shipping (20%). The next six examples also pertain to Hall and Ravinia counties.

Community Translation: A more diverse population distribution, occupational array, or economic setting can sustain more community types than a similarly sized but otherwise more uniform population distribution, occupational array, or economic setting.

For example, most people in Hall and Ravinia counties gather with people in their own occupational grouping. Farming families tend to have their own social circles, as do people that work in shipping, recreation, or tourism. Hall County has a significant number of people only within the farming social groups, while Ravinia County has social groups centered on all three occupations.

Occupational Translation: A more diverse economic or environmental setting can sustain more occupational types than a similarly sized but otherwise more uniform economic or environmental setting.

For example, the local newspaper in Hall County has a "help wanted" section that is divided into different job categories such as farmhand, construction, accountant, etc. The newspaper lists about twenty job types in an average week. Ravinia County's local paper has a similar help wanted section, and it lists an average of about thirty job categories in a given week.

Government Translation: A context that provides a higher diversity of situations for individual people and groups can sustain a higher diversity of governmental responses than a similarly sized but otherwise more uniform situational context.

For example, about once per year, people run through emergency practice drills and procedures in both Hall County and Ravinia County. The drills in Hall County include wildfire, tornado, and hurricane response. Depending on their specific location, Ravinia County residents perform these same drills, but also perform drills associated with floods, tidal waves, and pandemics.

Environmental Translation: An area with a more diverse physiography, climate, or occupational array can sustain more diversity in land uses than a similarly sized area with a more uniform physiography, climate, or occupational array.

For example, about 80 percent of the land in Hall County is used for crops, hay, or livestock pasture. The remaining areas are mainly in natural condition or used for residential, commercial, or public transportation purposes. About 60 percent of Ravinia County is used for agriculture, and another 20 percent is used for residential, commercial, or transportation purposes. However, unlike Hall County, Ravinia County also has a large area dedicated to a dual usage recreation *and* ocean wave/storm buffer. Just like Hall County, Ravinia County has many roads for public transportation, but it has additional transportation areas of restricted-use waterway, where shipping safety is the primary focus.

Cultural Translation: A more diverse environmental or economic setting can sustain more cultural groups than a similarly sized but otherwise more uniform environmental or economic setting.

For example, farmsteads in the area of Hall and Ravinia counties tend to be passed from one generation to the next, and the culture in farming regions is rather stable, family oriented, and tied to the bounty of the earth. People that work in shipping also tend to be family oriented, but many of them have had extensive international travels and they tend to have a more global perspective. People from shipping families also tend toward a love and value for the sea, and a humility toward the power of nature and the ever-present risk of seafaring. People in the region who work in tourism and recreation tend to have an outgoing, playful, and entrepreneurial spirit. In Hall County, the farming region subculture is predominant, whereas all three of these subcultures are well represented in Ravinia County.

Physiological Translation: A more diverse environmental, economic, or occupational setting can sustain more abilities and body types than a similarly sized but otherwise more uniform environmental, economic, or occupational setting.

For example, a majority of people in Hall and Ravinia counties remain within the economic sector that their own family participates in, although due to differences in interests and abilities, there are always some who leave the economic sector they were born into, or who move into or out of the region. Along with this, there is a tendency for self-selection and competitive selection to occur for each person on the way toward a career path. This selection is based upon aptitudes, interests, and physical abilities. People that gravitate toward or remain in farming tend to be patient, uninterested in chaotic situations, and physically strong. People that gravitate toward tourism and recreation tend to like lots of change, meeting new people, getting ahead, and trying new things. People that find their way or remain in shipping are often men who prefer to broaden their horizons and be in risky situations, who want to have a clear-cut endpoint to their tasks, and who like to have their bodies put to the test.

A third factor that tends to correlate with biodiversity is the level of environmental support. When environmental support for a given lifeform is high, that lifeform tends to be represented by more species and associated natural community types (community types based on that lifeform). When environmental support for a given lifeform is low, that lifeform tends to be represented by few species and few associated natural community types. Bear these simple concepts in mind when reading the next pattern, which has wording that may seem complicated by relationships between hierarchical levels of biology.

Pattern 39: When the environment of an ecosystem is very supportive of a species class, that ecosystem can sustain a higher diversity of species within that species class, and it can sustain a higher diversity of natural community types based on species from that species class, in comparison to an ecosystem with a less supportive environment for that species class.

Essential Translation: When there is stronger environmental support for a given form, a complex system can sustain a higher diversity of members of that form, and it can sustain a higher diversity of subsystems which are based upon that form, in comparison to a complex system provided with less environmental support for that form.

This pattern can now be translated into the context of human systems:

Economic Translation: When there is stronger economic and environmental support for a given economic sector, that sector can contain and sustain a higher diversity of economic patterns and strategies compared to an economic sector with less economic and environmental support.

For example, differences between two regions—Pleasant Province and Forest Province—highlight the influence that support can have on diversity.

Pleasant Province has a physiography of wide fertile valleys and a relatively gentle climate with plentiful rainfall. It has 40 cities with populations of 50,000 or more. Pleasant Province has an attractive landscape of valleys, wooded hillsides, and seasides, but scenic vistas are often lacking. It has laws that protect the environment from degradation, but it has avoided using its funds to protect or purchase much of its scenic land or open space. Pleasant Province has no tax breaks for small companies and no special programs to promote startup businesses. The recreation and tourism industry in Pleasant Province is relatively small. It includes about 40 public agencies and private companies. Each of the agencies and companies has a slightly different type of funding or revenue stream—in broad categories, approximately 4 agencies draw their income from provincial public support, 7 draw income from local public support, 21 gather income from client fees, and 8 companies rely on product sales.

Forest Province has a mix of grasslands, forested mountains, and canyon lands. It has a more intensively seasonal climate with colder winters, hotter summers, and occasional water shortfalls. The province contains two large metropolises and another 10 cities with more than 50,000 people.

Over several centuries, Forest Province has used its funds and legal system to protect and purchase much of its most scenic open space. Some of its primary travel and sightseeing hotspots include its two largest cities, its rocky and wild seashores, its dry canyon lands, dozens of miles of untamed scenic rivers, and its forested mountains. Its two big cities have planned shopping and attraction centers that draw hundreds of thousands of people every year. Forest Province gives tax breaks to small companies, and it has minimal governmental bureaucracy, along with streamlined permitting and licensing processes. It also has special programs to promote startup businesses. Forest Province has a very strong recreation and tourism industry. Not only is Forest Province's recreation and tourism industry large, it is economically diverse. There are between two hundred and three hundred unique public agencies and private companies that serve this industry. Each of the agencies and companies has a slightly different type of funding or revenue stream—in broad categories, approximately 12 agencies/companies draw their income from provincial public support, 17 draw income from local public support, 185 gather income from client fees, 46 companies rely on product sales, and 4 companies rely on revenue from advertising tourism destinations.

Although Pleasant Province hasn't taken major steps to support to its recreation and tourism industry, is has taken many steps in support of colleges and universities. It has a long history of using public revenues to support urban infrastructure, and to support all levels of education. Institutions of higher learning, for example, are given special tax status. Its citizens tend to be well educated, and the internal demand for secondary education is high. The province additionally attracts many students from abroad. Each of the forty cities in Pleasant Province hosts at least one college, university, or training center. From small training schools to large universities, Pleasant Province contains a total of 135 institutions of higher learning. Primary funding mechanisms for these institutions include nine types: provincial public sources, local public sources, private corporations, tuition, endowments, investments, donations, product sales, and volunteer contributions.

Forest Province has a long history of supporting transportation and utility infrastructure projects—but this is especially in association with sectors such as its agriculture and logging industries, rather than for the benefit of urban areas. The province does provide public funding for education, but the level of funding is not especially strong relative to other priorities. For example, it has no special tax status for institutions of higher learning, which number 48 within its borders. These institutions range from small training schools to large universities. Primary funding mechanisms for Forest Province's institutions include six types: provincial public sources, private corporations, tuition, endowments, product sales, and volunteer contributions.

Comparisons between Pleasant and Forest provinces provide the basis for the next two examples as well.

Community Translation: When there is stronger economic and environmental support for a given economic sector, that sector can contain and sustain a higher diversity of human community types and human community interconnections, in comparison to an economic sector receiving less economic and environmental support.

A researcher studies the focal interests—or commonalities—around which human friendship and activity groups develop within Pleasant Province and Forest Province. In Pleasant Province, he finds that about 40 percent of the communities are formed principally around an education or intellectual commonality. These commonalities include stage levels such as incoming freshman or graduate students, interests such as biochemistry, and values such as religious affinity or political view.

In contrast, only 22 percent of the communities in Forest Province are formed principally around education or intellectual commonalities. In addition, when the researcher categorizes the education or intellectual commonality groups into specific *types* (such as freshman status groups, professor groups, or law interest groups), he finds that there are more in Pleasant Province than in Forest Province, which have 60 types and 40 types, respectively.

Afterward, a second researcher is inspired to study friendship and activity group commonality in the provinces relative to recreation, which includes activities such as kayaking or endurance climbing. In line with her predictions, the higher numbers are in Forest Province. Not only does Forest Province have over twice the ratio of recreation group commonalities within its population, but when all the groups are categorized into specific types, there are about twice as many recreation-based group types in Forest Province as there are in Pleasant Province.

Occupational Translation: When there is stronger economic and environmental support for a given economic sector, that sector can contain and sustain a higher diversity of businesses and professions in comparison to an economic sector receiving less economic and environmental support.

For example, there are approximately 350 total fields of study that can be pursued at the many institutions of higher learning in Pleasant Province. These fields have a wide span that includes 25 languages, 20 types of biological science, and 10 types of history. In Forest Province, one could potentially pursue about 250 fields of study, including 18 languages, 15 types of biological science, and 5 types of history.

A fourth factor in nature that tends to correlate with biodiversity is the age of an area. When nature in a particular place has been allowed to operate free from significant disruption, the number of lifeforms in that place can increase. Stated another way, if there are two very similar places and one is older, the older one will tend to have a higher number of species. The gain in species with age might plateau at some point, rather than rising indefinitely. Thus, it is possible that a comparison in species numbers between a very old place and a very, very old place would not result in a clear pattern.

Pattern 40: Ecosystems that have existed for longer period of time, or have existed for a longer period of time in a particular location, generally sustain more biodiversity (diversity of individuals, species, species groups, and natural communities) than similar but otherwise younger or more recently established ecosystems.

Essential Translation: Complex systems that have existed for a longer period of time, or have existed in a particular context for a longer period of time, generally sustain more diversity of components and subsystems than similar but otherwise younger or more recently established complex systems.

This pattern can now be translated into the context of human systems:

Economic Translation: Economies that have existed for a longer period of time, or have existed in a particular context for a longer period of time, generally sustain more diversity of sub-economies and economic patterns than similar but otherwise younger or more recently established economies.

For example, the city of Carrolton began as a farming hamlet. Later, religious missionaries settled there. When a regional wave of industrialization occurred, Carrolton happened to be strategically located, and within 70 years, it grew rapidly into a city with a population of about 500,000 and an economy based on metal smelting and manufacturing.

After another 300 years, Carrolton's population size had not changed much, but its economy had become more diversified. Right after its initial boom, smelting and manufacturing generated two-thirds of the city's income. But after 300 years had elapsed, smelting and manufacturing accounted for only a quarter of the city's income, and other economic sectors had also become very important. This included banking, scientific research, and healthcare. Changes that occurred in Carrolton through this 300-year period are the subject of the next six examples as well.

Community Translation: Human communities that have existed for a longer period of time, or have existed in a particular context for a longer period of time, generally sustain a higher diversity of human groupings than similar but otherwise younger or more recently established human communities.

For example, soon after the wave of industrialization built the city of Carrolton, if one wanted to join a sports team, there were three sports to choose from. After 300 years had elapsed, there were eight different sport leagues that one could choose to participate in. Furthermore, social networks that were originally tightly structured around the demands of the manufacturing economy had diversified to include a broader range of ages, livelihoods, and interests.

Occupational Translation: The occupational diversity of human groupings that have existed for a longer period of time, or have existed in a particular context for a longer period of time, is sustained at a higher level than the occupational diversity of similar but otherwise younger or more recently established human groupings.

For example, soon after the wave of industrialization built the city of Carrolton, about three quarters of adults employed outside the home were working in low- or semi-skilled physical labor in smelting or manufacturing. After 300 years had elapsed, the workforce was still comprised of about 33 percent low- or semi-skilled physical labor, but another 25 percent worked in highly skilled trades and labor, and another 20 percent did office or clerical-type work.

Government Translation: Human groups that have existed for a longer period of time, or have existed in a particular context for a longer period of time, generally sustain a higher diversity of governmental or behavioral responses than similar but otherwise younger or more recently established human communities.

For example, soon after the wave of industrialization built the city of Carrolton, expectations for a typical man were to complete the tenth grade in school, and then become a worker. Being a worker normally required showing up on time for an eight- to twelve-hour work shift, and being ready to work physically hard throughout that shift. If he did this and followed other relatively clear codes of conduct stipulated by the local dominant religion, he had a high likelihood of financial and social success.

After 300 years had elapsed, there was a wider span of behaviors that could lead to success for an average male. Some young men continued past the tenth grade and went through a lengthy secondary education process. Along with technological changes, some men could work more from an office, home office, or on-call setting. Further, the social norms had diversified somewhat, and the way men spent time was less standardized. Men in some families and communities focused more on hard work, others on family time, and others on spiritual interests.

Environmental Translation: Human groups that have existed in a particular place for a longer period of time generally sustain a higher diversity of environments and land uses than similar but otherwise more recently established human groups.

For example, Carrolton was essentially built up from a village into an industrial city within a few decades. In its early industrial form, the city had standardized and somewhat regimented land uses—neighborhoods of similar-looking row houses and apartment buildings were set on a grid of streets that surrounded an industrial downtown with a few stores.

After 300 years had elapsed, the land uses and appearances were much more varied. In four locations, whole blocks of homes had been removed in order to create public parks. Neighborhoods had diversified to some degree—some neighborhoods contained community gardens, some had a very clean kept appearance, and some contained recently constructed churches. Dozens of first-floor apartments had been converted to local storefronts. Certain roads had become far busier, while a few others had been completely closed down to vehicle traffic in order to accommodate more pedestrians. In addition, the downtown emphasis was no longer just upon manufacturing; there were also many office buildings, theatres, and restaurants.

Cultural Translation: Human groups that have existed in a particular context for a longer period of time generally sustain a higher diversity of cultural values than similar but otherwise more recently established human groups.

For example, soon after Carrolton was first built up into an industrial city, there was a somewhat uniform common culture, which was based on being dependable and on the rhythm of the industrial workday. For most people, the value of life was tied to their hard work, the products their factory would produce, the survival of their families, and their engagement in one of two religious groups.

After 300 years had elapsed, the culture was more varied. Many individuals still had very similar values and traditions that were unchanged from 300 years prior, but many others held values associated with other types of work and lifestyles. For example, among the people who worked in banking, there tended to be more value placed upon shrewdness and income generation. Among the research profession, there tended to be a higher value for creativity. And among the healthcare profession, public welfare was held as highly important. Furthermore, most residents of the city still associated with one of two religious organizations, although the particular tone and style within individual houses of worship had diversified along with the city's general population.

Physiological Translation: Peoples that have existed in a particular context for a longer period of time generally sustain more diversity of ethnic or genetic groupings than similar but otherwise more recently established peoples.

For example, soon after Carrolton was first built up into an industrial city, 99 percent of the population represented one ethnic group from regional villages that sent their extra hands to the city for work. About 200 years later there was a worker shortage, and members of another, more distant ethnic group were encouraged to move to Carrolton and work in smelters and factories. Some of these individuals remained permanently, settling into a particular neighborhood enclave. With about 20,000 residents in this enclave, this second ethnic group eventually made up about 4 percent of Carrolton's population.

Diversity is important for complex systems because it can relate to all sorts of internal processes and it can improve resilience. Patterns 37 through 40 suggest that diversity levels within a system aren't random. The levels of diversity that can be sustained within a system vary by the size of the system, heterogeneity within the system, supports provided by the system, and maturity level of the system.

Chapter 17 Innovation and Management

Now we have 40 patterns that help to demonstrate ways that complex systems operate. By knowing how complex systems operate, we are in a better position to design them sustainably. But making sustainable plans won't be enough. Even if initial designs are good, we will eventually face the unexpected, the unforeseeable, and the unknown. We will face new challenges, conflicts, and stress points. Maintaining system sustainability will require management-level decisions. Even in these uncertain and complicated situations, we can look to nature to provide guidance.

In the most general sense, a "problem" can be any situation in which the goals or aspirations for a system are being undermined, put out of reach, or stymied. When problems arise, maintaining sustainability goals requires problem solving, conflict resolution, goal achievement, adaptation, or addressing challenges—which we can lump all into the single term of "innovation." An *innovation* is just a new way of doing something. When a system is faced with a problem, innovation will be needed to maintain a status or reach an objective.

Nature *abounds* with examples of successful innovations. Apparently through trial and error, plants, animals, and their communities have "figured out" how to survive all kinds of amazingly difficult conditions, such as extreme heat, high pressure, high salinity, flooding, extreme cold, low light, low energy, low oxygen, water deprivation, nutrient deprivation, and cyclical extremes.

For example, plants need water to survive, but they have developed many amazing innovations (which are often termed "adaptations" in biological sciences) that allow them to live where there are long periods with hardly any rain. The innovations/adaptations make use of strategies such as water storage, water conservation, and weather timing. Here are some examples of these plant adaptations:

1. Dozens of arid-land plant species effectively collect water during moist periods and store it in their trunks, leaves, branches, or roots. For example, many species of barrel cactus (and in fact, most cacti) store water in their stems and rounded trunks. Huge African baobab trees store water in trunks, and the tiny African living stone plants store water in their leaves. Some agave plants in the Americas also store water in their thick leaves.

2. Another fundamental adaptation to scarce water is to have very deep root systems. Deep roots can reach down into lower moisture and groundwater layers even when rainfall is absent. The velvet mesquite plant found in deserts of the American Southwest has roots that can extend more than 50 feet downward. Other North American plants of deserts and dry lands with very deep root systems relative to their size are Colorado pinyon pine, sagebrush, tarbush, and desert bitterbrush.

3. Some plant species have especially hairy or fuzzy leaves that help to deflect direct sunlight and hot, dehydrating winds. The hairs thereby reduce the amount of water evaporating out of the leaves. Examples of these desert and arid grassland species are North American rabbitbrush, brittle brush, scarlet globemallow, woolly plantain, littleleaf rhatany, and winterfat.

4. A similar adaptation is displayed by plants that have a waxy coating or film on their leaves. Since wax tends to repel water, water cannot escape out of the leaf, and so the wax reduces the amount of water lost to evaporation. Some of the plants that utilize this water conservation strategy are the North American creosote bush, plains yucca, Utah juniper, and Sonoran scrub oak.

5. Another plant strategy is simply dropping leaves off during dry periods. This allows them to enter a sort of dormancy without much need for water. A plant that drops its leaves cannot grow much, but if it grows its leaves back when rains return, then at least it can survive the driest periods. Some species with this adaptation are the North American coyote gourd, bigroot, desert ironwood, and triangle bursage.

6. A slightly more extreme twist on the leaf-dropping adaptation is performed by short-lived species (sometimes referred to as "annual" species). They survive as dormant seeds that only grow for a few weeks or months after soaking rains. Once they mature, they quickly set seed and die, leaving behind a new generation of seeds that await the next wet period. North American desert annuals that utilize this adaptation are the Arizona poppy, purple mat, and desert sunflower.

7. Another ingenious way of surviving desert conditions is a type of biochemical adaptation. In order to conduct photosynthesis and make energy, most plants of humid regions collect carbon dioxide during the day through tiny openings in their leaves. However, as they gather this carbon dioxide, they unintentionally lose a lot of water from their leaves. Losing water in humid regions is not such a problem, but in arid regions, it is life-threatening. So numerous plant species have invented a way to avoid this evaporative water loss. Instead of gathering their carbon dioxide during the hot daytime, they gather it at night while temperatures are low and humidity is high. Thus, they don't lose very much water in the process of gathering the carbon dioxide. They store the carbon dioxide until the sun rises when it can be used for photosynthesis. The technical term for this nighttime gathering of carbon dioxide is crassulacean acid metabolism (CAM). Some of the

species that exhibit this chemical innovation are pineapple, purslane, Mojave yucca, and most cacti.

There are even more adaptations that plants utilize to survive desert conditions. If we investigated, we would also discover interesting innovations and behaviors that animals utilize for desert survival. We could look at other extreme conditions too, such as high salt levels, and we would find similarly ingenious adaptive mechanisms that species have for living in these conditions.

In fact, wherever we want look across planet Earth, nature has often found a *multitude* of adaptations for coping with common problems. Thinking back to all those plants living in deserts and arid lands, which one of the adaptations should be considered the most valuable? From a common sense perspective, the answer would probably be the first one, water storage, right? What could be a more fundamental solution than storing water in a place where it is scarce? But then again, one could argue for the importance of having deep roots. Surely, no survival strategy in deserts could be superior to deep roots and their ability to reach water that is being held deeply below the soil. Still, plants with hairy leaves might have found the smartest solution because growing little hairs on leaves seems like it would require the least effort. On that note, maybe waxy leaves are superior to hairy leaves because a waxy coating is only a thickening of a surface wax layer that would be there anyway. Then there's the nighttime accumulation of carbon dioxide. It seems more technologically advanced than all the others.

Which desert plant adaptations are really best? We could go on and on debating. But having this long debate would suggest we were missing a fundamental point. Nature has already settled the debate, hasn't it? Nature has already told us that in fact, all of these solutions have merit. The proof is that year after year, all of these survival solutions are being used out on the ground where the rubber meets the road, and where poor solutions don't survive long. In different places and times—sometimes singly, and sometimes in combination within a species—each innovation for solving the water scarcity problem is used and passed along for another year and to another generation of plants. Solutions without merit would never survive these dry and competitive conditions.

Were we to look at any other common problem faced by life-forms, we would find a multitude of innovations across the planet. As interpreted here, nature is showing us that even the most difficult problems often have more than one workable solution.

What else does nature indicate about innovation and problem solving? Plant adaptations to dry situations demonstrate a second principle. There are many, many plant species that inhabit deserts around the world. These species, each with unique forms and adaptations, come from many different *plant families*. The concept of *family* in the biological sciences is somewhat different from the concept of family in the human world. A biological family (also known as a *taxonomic family*) is a conceptual hierarchical grouping of organisms that are somewhat similar and usually closely related.

Some common examples are the cat family (which includes housecats, bobcats, lions, tigers, cheetahs, leopards, etc.), the dog family (which includes dogs, wolves, coyotes, foxes, jackals, etc.), and the orchid family (which includes hundreds of orchid plant species from around the world). All the species within a single family are believed to share common ancestors prior to evolutionary events that allowed the development of separate species (see discussions prior to Pattern 27 related to Protections).

Thus, in theory, at one time on Planet Earth, a single type of cat species or cat-like species lived and eventually evolved into all the modern cat species currently alive. The same concept would hold true for other families as well—whether the dog family, the orchid family, the owl family, the sunflower family, etc.

Back to the innovation and problem-solving theme, the species that are adapted to survive in extremely dry situations represent *dozens* of families. If we look at the desert survival strategy of water storage, the baobab is in the mallow family, living stones are in the ice plant family, barrel cacti are in the cactus family, and agaves are in the asparagus family. If we look at the survival strategy of hairs to deflect the sun's radiant heat, we see that this, too, has been adopted by numerous species from numerous families—such as rabbitbrush and brittle brush in the sunflower family, scarlet globemallow in the mallow family, woolly plantain in the plantain family, littleleaf rhatany in the rhatany family, and winterfat in the goosefoot family. Waxy leaves are utilized by plants in more than one family—including the bean-caper family, the agave family, the cypress family, and the beech family. Several families are also represented by plants with deep roots, drought-initiated dormancy, water-initiated sprouting, nighttime carbon dioxide absorption, etc.

If plants from different families are using the same type of innovation/adaptation to survive, it usually means that they have discovered the same (or similar) adaptation independently, along their own unique historical and biological pathways. Biologists call this phenomenon *convergent evolution*.

Convergent evolution is when two essentially unrelated species develop a similar adaptation. There are thousands of examples of convergent evolution in nature. Unrelated creatures have "figured out" how to fly with wings, how to hide their food, how to camouflage themselves, etc. In these cases, multiple and independent unique pathways are leading to the same or very similar adaptations. The solutions don't all look *exactly* alike—we are, of course, talking about different species from different families. But they are effectively exhibiting the same strategy and innovative technique, just having gotten there from different starting points.

Combining this with the previous innovation finding, not only does nature often find multiple workable innovations or solutions to meet a particular challenge, it is also capable of developing those innovations independently along multiple pathways. It is suggestive of a rather unrestricted, open process for innovation, problem solving, and goal achievement.

Of course, having open capability for innovation does *not* mean that anything goes. As we discussed much earlier, sustainability cannot be achieved unless *functionality* is achieved first. Notice that the innovations that plants are using to survive dry conditions—storing water, deep roots, hairy leaves, etc., are at least potentially functional solutions. None of these innovations is based on having no water at all. Such a design would be nonfunctional—plants cannot conduct their chemical and physiological processes without at least some water.

Similarly, we see many animals flying with wings, but we don't see any animals flying around by flitting their eyelids or wiggling their toes. Again, at a minimum, strategies and innovations need to be functional, and then the focus can turn to whether they can be sustained.

Also, the fact that several strategies and solution pathways may be available does not mean that each one is equally functional or equally sustainable. Yes, there are many successful strategies that plants use to survive dry conditions, but each species generally uses a different combination of these strategies, and due to climate and soil differences, some of these strategies are found more often in some deserts but less often in others.

These "multipath" tendencies in innovation, goal achievement, and problem solving also seem to be at work at other hierarchical levels above and below the species level. For example, it is not hard to find cases of individuals from the same species addressing the same need with a different strategy—such as when only some redpoll songbirds migrate southward while others remain in residence.

It is also not hard to find the same sort of strategy or gross form employed by very different ecosystems and natural communities on different continents. For example, at least until a few centuries ago, the tropical and subtropical savanna ecosystems of Africa and the temperate savanna ecosystems of North America occurred in very different climates and under the influence of different species groups; yet both contained vast expanses of grassland, scattered trees, large herbivores, and large predators.

> Pattern 41: Different offspring, individuals, species, species classes, natural communities, and ecosystems are often capable of two, three, or a multitude of successful and sustainable responses to one particular type of challenge.

Essential Translation: Sustainable complex systems are often capable of two, three, or a multitude of successful and sustainable responses to one particular type of challenge.

This pattern can now be translated into the context of human systems:

Economic Translation: From the local to the individual levels, there are often two, three, or a multitude of successful and sustainable responses to one particular type of economic challenge.

For example, severe floods sweep through large portions of an urbanized region, causing major building and infrastructure damage, and upending many local businesses.

Three men who are independently involved in home construction and repair believe that this represents a good opportunity to launch their trade at a larger scale. Given their backgrounds and interests, they each strategize on how to approach this. One pursues custom home construction. The second pursues rapid prefab home sales, assembly, and installation. The third markets his services for rapid repair to badly damaged but salvageable homes. Within eighteen months, all three businesses are turning a profit. Over the next few years, each of the men find that their businesses can barely keep up with demand, and profits are beyond their expectations.

Community Translation: There are often two, three, or a multitude of successful and sustainable responses to one particular type of community challenge at the individual, family, or regional level.

For example, several dozen families flee political persecution and settle in a small city within a neighboring nation. Having been uprooted with little preparation time, these families begin life in their new surroundings without a clear sense of place or community. One of these families becomes involved with youth sports and coaching along with their children's sports participation. A second family finds common ground and a sense of place while volunteering for local environmental rehabilitation projects. A third family gains a sense of community by starting a small business and becoming active within the local commerce club. Within a few years, these three families have integrated quite strongly into their community.

Occupational Translation: Individual people are often capable of two, three, or a multitude of successful and sustainable business or career choices.

For example, Bridgette decided in her teens that she wanted to become a nurse. She went through nurse training and was eventually able to find a good job in her home town. After five years, she learned that her job location was going to be transferred to a larger town 45 miles away. She did not want to move or commute a long distance, and so she made a difficult choice to quit her nursing job. Soon afterward, she obtained a job as an office clerk. The pay in her clerk position was good, and she liked her coworkers. She was successful as a clerk overall, but eventually a new office manager was hired. The new manager had very strong opinions and was not trusting of colleagues. Eventually, tension in the office led Bridgette to resign. About one year later, Bridgette had the opportunity to apply for a position as the Village Manager. She applied, interviewed, and was eventually hired. She worked in this position until she was nearly 70 years old, receiving 18 pay raises, and much praise from village residents.

Government Translation: Individual people are often capable of two, three, or a multitude of successful and sustainable choices or pathways in response to their situation.

For example, Peggy, Sharon, and Tyler all grew up in the same city and attended the same middle school. Each of them struggled greatly in the standard academic atmosphere. Peggy didn't learn abstract ideas easily. Sharon had significant trouble with symbolic and spatial cognition. Tyler was very quickly bored and distracted. All three tended to perform poorly on tests, and all consistently received low grades. Independently of one another, all three considered dropping out of high school, but each was encouraged by family and teachers to remain. Peggy was close to quitting school altogether when she learned that she had been accepted into an alternative high school that emphasized hands-on skills. Sharon did choose to leave high school, but did so with an independent study plan, and she was eventually able to receive the equivalent to a high school diploma by passing three tests. With the encouragement of an older sister and a younger brother, Tyler elected to stay in school, where he essentially toughed it out and received grades that were just high enough to earn his diploma after four years.

Environmental Translation: From the regional to the very local level, there are often two, three, or a multitude of successful and sustainable land uses.

For example, three small towns sat outside of the urban fringe of a large city. When regional planners projected that the urban fringe footprint would reach all three towns in about twenty-five years, each of the towns began to plot out its future land-use patterns. One of the towns adopted a simple road-grid layout, with zones for commercial usage along main roads, and homes along smaller roads. A second town adopted a "town center" approach, with a small bustling downtown surrounded by dense housing. Beyond the dense housing, the town planned for scattered homes, and then finally a ring of green space. The third town planned for five hamlets around a central business district. Each hamlet was to have its own elementary school and be separated by green space, effectively creating a sense of five villages. Fifty years later, all three towns sat well within the overall urban footprint and had developed very similarly to the form called for in the original plans. Each of the towns (having become more like cities) attracted certain types of residents and businesses, and from a development and taxing perspective, each town viewed itself as successful.

Cultural Translation: Individuals, households, clans, and regional human groups are often capable of two, three, or a multitude of successful and sustainable cultural values and forms.

For example, there are three villages along a 50-mile stretch of coastline. The villages have similar populations sizes of about 3,000 each. Even though the villages were founded many hundreds of years prior by members of the same ethnic group, they all have cultural differences. Some of those differences are exemplified by their marriage partner system, burial traditions, and emphasis during ceremonies. The western villagers practice polygyny, burial at sea, and emphasize ancestors in their ceremonies. The central village practices cremation and scattering of ashes at sea, and only allows its chief to have multiple wives. Ceremonies in the central village include respect for ancestors, but there is a stronger emphasis upon the connection to animal spirits. The eastern village allows polygyny, but it is generally discouraged relative to monogamy. The eastern villagers practice seaside burial and have more esoteric ceremonies that emphasize the individual path. Major conflicts between the villages are very uncommon, and they have a long history of mutual tolerance of the others' beliefs and values.

Physiological Translation: Human individuals and groups are often capable of two, three, or a multitude of successful and sustainable physiological or genetic responses and expressions.

For example, Sam, David, and Ben grew up in the same neighborhood and were best buddies on the playground. Each of them also happened to be naturally gifted athletes. As children, they each played and enjoyed a variety of sports. In high school, however, they were individually encouraged by their families and coaches to narrow their focus. Sam liked tennis, and his hard work at practice eventually won him a college scholarship. David pursued soccer, and in his junior and senior years, his abilities excelled. He obtained a partial college scholarship, and eventually went on to coach soccer and run a soccer league. Ben turned toward dance, where he practiced with great intensity. Eventually he won parts in several musicals. Staying friends through adulthood, Sam, David, and Ben would joke that with enough practice, any one of them could have become the best tennis player, soccer player, or dancer.

> Pattern 42: Different and unrelated individuals, species, species classes, natural communities, and ecosystems are often capable of independently arriving at a single strategy or solution along very different pathways.

Essential Translation: Different and unrelated complex systems are often capable of moving independently along very different pathways to arrive at virtually the same successful and sustainable strategy or solution.

This pattern can now be translated into the context of human systems:

Economic Translation: A successful and sustainable economic strategy or solution can often be reached from along two or more very different pathways.

For Example, Dave and Eileen Jordan were young parents who hoped to send their two children to college, but their monthly savings was well below the amount they needed for this. They devised a plan to increase their income. Dave was paid hourly, and by putting in extra time at work, he could earn more. In addition, Eileen was able to find part-time employment from home. Then when their oldest boy was a teenager, he obtained part-time jobs on weekends and over summer. This income enhancement plan was an often difficult and slow process, but by the time their children finished high school, the Jordan's had accrued enough to pay for basic expenses at several state colleges. Meanwhile Jim and Debbie Weiger lived on the other side of town and faced the same prospect of being unable to send their children to college. They also embarked on a savings plan, but instead of working more hours, their plan was based upon frugality and careful investments. They combed their monthly expense list and determined what could be eliminated or downsized. At the same time, they began managing their savings very carefully with the objective of growth. They felt frustrated sometimes at the low spending ceiling they had to set, but by the time their older daughter was seventeen, they already had enough money to send all three of their children to college.

Community Translation: A successful and sustainable strategy or solution pertaining to a community can often be reached from along two or more very different pathways.

For Example, Tim's family was relatively well off, and he grew up in a large house set on a large lot in an expensive neighborhood. Tim went to a private college, where he met his eventual wife, Lauren. Lauren's younger sister, Stacy, dated and eventually married Joel. Were it not for their wives being in the same family, Tim and Joel would probably have never met. Joel grew up in a very poor part of the same city. In Joel's neighborhood, crime rates were high, and much more emphasis was placed on survival than on achievement. Tim and Joel saw each other at numerous family gatherings, and were wary of each other for the first few years, seeing the other as "one of them." However, over time, they each found that they had lots of common interests and attitudes. By the time they each had young children, they got along very well and considered the other to be one of their own family members.

Occupational Translation: A successful and sustainable occupation or business sector can often be entered from along two or more very different pathways.

For Example, a business strategy consulting company has three districts, each headed by a manager with a different background. Jason, the first district manager, began his career in the military. There he developed a sense of goal setting and asset coordination, which influences his management style. Ophelia had a graduate degree in psychology. Her nuanced understanding of individual and collective motivation allowed her to steer second-district work efforts and production in unique ways. Jack was the third-district manager. He had a training and work background in public speaking and marketing. His managerial style included lots of contact with his employees and repetition of established values.

Government Translation: A successful and sustainable strategy or solution pertaining to behavioral management, rules, or laws can often be reached from along two or more very different pathways.

For Example, South Corp manufactures precision parts for industrial machinery, but they have three tough competitors. The competition drives them to continually innovate in order to improve product quality and lower product cost. One of their quality innovations is to provide some parts at no cost if they are allowed to later retrieve the part and inspect its wear and tear. In several cases this has led them toward new materials and manufacturing processes. Another innovation relates to cost, where workers are able to receive a net pay increase if they produce a product in less total time or with less total cost and without a compromise in quality. East Corp produces high-quality kitchen appliances, and is in an atmosphere of strong competition from two companies that make similar items. East Corp makes a continual effort to innovate in order to improve quality and reduce cost. One of East Corp's quality innovations is to give out free appliances for two years, and then to retrieve them for wear and tear studies. Another innovation is to pay workers more if they can accomplish the same job in less time. These two innovations were developed in-house with no knowledge of essentially the same practices in place at South Corp.

Environmental Translation: A particular successful and sustainable land use or land cover can often be achieved from along two or more very different pathways.

For example, one tribal group practices a semi-shifting agriculture that utilizes tree clearing followed by about two decades of planting crops. Their plantings combine a type of millet, a bean, and a sprawling vegetable. They have found that by planting all three species together, their yields are maximized while their work requirements are minimized. Furthermore, their ability to successfully farm any one plot is extended by this combined planting method. About 5,000 miles to the east, a tribal network from a different ethnic group also practices agriculture. Their crops are grown in floodplains and irrigated grasslands. Even though the context is different, they practice the same interplanting technique—with millet, a type of bean, and a sprawling vegetable. The two groups have no history of contact, and apparently developed the interplanting technique independently of each other.

Cultural Translation: A particular successful and sustainable cultural practice or belief can often be founded independently two or more times, or can be developed from along two or more very different cultural trajectories.

For example, a researcher finds dozens of examples of harvest festival traditions among local and regional cultures across the planet. The researcher is able to verify that most of them go back at least 100 years, and evidence suggests that most of the festival traditions are likely to have originated independently of the others.

Physiological Translation: A particular successful and sustainable individual stature can often be reached from along two or more very different pathways; and a successful and sustainable racial or ethnic genotype can often be reached from along two or more ancestral lines that are not closely related.

For example, in a study of human physiology, patterns of average height, body fat content, nose shape, and skin tone are mapped across the planet. Results indicate that even though many groups are not closely related in terms of genetics and ancestry, the people in higher latitudes often share similar body characteristics, as do those of middle latitudes, and those of lower latitudes.

Now that we have looked at aspects of innovation and problem solving, we will take a look at the management of natural systems, and what it can teach us about the management of complex systems in general.

Biologists—including land managers, land stewards, ecologists, foresters, and all manner of conservation workers—are often put in the position of trying to meet an environmental or conservation objective. The objective might be to reforest an area, to boost populations of a rare animal, to maximize diversity of native species, or to eliminate/reduce the presence of non-native and invasive species.

Some of the actions available to biologists include planting trees, planting wildflowers or various other herbaceous plants, scattering seeds, introducing or reintroducing species, breeding plants, breeding animals, thinning trees, removing forest, selective weeding, vegetation clearing, animal culling, animal relocation, plowing, altering soils, flooding, draining, and burning. This management work is often conducted in tandem with types of monitoring or research.

Biologists charged with making management decisions have to determine not only *what* to do, but *how*, *where*, and *when* to do it. They generally have to base their decisions on a variety of complicated factors, including multiple program objectives, existing conditions on the ground, available scientific resources and information, accumulated experience, geographic context, and institutional support or constraint. Because these factors vary so much from place to place, time to time, and situation to situation, management decisions vary a great deal.

In spite of these situational differences and the great variety of management activities used to address them, there is a general pattern that emerges concerning these management activities: the actions that tend to be the most efficient and easy to sustain are those that occur at hierarchically higher system levels, as opposed to lower system levels.

For example, managing an ecosystem tends to be more efficient than managing all the natural communities in that ecosystem; managing a natural community tends to be more efficient than managing all the species classes within that natural community; managing a species class tends to be more efficient than managing all the species within that species class; managing a species tends to be more efficient than managing all the individuals of that species; and managing an individual tends to be more efficient than managing all the offspring of that individual.

Perhaps an example would help clarify what higher-level actions are in practice. Purple loosestrife is a European wetland herb with showy purple flowers. Purple loosestrife seeds were brought to North America sometime in or before the early 1800s. The plants spread gradually at first, and then proceeded to spread rapidly across much of North America. When purple loosestrife enters new wetlands, it can form very dense colonies within a few decades. Because it displaces many native species and can even impede water flow, it is considered invasive and undesirable within North America.

If purple loosestrife is found in a wetland, a person can attempt to eradicate it by attacking individual plants or plant clusters. Removal methods can include cutting, pulling, and spraying with herbicide. Unfortunately, all of these methods require lots of effort per plant. Cutting or pulling individual purple loosestrife plants (while carefully avoiding damage to the desirable native plants) in a wetland is extremely difficult work. The mud, deep water, insects, and physical strain of bending and pulling roots all contribute to the necessity of a very large human effort for just a small amount of loosestrife control. Furthermore, it is very difficult to cut all the stems and extract all the root pieces of these plants from the mud—and if stems or roots are left intact, the plants will generally come right back within a few years. Also, once purple loosestrife has found its way into a wetland, the seeds become scattered throughout the soil, and they will grow following any plant removal.

Reduction of purple loosestrife plants via herbicide can be a bit easier, but it is somewhat expensive, and it is difficult to avoid spraying herbicide on desirable native plants. Furthermore, when purple loosestrife plants are killed with herbicide, their seeds remain in the mud, and they are apt to grow back within a few years. In summary, unless there are only a few of them, efforts to attack the individual purple loosestrife plants tend to be expensive, arduous, and only temporarily effective.

Another method of purple loosestrife control could be to ban this species' import into North America. This effort might require some legal negotiation and governmental effort. However, the effort needed for a species ban would probably be very, very, very small compared to the effort required to remove the plants once they are widely established in North America. Of course, purple loosestrife is already in North America and cannot really be banned. However, the point is that *if* purple loosestrife had been banned in the first place, that banning effort would have required a relatively low expenditure of energy, and as long as the ban were effective, its benefits would be ongoing.

Yet another method of purple loosestrife control could be to identify European insects that feed on the plants, and release those insects into North America. This requires a dedication of scientific resources, but if such insects can be identified and their introduction to North America is successful, they could greatly reduce the densities of purple loosestrife plants. This latter approach—the predatory insect approach—was actually conducted in the late twentieth century. The results have been relatively effective, with ongoing and apparent purple loosestrife reduction wherever the insects have colonized.

The case of invasive purple loosestrife control is a good example of the of attempts to exert influence at the individual level (via pulling, cutting, or herbiciding) versus at the species level (banning the import of purple loosestrife), or versus the ecosystem level (if the import ban is applied to all species from other continents), or versus the natural community level (introducing predatory insects). In these cases, actions taken at the lower (i.e., individual) hierarchical level are less efficient and more difficult to sustain than actions taken at the higher hierarchical levels—species, ecosystem, or natural community.

Regardless of the hierarchical level that is being managed, wild creatures and their communities are complex systems. As discussed previously, complex systems often have thousands and thousands of components (such as different creatures, soil, water, rocks, weather, etc.), and can be influenced by a multitude of feedback loops and independent forces. In addition, nature is full of creatures that have free will. The presence of thousands of components in combination with free will means that at any given moment, the system is not completely like any other system that has ever been before.

Because of this, the biologist is always facing a situation that is to some degree novel, and will *always* continue to be novel as unexpected events continue into the future. The novel aspect means that a biologist is never very comforted with a guaranteed formula for success. Add in the fact that complex systems cannot be controlled anyway (at least not in the firm way that a kitchen faucet can be controlled), and the biologist is never really positioned to completely understand or exert authority over their system of interest.

It can be humbling to be in this position, but efforts taken from this position are not a complete loss. Progress towards a goal can be made, but one of the keys to this progress is for the biologist is *to be an avid student*. If the biologist makes the best management choice based on the available information, and is willing to be a student of the system afterward, then he or she can learn even more about the system and how it responds to a particular management choice. That learning can be incorporated into the next management decision, after which the biologist will have another opportunity to learn even more, and to incorporate that learning into the next management decision, after which the biologist will have another opportunity to learn even more, and so on and so on.

In the realm of ecology the term *adaptive management* is often used for this process of cycling repeatedly through action, learning from the results of that action, and taking new action that accounts for what has been learned. To use adaptive management is to assume the role of a steward rather than a dictator, and then to continually review the system's conditions and responses, and to learn from that before taking additional actions.

It is a sort of iterative and repeating cycle—observe, then learn, then formulate a plan, then act, then allow the system to respond, then observe, then learn, then reformulate a plan, then act, then allow the system to respond further, etc. Sometimes, the biologist doesn't even know what the system is going to be like if management is successful. This might seem like a frightening position to work from—not knowing what form success will take—but some people find it interesting, and not knowing what is true or false is perhaps better than "knowing" that something is true when indeed it is false.

> Pattern 43: When managing natural systems for a particular objective or set of objectives, the most efficient and easily sustained strategies are often those that are implemented at the higher of any two alternative hierarchical levels.

Essential Translation: The most efficient and easily sustained strategies when managing complex systems toward an objective(s) are often those that are implemented at the higher of two alternative hierarchical levels.

Note that Pattern 43 only suggests the ease and efficiency of acting at higher hierarchical levels. It does not guarantee that any particular strategy will be effective (functional). The selection between alternative solutions would presumably include the consideration of effectiveness along with hierarchical level.

This pattern can now be translated into the context of human systems:

Economic Translation: The most efficient and easily sustained strategies when managing economies toward an objective(s) are often those implemented at the higher of two alternative hierarchical levels.

For example, two small nations believe that they could and should be engaged in more international trade. The first nation pursues a trade enhancement policy by sending teams of government representatives to negotiate trade deals with foreign companies and other governments on behalf of its own corporations. The second nation builds two new highways connecting it to foreign countries, and expands one of its smaller seaports to accommodate larger ships. Highway and port construction are somewhat expensive, and in the first five years, the second nation spends far more on its trade program than the first nation does. However, within about twenty years, both nations have enhanced their international trade levels considerably. Then, both nations go through periods of government budget cutting. When this happens, the first nation downsizes its trade enhancement program, and its international trade levels drop significantly. The second nation also reduces its government spending, but with the highways and expanded port still in place, there is no drop in international trade.

Community Translation: The most efficient and easily sustained strategies when managing community interactions toward an objective(s) are often those implemented at the higher of two alternative hierarchical levels.

For example, Shelly was divorced, and she moved with her two young children into a new neighborhood. She was pleased with the neighborhood in several regards, but with her young children attending public school, she gradually realized that the adjacent neighborhood they had to walk through to reach the school wasn't very safe. Furthermore, right in her own neighborhood, her children had begun hanging around with other children that had crude and discomforting behaviors. In balance with her job, Shelly found herself walking with her children to and from school, and keeping a constant eye on them whenever they played outside.

After about eight months of this she felt worn out. She weighed options such as driving her children eight blocks back and forth to school twice each day, having a talk with the neighboring children or their parents, and keeping her children inside all day. After careful consideration, she decided to move to a different neighborhood. She scoped out several areas carefully. She found a neighborhood that she believed was indeed friendly and had a very safe walk to school. Her move was very hard work and the house she moved into wasn't perfect, but over the next several years, she spent almost no time worrying about her children's safety or the playmates they found.

Occupational Translation: The most efficient and easily sustained strategies when managing occupational systems toward an objective(s) are often those implemented at the higher of two alternative hierarchical levels.

For example, a national government has several lines of evidence that there aren't enough doctors practicing within its borders—including long lines at the doctor's office, high prices for healthcare, unfilled doctor positions in hospitals, etc. For five years the government funds a small program trying to convince young people to enter the medical profession. However, this program seems to have little positive effect. Then the government adopts a new strategy by creating an alternative to its medical licensing system. In the new apprentice system, people can be paid to work their way from an entry-level lab technician all the way up to a doctor level. They are required to take courses intermittently along the way to enhance their training, but these classes are more affordable to people who are have been working intermittently. With this program in place for ten years, the number of doctors increases slightly, and the government projects that the total number of doctors will triple during next ten years.

Government Translation: The most efficient and easily sustained strategies when managing human behavioral response and self-governance toward an objective(s) are often those implemented at the higher of two alternative hierarchical levels.

For example, a grade school principle has grown weary of what seem like an excessive number of fights and injuries out on the playground. He has talked to the children again and again, removed privileges, and even suspended children on a few occasions. But the problems continue. Finally in discussions with a playground monitor, he comes to believe the children are cramped and fighting over the rights to equipment and play space. He then divides the recess into an early group and a late group, and put his two roughest boys in separate groups. He is able to obtain regional funding for one new piece of playground equipment (actually a duplicate structure of a very popular slide). Then he pursues grant money to purchase a lot adjacent to the school. Three years later, he does win the grant to buy the lot, and he expands the playground. But even before that, the number of fights, injuries, and discipline situations has dropped by about two-thirds.

Environmental Translation: The most efficient and easily sustained strategies when managing land use systems toward an objective(s) are often those implemented at the higher of two alternative hierarchical levels.

For example, a farmer battles the same annual and perennial weeds year after year in his large vegetable beds. He determines that as he ages, he will be unable to sustain the battle as easily. He then divides his vegetable fields into six plots. He purchases three goats, and pens the goats in one plot until they have chewed the weeds down to the dirt and left their fertilizing manure in the field. As he moves the goats to the next plot, he plows and plants the grazed-down field. After operating on this management cycle for several years, he has found that his profits have remained stable, yet his level of required effort has been reduced.

Cultural Translation: The most efficient and easily sustained strategies when managing cultures toward an objective(s) are often those implemented at the higher of two alternative hierarchical levels.

For example, administrators at a public school system are concerned and alarmed after numerous narcotic-related incidents and two drug overdose deaths among their students. After interviewing many teachers and counselors, they determine that few students are likely to be actual drug addicts, but, that drug use has become esteemed and fashionable in the local schools. The administrators institute a new policy in which anyone caught in a drug-related incident is expelled from school. They are allowed to return to school however, if accompanied by a parent through their classes for five consecutive days. The fashion value that many students tie to drug use disappears over the next year.

Physiological Translation: The most efficient and easily sustained strategies when managing human minds and bodies toward an objective(s) are often those implemented at the higher of two alternative hierarchical levels.

For example, a health insurance company has an incentive program for its most frequently ill customers. The company will lower the monthly premium for those customers who attend a lifestyle program, which the insurance company pays for. The lifestyle programs are regimented toward particular amounts of sleep, waking time, foods, exercise, and hours of sedentary behavior. Four years into this program, the insurance company determines that corporate net loss from insurance holders who have joined the program, including an accounting of lower premiums, insurance payments to doctors, and membership dues for lifestyle programs, has seen a net decrease of 40 percent.

> Pattern 44: When managing natural systems for a particular objective or set of objectives, an often effective way to make progress in the face of multiple past, present, and future uncertainties is to repeatedly observe and learn in order to actively inform the reevaluation and revision of management strategies.

Essential Translation: When managing complex systems for a particular objective or set of objectives, an effective way to make progress in the face of multiple past, present, and future uncertainties is often to repeatedly observe and learn in order to actively inform the reevaluation and revision of management strategies.

This pattern can now be translated into the context of human systems:

Economic Translation: When managing economic systems for a particular objective or set of objectives, an often effective way to make progress in the face of multiple past, present, and future uncertainties is to repeatedly observe and learn in order to actively inform the reevaluation and revision of management strategies.

For example, in one nation, there is a ministry of wages and a ministry of taxes. These agencies are designed to work closely together with a mission of promoting national economic vitality. Some of the measuring sticks these ministries use to gage economic vitality are the widespread affordability of food, housing, and energy; a widespread ability to save money; and tax revenue. The ministries spend much of their time and energy trying to optimize tiered wage levels and tax rates (both of which they are allowed to control), in order to develop the most optimal combination of per capita resource affordability, per capita savings ability, and government tax revenue. These ministries were guided for several decades by the advice of theoretical economists.

Under their guidance, economic vitality was good in some years, but there tended to be chronic problems with per capita savings and with tax revenue.

Eventually the ministries simply began intensively reviewing past economic performance data, and using those trends to adjust the ratios and brackets they had control over. They eventually settled into a range of wages and tax brackets that seemed to provide relatively widespread and consistent national economic vitality. However, the ministries continued to closely monitor dozens of economic indicators and make occasional adjustments. Workers within the ministries learned by experience that the semi-stable national economy goes through cycles or swings—sometimes for reasons uncertain, and they always needed to be ready to make adjustment to keep the economic system as optimally aligned as possible.

Community Translation: When managing communities or community interactions for a particular objective or set of objectives, an effective way to make progress in the face of multiple past, present, and future uncertainties is often to repeatedly observe and learn in order to actively inform the reevaluation and revision of management strategies.

For example, a community organizer is trying to help reduce crime levels and improve safety in a somewhat rundown city neighborhood. Her first effort is to help establish a neighborhood watch program, which she knows has been effective in some other neighborhoods. This begins with lots of energy and interest, and once the neighborhood watch program is up and running, crime seems to decrease slightly. However, participation in the program seems to drop gradually during the next year, and by the end of the year, it doesn't seem to be making much difference. The organizer's next effort is to get anti-crime cooperation and buy-in from two local churches, and thereby motivate the local community to work together against crime. This too seems to attract lots of attention for about a year, but again the enthusiasm seems to fizzle out as people's lives are overtaken by other events. After hearing success stories from a colleague in another city, the organizer attempts to initiate a neighborhood beautification program that promotes community action in cleanup projects. This raises only a bit of interest among the residents, although there are small groups of youths who are assigned to work with her in this beautification program.

On one particular afternoon she is very frustrated that instead of cleaning, several youths keep fooling around by making odd-shaped piles or designs out of trash and debris. But after more consideration, she asks them to create art from the trash. After watching the youths respond to this, she tries shifting the cleanup work into an art and beauty program for teens and youths. Over the next two years, this draws a great surge of interest. Small grants and teen labor assistance pave the way for the introduction of art boards, mural walls, and performance arenas. With growing and enthusiastic participation, residents spend more and more time out of their homes and working together. Over the next few years, more and more people get involved—cleaning up junk piles, planting flowers, painting curbs and porches, creating art, giving performances, etc. Eight years into the neighborhood art and beauty program, crime rates have dropped by half.

Occupational Translation: When managing occupational groupings for a particular objective or set of objectives, an effective way to make progress in the face of multiple past, present, and future uncertainties is often to repeatedly observe and learn in order to actively inform the reevaluation and revision of management strategies.

For example, there is a program that assists former active duty military veterans in finding new employment. The program manager requires each veteran to take a series of very thorough skill and aptitude tests. After the test, the program's caseworkers narrow down the job types that they will try to match with each veteran job seeker. The caseworkers place a heavy emphasis on the tests to properly match people with their job skills. At the same time, the program gathers basic follow-up data about the job placements they make—how long they last, whether they result in promotion, and how satisfied the veteran job-holders are.

In reviewing the program data, it gradually becomes apparent to the program manager that the matches that are most satisfying, enduring, and successful for the veteran are not necessarily the perfect skill matches. Instead, it seems that the jobs that provide the best matches, regardless of skill type, are ones that place the most emphasis on teamwork, the ones that have the most critical value for the organization, and the ones that happen to employ the highest number of other veterans. Upon realizing this, the program manager shifts emphasis somewhat away from individual job skills, and toward conditions at the place of employment. Over time, returning data indicates a general improvement in outcomes following this change in emphasis.

Government Translation: When managing human behavior or response for a particular objective or set of objectives, an effective way to make progress in the face of multiple past, present, and future uncertainties is often to repeatedly observe and learn in order to actively inform the reevaluation and revision of management strategies.

For example, a divorced father of three is repeatedly frustrated and embarrassed by the behavior of his children. Despite frequent scolding, he finds them to be undisciplined, and impolite among the company of his friends. He is angry, and his inclination is to scold them even more severely and begin disciplining them very harshly. However, before he reaches this point, out of curiosity, he begins watching the parents of the most disciplined children he knows. Along with this, he quietly watches the way that those parents and their disciplined children interact. He then gradually shifts his strategy. At home, he starts to instill a more disciplined schedule in regard to bedtime, homework time, free time, etc. Then, he makes a point to individually thank each one every few days for something considerate that they did. At the same time, he tries to raise the bar for the type of behaviors he expects from them, pointing out to them even small things that he finds inappropriate. He recognizes that this new regimen requires lots of work on his part, which he doesn't like, but after only five weeks, he notices that incidents of the most unwelcome behavior are less frequent. After one year elapses, he realizes that not only are his children behaving better, but he doesn't seem to work so hard to achieve that outcome. Over several years, he finds that he can be slightly more lax about schedules because the children have come to hold internal expectations of discipline without much outside coaxing.

Environmental Translation: When managing land uses or the environment for a particular objective or set of objectives, an effective way to make progress in the face of multiple past, present, and future uncertainties is often to repeatedly observe and learn in order to actively inform the reevaluation and revision of management strategies.

For example, a development company proposes to construct a resort on a coastal island. The provincial government, which has authority to permit or deny the project, requires that the company conduct an environmental suitability analysis. The development company's consultant completes the analysis and concludes that 20 percent of the island is unsuitable for building due to either beach erosion, dune shifts, wave action, flooding, or storm surges. However, according the consultant, the remaining 80 percent of the island can be safely developed. Development is then permitted and initiated three years later.

However, about two years after building has begun, a powerful storm destroys much of the partially constructed resort. The project developer cleans up and gets back on track, but within one year, the provincial government finds out that the developers have been attempting to block dune formation across several of their new roads. At this point, the provincial government assesses the consultant's study and determines that it contained incorrect conclusions. The provincial government halts further construction and hires their own consultant, who recommends that a series of weather, water, and land monitoring stations be laid out across the island. Left with an alternative of scrapping the project entirely, the developer complies with the monitoring recommendation.

Twelve years of monitoring and statistical analysis suggests that only 25 percent of the island can be developed without constraint. Another 35 percent is found to be developable with certain safeguards or special practices in place, such as building stilts, weather resistant building material, and emergency pumps. The remaining 40 percent cannot be safely built upon. The project is then renegotiated, redesigned, permitted, and constructed. During the building period and over the next twenty years, provincial officials pay for ongoing collection of the weather, water, and land monitoring data in order to develop a better understanding of environmental changes on this type of island, and to inform them should there be additional unforeseen problems to address.

Cultural Translation: When managing cultural patterns or values for a particular objective or set of objectives, an effective way to make progress in the face of multiple past, present, and future uncertainties is often to repeatedly observe and learn in order to actively inform the reevaluation and revision of management strategies.

For example, in one region with numerous lakes, fishing enthusiasts and retailers recognize that interest in their sport seems to be on the decline. In an attempt to encourage a renewed interest in fishing, these advocates take action. First, at the recommendation of a marketer, they print brochures extolling the value of fishing. They send one brochure to every home in the region. Although the advocates assume the brochures will boost interest in fishing, they seem to have little impact. So, they abandon the brochure concept for a different recommendation, which is to create a radio ad encouraging fishing for families. Again, despite expectations, this effort does not seem to change the downward trend in fishing.

Then the advocates try holding a fishing derby in one of the large local lakes. The derby seems to be an exciting and well-attended event, but most of the attendees are already dedicated fishermen and women. However, sensing a bit of momentum after holding the derby for two years, advocates strategize to sign up local social leaders, athletes, and public figures in the derby. This generates far more attention in the community, and the derby is attended even more heavily.

Despite this success, the fishing advocates still think there is more room for improvement. Eventually, the fishing advocates are able to convince the local schools to release two hours early on the day before the derby, officially in preparation for the derby. Once school lets out early for the event, participation in the derby grows significantly, and regional public interest in fishing during the rest of the year seems to increase. The derby is held every year thereafter, and while it is consistently well attended, the promoters continually try to find new ways of promoting it even further.

Physiological Translation: When managing other people in relation to a particular physiological or population objective or set of objectives, an effective way to make progress in the face of multiple past, present, and future uncertainties is often to repeatedly observe and learn in order to actively inform the reevaluation and revision of management strategies.

For example, Michelle is married and has fraternal twin boys. While her husband is at work, she spends lots of time homeschooling the boys. She had been following a homeschooling teaching schedule for about two years when she came to the conclusion that it wasn't going as well as she hoped. Though she felt unsatisfied, it was hard for her to put a finger on exactly what was wrong or how it should be going instead.

After talking it over with her husband, Michelle decided that the homeschooling goals should be for her to teach without expending extreme effort, and for the boys to learn, to be happy, and to behave well. At the suggestion of some other homeschooling parents, she tried making some revisions to the lesson plan. With this, the boys seemed to be picking up math a little better, but their abilities in reading and history didn't improve, and they weren't especially well behaved, nor did they seem very happy.

Then just before bedtime, Michelle began asking them how well they liked each day overall, and what the best and worst parts of each day were. She also took daily notes on their responses and on her own thoughts of how well they seemed to learn and behave. Michelle then decided that she would try to put the boys' fascination with playdough to work. She began having them form letters and words out of playdough. Then she asked them to make historical events and figures out of playdough.

After two weeks of this, her boys seemed to be picking up reading and history concepts more quickly. Through her notes she began to see that the boys were most happy when they were able to gather with other homeschool children, and they were least happy when trying to study and memorize information. With this insight, she initiated more frequent gathering with other homeschool children. While watching all the children interact, she had the idea of the children acting as the teachers to each other. Her boys were highly enthusiastic about teaching their knowledge to other children. In preparing to "teach," her boys seemed unbothered by having to review or memorize information.

All the while she continued to ask them how they felt about each day. She could see that their enjoyment of most days had improved. However, she still felt that their typical day had a level of satisfaction missing. Recognizing that the one-and-a-half-hour free time in the afternoon was often the best part of their day, she adjusted the schedule to include two 50-minute playtimes—one in the morning and one in the afternoon. Following this change and the other changes that she had already made, she felt that most days went very well, and the children were learning, happy, and well behaved—all this without having to drive herself crazy making it happen.

That makes forty-four patterns relating to all sorts of complex system features. These patterns address topics such as general dynamics, hierarchical influence, interdependence, differential impacts, wealth and success, waste, successional change, competition and system coordination, timeframes and time lags, group cohesion, group decision making and representation, expression, Protections, group formation, behavioral accountability, resilience, invasion, diversity, innovation, and management. There are many, many more patterns in natural systems, and many other ways that nature can be viewed and modeled. However, it seems that these forty-four patterns can provide at least some perspective for addressing the condition of many complex systems, structures, and institutions. Do you remember the school planned for the village of Landesby? We wanted to try and understand, beforehand, what level of sustainability might underlie different school designs or the school as a whole. Let's go back and take a look now at potential plans for the school, and see if the patterns that we found in nature provide us with insight.

PART IV

APPLICATION

Chapter 18 Return to the Village School

Landesby is semi-rural, with several thousand residents and several hundred children. The children are being raised in very traditional patterns, and they spend much of their time learning by doing. This includes learning the village's religion and customs as well as generally learning a craft or vocation.

The school was proposed by people who thought the children should have a more intellectually rigorous upbringing that would teach them reading, math, history, science, and more. This, they believed, would be good for the children and for Landesby, because the children would become more capable of solving problems and making good decisions. But there were also school opponents. They believed that the village and children were already doing well, and they didn't want to see resources directed to something that wasn't really necessary.

In order to help evaluate the viability or sustainability of this proposed school, we conducted a functionality analysis and a sustainability analysis (see Chapter 5). Our attempt to project sustainability into the future resulted in about forty-eight issues of support or concern. Our brainstorming responses to the forty-eight issues resulted in a fifty-seven-item list of alternative or potential to-dos. When we went further and studied how the school might affect its own surroundings through a compatibility analysis with fifty-seven to-do items plus ten inputs and outputs (sixty-seven total items), we ended up with 128 items. Our analyses provided lots of insight into the potential workings of a school or the types of support it could require.

But despite lots of analysis, we were left with several large uncertainties. Our analysis seemed to be based on lots and lots of predictions and best guesses. How well will our plans work if those predictions and best guesses prove false? Probably not too well. We also couldn't really manage the to-do list associated with sustainability. It was *too much* to do, given the likely funding or support anyone would have for establishing a school. Further, the threads of impact upon other local and regional complex systems were far more complicated than we could grapple with. Assuming we are allied with the school advocates, even if we find that the school might have a degrading impact upon Landesby's cultural or economic survival, will we forge ahead with the school anyway? We might recognize that something about the school isn't quite right, but we might be left with the uncomfortable options of giving up entirely, or just accepting the likelihood that the school will spur the loss of sustainability within some other important systems.

In short, the reductionist approach to complex system design is extremely informative, but it takes lots of work, generates results based on some false assumptions, suggests a level of activity or response we won't be able to provide, and puts us in the position of inadvertently weakening or destroying other complex systems.

This set us on a journey to seek something else—a framework for complex design wherein we don't have to have knowledge and control over every event and cause-and-effect relationship. Can the patterns we have found so far be used to help guide the school design?

Let's imagine that the school advocates have looked over our analysis, including the extensive list of fifty-seven alternatives plus ten inputs and outputs (refer to Chapter 5, Section 5.12 for the complete list), and they have asked our design team to carefully review sixteen of them. These sixteen alternative items involve a wide range of topics:

1a. **Provide a secure and comfortable building in which the school can operate.**

1c. **Site the school building away from areas undergoing environmental change.**

1f. **Reuse building materials** from buildings that are being torn down in order to save money.

2c. **Advocate for the establishment or funding of national or regional post-secondary institutions.**

3d. **Hire teachers from the same clan or group.**

3e. **Hire a school manager or principal with a decentralized management style, or one who is likely to align well with teacher values.**

3f. **Praise the teachers for school successes** whenever possible.

4f. If there is a government assimilation program, **encourage the villagers to abandon their native language** for the national language.

5g. **Help the villagers fight high tax rates or to find tax loopholes.**

6a. **Pay the teachers well** to prevent them from being distracted by another job.

6g. **Establish the school to be as independent and non-reliant as possible** upon the national government.

7c. **Make sure the children have plenty of time for play and games while in school.**

8e. **Encourage the village children to leave the country to find work** once they are educated, if the national economy holds no prospects for them.

9g. **Encourage the villagers to enter negotiations with their enemies.**

10d. Depending upon the size of the building, we might need to **hire a janitor or groundskeeper.**

11a. **Children will be educated** in areas of reading, history, math, science, etc.

Our team agrees to carefully review these sixteen alternative items. But our team has also just learned about the forty-four sustainability patterns, which gives us a tool to look at each of the sixteen items in a new way. By comparing and contrasting each of the sixteen alternative items to each of the forty-four sustainability patterns, we should be able to generate a sustainability profile for each alternative. Our team members are eager to find out what these profiles look like.

18.1 Pattern Review

Before entering into the review of the sixteen items, the essential translation of each pattern will be restated and discussed, especially in general relation to a school.

> **Pattern 1** essential translation: Sustainable complex systems have ongoing, changing internal processes and states that are important for system operation but often do not represent, or do not necessarily represent, net system shifts or long-term permanent changes.

This pattern means that sustainable complex systems are always in motion internally, but in overall form they are not usually changing that much through time. According to this pattern, the school will need to be internally dynamic but somewhat stable externally. This could mean students are coming and going daily through the years. It could mean that teachers or teaching formats can change, or that books and lessons can change, but the school's goals and general topics being studied are less likely to change as often.

> **Pattern 2** essential translation: Sustainable complex systems may be largely stable in overall form, but they will tend to shift gradually as they refine their internal and external relationships, or adjust to shifting internal or external conditions. While there may be additional wandering shifts that have a more random or oscillating component, the shifts in this case are directional and in effect, may appear purposeful.

This pattern means that sustainable complex systems have a stable overall form, but that form is not a static one. Overall form *does* shift in response to the need to continually create a balance between all forces. We should recognize that the school's general form can be specified, but it cannot be tightly constrained, and thus unable to adjust or evolve. For example, it might be set up for five grades, but if this is found to be unstable or insufficient, then changing the number of grades is warranted. It might mean that the budget is set at a certain initial amount that changes over a few decades as conditions around the school and village change.

> **Pattern 3** essential translation: Sustainable complex systems display a general pattern in which hierarchically lower levels have a relatively weak influence upon hierarchically similar or higher levels; and in contrast, hierarchically higher levels have a relatively strong influence upon hierarchically lower levels. This means that localized impacts upon any system might have little impact upon its relatively higher hierarchical levels, and significant impact upon its relatively lower hierarchical levels.

This pattern means that sustainable complex systems are "ruled" by larger forces, not by a few small parts. It suggests that if the school is to work properly, it can be heavily influenced by things at a wider scale or higher level. Examples of this might include the status of Landesby (e.g., if it declines or needs to move), the nation (assuming there is one), or a board that steers the school. In contrast, the school should not be overly influenced by small numbers of its constituents, such as a particular textbook, a few atypical students, one or two teachers, etc.

Recall the seven modalities of economy, community, occupation, government, environment, culture, and physiology. Notice that a school is not just one of these modalities. It includes all seven. In this case, trying to break down the school into separate modalities for analysis would probably cause more confusion than anything. Therefore, we will look at the school design versus the essence or "spirit" of the patterns. However, the analysis we perform cannot ignore the fact that the forty-four patterns often have unique combinations of modal translations.

> **Pattern 4** essential translation: Entities or subunits that constitute part of a significantly larger system usually have to be behaving similarly and repeatedly, or simultaneously, in order to have a significant impact upon the larger system.

This pattern means that sustainable complex systems can be swayed by their internal parts/beings if those parts or beings are collectively behaving in a directional manner. This could mean that the school should be able to be influenced by proportionately large groups of students, or by numerous teachers or parents. In other words, the school would fall outside of a sustainable complex system design if it were very unresponsive to the apparent views or interests of majority numbers of involved interest groups.

> **Pattern 5** essential translation: Individuals in sustainable complex systems are only able to survive and thrive with a combination of their own effort and a supportive environment.

This pattern means that individuals have to expend effort on their own behalf if they are going to make it in life. But it also means that this effort has to occur in a context where there can be some payoff for expending that effort. Perhaps the school's design should include an expectation of individually motivated action, and of support for each individual. Individuals that expend effort on their own behalf should have some reasonable opportunity of benefitting from their effort, and individuals that don't expend effort on their own behalf should not be provided with significant reward for that behavior.

> **Pattern 6** essential translation: By partaking in their normal activities, individuals and groups in sustainable complex systems actively help to create and sustain the environment that they live within.

This pattern means that appropriate or preferred conditions within a sustainable complex system are often upheld by the beings that inhabit the system. This could mean something to the effect that the school is regenerated in part by the students or the teachers, or that students and teachers help to maintain the village, or visa versa.

> **Pattern 7** essential translation: Individuals and groups depend on other individuals and groups to maintain conditions that they operate under.

This pattern means that individuals and sustainable complex systems rely upon each other for survival and maintenance. This interdependence is sometimes subtle and sometimes more evident over greater lengths of time. It means the school design cannot ignore behaviors that are going to damage the environment that others depend upon. It means that the more the school can contribute to a better environment for others, the more it is generally contributing to sustainability.

> **Pattern 8** essential translation: The effect that individuals and groups in sustainable complex systems have upon the environment that they help to maintain varies mostly by the type of role they play within that system; that environmental effect varies little between individuals and groups that act within the same system role.

This pattern means that while all individuals and groups uphold their environment, the way they uphold it depends upon their role in the system. Thus within the school there needs to be an allowance for the different roles and types of contributions that each teacher and student are apt to make. In addition, there should be an allowance for the different roles that the students will play when they become adults. "Equal" contributions should not be expected from every person.

> **Pattern 9** essential translation: Individual adults in sustainable complex systems typically have different levels of material success. In a given region or location, the median material success ratio between the more successful and less successful might be about 2 to 1 or less. The success ratio between the quite successful and quite unsuccessful averages somewhere around 5 to 1. The material success ratio between the extremely successful and extremely unsuccessful averages somewhere around 20 to 1. The case of the "extremely unsuccessful" individual is likely to be more common than the case of the "extremely successful" individual.

This pattern pertains generally to material income, not forms of wealth. It suggests that limited income inequality is the norm. For the school, this would probably mean that teacher, administrator, and janitor pay are best held within ratios of about 2-to-1. Furthermore, under the assumption that all involved live in Landesby, the pay that schoolteachers or administrators receive should be relativized toward village income level and probably kept within a 5-to-1 ratio.

> **Pattern 10** essential translation: Individuals in sustainable complex systems primarily accumulate material wealth in order to address significant and predicable or likely shortfalls in resources. They also accumulate improvements or intangible "wealth" in their community interactions, their cultural knowledge, and their physical/mental adeptness.

This pattern means that individuals don't accumulate economic or material resource wealth beyond that which is needed to survive possible resource shortfalls. However, individuals accumulate less tangible types of wealth relating to skills, knowledge, and interconnections. Perhaps the school design and operation should avoid creating or relying on situations of individually stockpiled material wealth. It should also be supportive of situations in which individuals accrue skills, knowledge, and intangible connections.

> **Pattern 11** essential translation: Aggregates, or groups of individuals at different scales in sustainable complex systems, tend to obtain forms of wealth that boost the group's likelihood of persistence far into the future.

This pattern means that groups of individuals in sustainable complex systems can collectively amass tangible and intangible wealth that tends to promote the group's persistence. Given this, collective wealth and posterity should be enhanced by the school if possible, not eroded.

> **Pattern 12** essential translation: In sustainable complex systems, material waste exists for a limited duration until useful acquisition or transformation. Thus, material waste does not accrue in sustainable complex systems. This pattern means that sustainable complex systems don't produce internal or external net waste.

Under this pattern, waste in sustainable complex systems is short-lived because it is repurposed or transformed into positive or neutral components. Perhaps no net trash should be produced by the school. If the school produces waste, it should be reused or transformed for another purpose.

> **Pattern 13** essential translation: The first stage of four stages in the development of a still unformed sustainable complex system is often marked by relatively little interaction, competition, or cooperation between the first components to become established in that system.

This pattern means that sustainable complex systems are initiated with a disorganization stage wherein all patterns of interaction tend to be weakly developed. Establishment and form in this stage carry an element of opportunism.

On behalf of the school, it is likely we would need to be understanding and respectful of the role that this successional stage plays. In this situation, Stage 1 of succession is probably akin to having a school brainstorming or contemplation stage, in which all sorts of ideas are entertained (look at the wide array of school design ideas brought forth in Chapter 5). Some of these ideas may last and some may not. However, in this initial design stage, it is important to avoid both dismissing unusual ideas and locking in on favored ideas. If we properly honor Stage 1 of succession in the design process, we will be open-minded toward new ideas and understanding that this initial stage can be messy and unorganized.

> **Pattern 14** essential translation: The second of four stages in the development of a still rudimentary sustainable complex system is often marked by increasing competition and little cooperation between the components that have become established in that system. Also, there is often an uncertainty as to what form the complex system will eventually take.

This pattern means that sustainable complex systems enter a second very competitive stage wherein various elements struggle for power and influence. Organization, coordination, and cooperation are generally not well developed in this stage.

In regard to the school, it is likely we would need to be respectful of the role of Stage 2 and be aware of the constraints this stage creates. In this situation, Stage 2 of succession is probably akin to having a full debate regarding the best school designs. There might be many creative, good, interesting, and/or coherent ideas for the school. But they cannot all go forward in one school (for example, the school can't simultaneously have five grades and no grade divisions). In this second design stage, it is important to be realistic and avoid being neutral for its own sake. If Stage 2 of succession is properly honored in the design process, we will look hard at the feasibility and practicality of each idea, and accept that some ideas are mutually incompatible.

> **Pattern 15** essential translation: The third of four stages in the development of a still forming sustainable complex system is often marked by decreasing competition and increasing cooperation between the components that are still established in that system. In this stage, patterns and form also tend to become apparent.

This pattern means that sustainable complex systems enter a third stage wherein competition has begun favoring certain elements and shrinking others. As this occurs, the system begins to develop identifiable form, and along with competition that is present, that form includes growing levels of coordination and cooperation.

We would need to be respectful of Stage 3's role in the school's planning. Stage 3 of succession could be akin to winding down the debate regarding the best designs and moving toward implementation. In this stage it is important to be able to let go of some of the ideas that are less appropriate or less sustainable (for whatever reason), and to forego ideas and potentials that are mutually incompatible. It will be important to move in a focused or relatively unified direction. There can still be conflicts at this stage, but there should be increasing clarity as we move through them.

> **Pattern 16** essential translation: The fourth and final stage in the development of a sustainable complex system is often marked by little competition, but significant coordination and cooperation between the components that occur within that system. This stage includes patterns and forms and the non-static continual activity and adjustment expected in all complex systems.

This pattern means that sustainable complex systems enter a fourth stage in which the system form has already been determined and developed, yet the need to continually achieve balance requires shifts and adjustments. There is still competition in this stage, but that competition occurs more at the local levels or in fine-tuning at lower hierarchical levels, rather than at gross form or upper hierarchical levels. Interactions in this stage are often relatively coordinated and cooperative.

Attaining this stage may be the goal of school planning, but it cannot be reached appropriately without first moving through Stages 1, 2, and 3. If we arrived at Stage 4, it would likely mean the school had been in operation for some time. To come to this stage, hurdles would have been cleared along the way, many kinks would have been worked out, and we might view the school as successful to some degree.

However, achieving this stage does not mean that challenges, conflicts, and problems won't arise. In fact, they necessarily will arise, but if those challenges are not extreme enough to topple the school project entirely, they can be managed within the general confines of the form and direction that has already been established.

> **Pattern 17** essential translation: Sustainable complex systems develop through a four-stage successional pathway in which progressively more time is spent in each stage. The higher the system level, the more skewed the system will be toward latter stages.

This pattern means that in complex systems, the first, disorganized stage of succession is shortest, and progressively more time is spent in each stage. For the school, our general goal is to reach Stage 4. But this stage can only be reached if the prior three stages have been approached and worked through. With complex systems, we probably don't overtly control how long each stage takes, but knowing relative stage lengths can help us to recognize if something is stuck or off course. If decisions are not moving past the contemplation stage, we are stuck in Stage 1. If we move through Stage 2 much faster than Stage 1, it might mean we are trying to rush through in an attempt to avoid tension or deny conflict. On the other hand, if Stage 2 is lasting many multiples longer than Stage 1, it might mean that Stage 1 was not given its full due earlier, or that we are stuck in conflicting mindsets and failing to resolve things. If the first three stages of succession have been approached, accepted, and worked through, then Stage 4, which can last much longer than the prior three stages, can unfold.

> **Pattern 18** essential translation: Sustainable complex systems go through their entire lifecycles, including their own Stage 4 level of development, in a relatively stable, refined, organizationally complex, and cooperative context.

This pattern means that complex systems stabilize more quickly in the context of hierarchically higher-level systems that are themselves stabilized. In order to understand Pattern 18, it can help to look back at Pattern 3, which essentially notes that higher levels of hierarchy can easily affect lower levels, whereas lower levels don't have the same strong impact upon higher levels. If sustainable complex systems spend most of their time in being stable, then the beings or other levels they contain can stabilize more easily.

As an analogy, if an ant is trying to balance on two legs on the back of an elephant, it will be much easier do if the elephant isn't moving too much. We should recognize that if a stable context for the school is chosen, the school will be able to reach a stable configuration more easily. Also, we should be aware that the school will probably provide one of the important life contexts for many children. The more stable the school is, the more it will enable the children to reach their potential, in general.

Pattern 19 essential translation: Shifting from a reference point of prior conditions, sustainable complex systems are in a continual pursuit of equilibrium with existing conditions.

This pattern means that sustainable complex systems are never in perfectly balanced configuration. It essentially means that idea of "finally making it once and for all" is not something that actually occurs within a physical complex system format. With this in mind, we would want to refrain from a rigid, locked school form. Doing so would prevent the school from achieving more balance. We would also want to recognize that some aspects of the school will be in a state of continual adjustment, rather than in a static state, and that it might be normal to feel like the school isn't fully where it "should" be at a given moment.

Pattern 20 essential translation: Sustainable complex systems undergo gradual or cumulative major events more often than they undergo instantaneous or concentrated major events, and thus gradual or cumulative major events tend to have more overall impact on those systems than instantaneous or concentrated major events. Here, *major event* refers to a large and long-lasting change in the condition or status of a system.

This pattern means that sustainable complex systems go through big changes that are more often gradual than punctuated. Even changes that appear rapid are often the outer manifestation of gradual preceding events. Indications that we understand this pattern would be patience, and actions and plans that assume gradual school change or progress.

Pattern 21 essential translation: Individuals within sustainable complex systems form groups when those individuals perceive that group benefits outweigh group costs. However, individuals or subgroups disband when grouping is not beneficial or when the group costs outweigh the benefits. In general, the larger the group, the less net benefit there is to maintain that group or to maintain it continually over time.

This pattern means that members of sustainable complex systems are selective about whether or not to aggregate together. This selectivity is based on the negative and positive aspects of grouping and ungrouping from each individual's vantage point. In regard to the school, we'd want to encourage group formation or collective action around collective needs or values. Assumption about which groups should occur, or who will benefit from the school, should be avoided. We should keep in mind that larger and larger numbers of people generally have less and less in common, at least from the perspective of grouping.

Pattern 22 essential translation: Sustainable decision making by groups of about 150 individuals or less can occur by a variety of modes. These include decision by an elder, dominant, or extremely knowledgeable individual; by several experienced or dominant individuals; by subgroups within the group; by majority rule; or by unanimous rule. Under more stressful or potentially life-or-death situations, the decision making might fall upon dominant individuals.

This pattern means that small groups aren't constrained to just one or two sustainable processes for decision making. However, in order for the decision-making process to be sustainable, it must accurately integrate information regarding group members and the choice at hand. This pattern would be relevant during any small-group decisions with the school. Such decisions should properly address the needs or views of the group members, along with the most realistic information regarding circumstances or environmental conditions. As long as this is done, small group decisions can be made by one person, a few people, or the whole group. Recall that herein, small groups are defined as 150 or fewer members, and that Landesby has several thousand inhabitants and several hundred children. Therefore, several of the school decisions would probably be large group types.

Pattern 23 essential translation: Larger groupings of individuals, i.e., about 150 or more, make sustainable group-level decisions through collective means. Collective group decisions tend to account for (or attempt to account for) the needs of all in the group; allow for the expressions that any in the group might make; and result in a choice that a given individual may or may not prefer. Collective decisions can take the form of majority rule, unanimity, various levels of consensus, etc.

This pattern means that sustainable decision making in large groups is generally constrained to collective rather than hierarchical or leader forms (but note that Pattern 24 allows for representation). This allows the decision to integrate information regarding the choices and perspectives of group members. Any school decision by groups, or affecting groups with more than about 150 members would need to be made collectively. Landesby has several thousand inhabitants and several hundred children. The most far-reaching or impactful decisions regarding the school would probably be considered large-group decisions.

Pattern 24 essential translation: Larger groupings of individuals (i.e., about 150 or more) can make group-level decisions through a relatively small number of decision makers. This can only occur when the individuals making the decision have exceptional access to information *and* they have a maximal understanding of the needs of their group.

This pattern means that sustainable large-group decision-making can include representation. Note however that true group representation can only occur when a group is very unified and when the representative aligns with that group. Also, representation is only necessary when decision makers have access to information that common group members don't. In designing and operating the school, we should determine if representation is appropriate. Also, if representation is being used, we would want to make sure it is applied correctly.

Pattern 25 essential translation: In sustainable complex systems, extreme competition for resources is ultimately avoided by suppressing extreme competition. This is done by reducing the density or numbers of competing individuals or groups of individuals until a balance point of moderate competition is achieved. At this level, individuals and groups have sequestered enough resources for significant, but not maximum, expression. Significant Expression requires that individuals and groups sequester resources for both a present and a future.

This pattern means that individuals and groups in sustainable complex systems take steps or become positioned to avoid extreme, chronic competition. This allows some of them to flourish. We might assume that if Landesby has been in place and stable for a long time, the villagers may have achieved a state of *Significant Expression*. Whatever we create or alter via the school, it should avoid denying groups and individuals the opportunity to suppress extreme competition and thereby experience *Significant Expression*.

Pattern 26 essential translation: In sustainable complex systems, extreme competition is suppressed via four Protection avenues: resource use, self-defenses, restraint, and barriers. This applies at the individual and group levels. Each of the four Protection types is fallible or practically limited, which often requires that a combination of the Protections be used.

This pattern means that individuals and groups in sustainable complex systems have four Protections, or ways by which to avoid extreme resource competition. If any of these Protections are removed, individuals and groups have to rely all the more heavily upon other Protections in order to retain their expression status. As the school is designed and operated, we would want to avoid creating or contributing to an atmosphere of weakened Protections for individuals and groups.

> **Pattern 27** essential translation: In sustainable complex systems, the diversification, or expansion of the number of self-replicating groups requires barrier separations between the groups. In general, the more complete the barrier or separation, the more pronounced can be the diversification between the groups.

This pattern means unique, sustainable human groups cannot form unless they can somehow separate long enough and fully enough from other groups. We would want to ensure that our actions relative to a school don't prevent the formation and differentiation or new groupings, if that is perhaps in the best interests of the children or villagers. Or, if there is some reason for us to try and spur diversification, we need to recognize the necessary role that barriers and separation play in that process.

> **Pattern 28** essential translation: In sustainable complex systems, maintaining the diversity of potentially competing self-replicating groups requires Protections. If only a degree of group Protections are lost between these groups, some diversity of groups can be maintained—the amount depending on the level of Protections that remain. If all Protections are lost between self-replicating groups, diversity of group types will become unsustainable, and the resulting final group will be based on numerous factors, including relative fitness.

This pattern means that two or more potentially competing or exchangeable human groups can only be sustained if they remain separated or protected from each other in some fashion. Separations or Protections can have a cultural, physical, genetic, behavioral, or other basis. A group that is disappearing is obviously unsustainable, but the disappearance of one group does not necessarily mean its larger system context is unsustainable. Knowing this, we would want to help sustain any Protections for groups that represent our cause or school purpose. Similarly, we would want to avoid inadvertently causing the diminishment of groups, especially if they represent our cause or interest.

> **Pattern 29** essential translation: In sustainable complex systems, governmental authority, or choice making authority, is local. Generally, the governmental authority for each individual is itself, or closely related or adjacent individuals.

This pattern means that the individuals living in sustainable complex systems direct their own behavior, although sometimes their behavior is guided by nearby individuals or relatives. So, individuals rarely "control" others, and if they do, the controller is a related or adjacent individual. The school design would need to include an expectation of individual choice. Setting up a system where individuals are routinely forced to make the choices that others provided would be contrary to this pattern.

Pattern 30 essential translation: In sustainable complex systems, government—or behavioral steering—is affected by accountability to four areas. The four areas are the physiological self; the environment; peers and associates; and competitors and predators. Behaviors that are in synch with the true quality and function of these four areas tend to result in an increased likelihood of receiving positive consequences. Conversely, behaviors that are out of synch with the true quality and function of these four areas tend to result in an increased likelihood of receiving negative consequences.

This pattern means that the behavior of individuals living in sustainable complex systems is influenced by their own bodily needs, their environment, their peers, and their competitors or attackers. Individuals who fail to account for these four areas experience a demise, while those that respond to these four areas tend to prosper. A school design should be accountable to these four areas, and it should help position individuals (or groups) such that they can also be accountable to these four areas.

Pattern 31 essential translation: In sustainable complex systems, the unfolding of government is often nuanced according to each individual situation. This includes nuance in the conditions that each group or individual faces, the accountability and consequences they are faced with, and how they respond behaviorally to that accountability.

This pattern means that the individuals living in sustainable complex systems are generally not faced with or influenced by the exact same situations. So, appropriate individual responses will tend to be unique. Even when outwardly faced with equivalent situations, each individual responds from a different historical disposition, and thus behavioral steering will be unique for each. The school should allow for some individual discretion in behavior.

Pattern 32 essential translation: In sustainable complex systems, there is an alignment between government and system function. That is, the pressure of accountability applied upon each individual and group of individuals tends to be aligned with the function of the systems within which they live.

This pattern means that individuals and groups in sustainable complex systems are often pressured to behave in ways that uphold the systems within which they operate. Regarding the school, we should try to make behavioral influences upon individuals be complementary to the needs of larger systems. A complicating issue can be that as we align people's behavior with system function, those very systems might be unsustainable, or those systems may be interfering with the sustainability of other systems. When evaluating a school plan, these more distant system elements may be unclear.

Pattern 33 essential translation: Sustainable complex systems are capable of responding resiliently to a given consequential disturbance by immunizing themselves. This occurs through repeated adjustments following prolonged or repeated exposure to the disturbance. The adjustments can take the form of resistance, avoidance, or incorporation. The development of immunity requires the expenditure of energy or the effective loss of energy by requiring behaviors that would otherwise not be preferable. Once immunity has occurred, depending upon the situation, the system will no longer experience the disturbance as a stress event, but as a normal or even necessary event.

This pattern means that sustainable complex systems are sometimes capable of making the adjustments necessary to survive repeating stresses or changes. However, these adjustments take time, and during that time, systems lose a degree of refinement and alignment. Also, after systems make these adjustments, they are often no longer as well adapted to their former, pre-disturbance conditions. Major disturbances could occur to the school or its context, or disturbances might be unfolding while the school is established. Our job would be to allow and assist the systematic response that is required for a given system to survive these changes.

Pattern 34 essential translation: The more a sustainable complex system is operating within its internal and external historic range of variability, the more resilience it will have in responding to continued or additional disturbances. The more a complex system is operating near the periphery or outside of its internal and external historic range of variability, the less resilience it will have in responding to continued or additional disturbances.

This pattern means that sustainable complex systems are more likely to survive a consequential disturbance if they are already operating within their typical or historical behavioral range. In operating the school, we would probably want to avoid pulling other nearby (presumably sustainable) systems outside their historical range of variability, because this leaves them somewhat more vulnerable to collapse or failure. And, if we notice that a system has recently been pushed out of its historic range of variability, we would want to be cautious about relying upon it too much.

Pattern 35 essential translation: When a "new" entity or group is introduced into a system, the tendency for the new entity or group to behave disruptively within the system is often associated with its tendency to behave with restraint, and with its similarity to the other occupants that are already present. The more restraint mechanisms that the introduced entity or group has, the less disruptively it will tend to behave. The more similar the new entity or group is to existing system occupants, the more likely it is that its introduction will be inconsequential and non-disruptive. The less similar the new entity or group is to existing system occupants, the more likely it is that its introduction will be consequential and disruptive. Consequential and disruptive events lower the likelihood that a receiving system will remain sustainable. Similarity between introduced and indigenous entities and groups in a system is just a special category of a system's historic range of variability.

This pattern means that sustainable complex systems are more likely to be disrupted by a new occupant that one—lacks restraints, and two—is unlike the existing occupants. In the school design we would want to avoid a dynamic in which one system is essentially being invaded by a new occupant. We should be aware of whether such an invasion is occurring to the school's context, because that could have disruptive repercussions for the school.

> **Pattern 36** essential translation: Individuals and other complex systems are generally made more resilient to disturbance by an inherent diversity of members or active components and/or ranges of potential behavior, and are made less resilient to disturbance if operating from a state of reduced membership or numbers of active components and/or reduced ranges of potential behavior.

This pattern means that sustainable complex systems are more likely to remain functional and intact when met with major challenges if they have a broader constituent base and/or a broader range of potential behaviors. We should understand the role of diversity in augmenting resilience, and we should try to avoid unnecessarily stymieing diversity.

In general, it is important to understand that the forty-four patterns operate in balance with each other. Complex balance in general result from a simultaneous combination of alignment and tension (push and pull). If any one pattern is singularly pushed to an extreme, balance among the patterns can be disrupted. For example, Pattern 2 suggests that complex system sustainability is not a fixed endpoint. However, if it is too unfixed, and too chaotic, then it isn't sustainable. And very unfixed or very chaotic behavior would work against Pattern 18, which notes that systems at each hierarchical level tend to provide a stable context for those below. Or, consider Patterns 14 and 15, which are the conflict and resolution stages of succession. If there is no conflict, a resolution cannot be reached. Pattern 33 states that a system can become immunized and resilient. This could never happen if the environment was always stable—as in Pattern 18. There are many similar interactions between the patterns that help them to create balance. Because of this, the patterns generally cannot be treated as independent units to be minimized or maximized.

The need for balance can even be seen within single patterns. For example, Pattern 21 highlights the relative value of group formation and group dissolution. The decision of whether a group should form cannot be based only on group costs, or only on benefits. Both have to be weighed, and then the decision of whether or not to group can be made. There can also be hierarchical balance and tension. The diversity that allows for increased system resilience at the higher level (via Pattern 36) can also destroy unprotected subgroups or subsystems (i.e., Pattern 28). Ideally, the forty-four patterns work together in balance with each other.

> **Pattern 37** essential translation: A larger system context can sustain more types of systems or subsystems than a smaller, but otherwise similar context.

This pattern means that sustainable complex systems can contain more diversity if they are larger. One of the primary causes of this pattern is that scale is needed to indefinitely hold complex components or subsystems. In relation to the school, we should recognize and respect the role that size and scale have in regulating diversity. In this case, diversity can be within or between cultures, ethnic groups, school subjects, economies, etc.

Pattern 38 essential translation: A system context that is more diverse can sustain more types of systems or subsystems than a similarly sized but otherwise more uniform system context.

This pattern means that sustainable complex systems can sustain a higher diversity of things if they have a more varied internal context. We would want to recognize and respect the impact that contextual heterogeneity has upon diversity. In this case, diversity can refer to subcultures, ethnic groups, school subjects, economies etc.

Pattern 39 essential translation: When there is stronger environmental support for a given form, a complex system can sustain a higher diversity of members of that form, and it can sustain a higher diversity of subsystems which are based upon that form, in comparison to a complex system provided with less environmental support for that form.

This pattern means that we can expect a higher diversity of the things that associate with economic sectors that are more strongly supported in a given sustainable complex system. While this economic twist is less obvious in the essential translation, it is apparent in the modal translations.

This pattern means that if the nation provides strong financial support for schools, then the general diversity of school types that can exist in the nation is increased. We should be aware that in turn, the expansion of potential school diversity means that people who create or oversee individual schools have more freedom of design in how their school operates.

Pattern 40 essential translation: Complex systems that have existed for a longer period of time, or have existed in a particular context for a longer period of time, generally sustain more diversity of components and subsystems than similar but otherwise younger or more recently established complex systems.

This pattern means that sustainable complex systems can contain more diversity as they age. The *age* of a system should be understood as the time that has elapsed since the origin of the system's formation, or since its last major catastrophic disruption, reconfiguration, or revolution. In designing and operating a school, we should recognize that gradual diversification can be a natural part of aging. We should avoid false assumptions of especially high diversity early in the school's operation or especially low diversity as the school ages.

> **Pattern 41** essential translation: Sustainable complex systems are often capable of two, three, or a multitude of successful and sustainable responses to one particular type of challenge.

This pattern means that sustainable complex systems can often successfully address a challenge in more than one way. Multiple solutions in a given situation may or may not be mutually exclusive. In designing or operating a school, we could embrace the possibility that any particular challenge, problem, or need *might* have more than one viable solution.

> **Pattern 42** essential translation: Different and unrelated complex systems are often capable of moving independently along very different pathways to arrive at virtually the same successful and sustainable strategy or solution.

This pattern means that sustainable complex systems can theoretically be situated very differently, and yet through adjustments can attain very similar goals or independently utilize the same successful strategy. Regarding the school, we would want to be aware that just because we or another group are differently situated does not mean we cannot effectively reach the same goal or utilize the same strategy.

> **Pattern 43** essential translation: The most efficient and easily sustained strategies when managing complex systems toward an objective(s) are often those that are implemented at the higher of two alternative hierarchical levels.

This pattern means that if one is attempting to shift or alter the behavior of a complex system, it is generally easier in the long run to do so from a higher level, rather than from a lower and more detailed level. Recall that Pattern 43 only addresses the efficiency and sustainability of management strategies, and that the effectiveness of strategies should also be weighed. In addition, any hierarchical level can generally be contrasted to either a higher level, or lower level. So judgment has to be utilized to determine whether a higher- or lower-level solution comparison makes more sense. Alignment of the school with Pattern 43 would require certain types of decisions to be implemented through the hierarchically higher of any effective alternative levels.

> **Pattern 44** essential translation: When managing complex systems for a particular objective or set of objectives, an effective way to make progress in the face of multiple past, present, and future uncertainties is often to repeatedly observe and learn in order to actively inform the reevaluation and revision of management strategies.

This pattern means that sustainable complex systems are often influenced by more factors than people can understand and anticipate, and so managing those systems requires repeated reevaluation of the system's status, and this information should affect subsequent management decisions. The school itself is a complex system. A proper understanding of Pattern 44 would mean that we monitor the status or progress of the school and the schoolchildren, and use what we learn to make adjustments.

18.2 Sustainability Profiles for the Alternatives

We can now review the sixteen alternative school items relative to each sustainability pattern. The school is in the planning stage, so we are reviewing *plans* or *potential* actions, rather than past events. Sustainability patterns can help indicate whether a plan is sustainable—and even whether the process used to form the plan or the thinking behind the plan—is aligned with sustainability.

As our design team combs through the sixteen alternatives relative to each sustainability pattern, it finds that some sustainability patterns are not applicable to the alternative—not germane. In other words, each alternative only "involves" certain sustainability patterns. Only the applicable alternative-pattern combinations are listed and discussed below.

Alternative 1a: **Provide a secure and comfortable building** in which the school can operate.

Alternative interpretation via Patterns:

> A well-constructed building is a form of collective wealth; this plan alternative would be aligned with Pattern 11.

> This plan indicates some understanding that it is important to have the school building provide a stable context for the children. In this respect, this plan is aligned with Pattern 18.

> If the building were large enough to contain all the children without very tight competition for space, this would be aligned with Pattern 25.

> This plan suggests an effort to serve people's physiological needs, which would be aligned with Pattern 30.

> The range of what individuals consider comfortable will vary. If people are to remain in a place for extended periods of time, there should be an expectation and allowance for different individual responses to its atmosphere or design in order to be aligned with Pattern 31.

> The contemplation of multiple solutions regarding accommodations during the decision-making process is aligned with Pattern 41 (see alternative 1d).

As interpreted, this alternative involves six of the sustainability patterns. The sustainability findings for this alternative are rather positive, but patterns suggest that maximum sustainability includes well-built facilities, large-enough facilities, and conditions that accommodate variation in the student's predispositions.

Notice that because the sustainability patterns are mostly abstract, they can apply to situations that might seem at first to be completely unrelated. For example, on the surface, providing a school building might have very little to do with competition, but a view of this alternative through Pattern 25 suggests that it actually could involve this topic.

Alternative 1c: **Site the school building away from areas undergoing environmental change.**

Alternative interpretation via Patterns:

> If the school building is in an area that is changing quickly, it is less likely that the school can hold an overall stable outer form. Therefore this alternative action is aligned with Pattern 1.
>
> If the school is sited in an area undergoing change or periods of chaos, it may be unable to emerge beyond Stage 2 of succession. The plan to find an area that can stabilize is aligned with Pattern 15.
>
> Siting the school building in an area undergoing change might keep school conditions in flux, which would tend to keep it in a Stage 2 successional form. This plan indicates some understanding that it is important to move on to Stage 3 rather than remain in the conflict of Stage 2 for a very long period. Having this plan or understanding is aligned with Pattern 17.
>
> This plan indicates some understanding that a destabilized context will hamper stabilization of the school, which is aligned with Pattern 18.
>
> Even areas that appear stable are, at the very least, still "catching up" to their full balance point. If this plan was made without that understanding, there could be an eventual conflict situation in relation to Pattern 19.
>
> This plan suggests an effort to respond to environmental realities, which would be aligned with Pattern 30.
>
> An area undergoing environmental change has probably already been pulled outside of its historic range of variability, making it less resilient to additional change, and more likely it could undergo a form of collapse. It is possible that this environment will continue to change unpredictably. This would constitute a destabilizing force for some time. By honoring the role that historic range of variability plays in resilience and stability, this plan is aligned with Pattern 34.
>
> This indicates we are open to, or capable of a range of choices in siting the school. The ability for a range of choices in response to disturbance enhances school resilience or its potential, in alignment with Pattern 36.
>
> This plan would probably help to avoid many problems such as damage to a building, emergency school closings, etc., that could be associated with placing the school in an area undergoing environmental change. It is aligned with Pattern 43 because it can be seen as a higher-level choice relative to potential building repair and school closures.

As interpreted, this alternative involves nine of the patterns. Sustainability findings for this alternative are strongly positive, but Pattern 19 suggests caution in assuming which areas may be subject to delayed environmental changes.

Alternative 1f: **Reuse building materials** from buildings that are being torn down in order to save money.

Alternative interpretation via Patterns:

> This would allow things to be reused rather than remain as waste items. This plan is aligned with Pattern 12.
>
> Reusing materials would negate the need for new resources, and is therefore a form of restraint. This would be aligned with Pattern 26.
>
> By responding to economic pressure to save money, we would be doing something that probably benefits the environment (see Pattern 12, Chapter 9). This would be aligned with Pattern 32.
>
> The contemplation of different possible solutions regarding building development is aligned with Pattern 41 (see alternatives 1b and 1e).

As interpreted, this alternative involves four of the patterns. Sustainability findings for this alternative are strongly positive.

Alternative 2c: **Advocate for the establishment or funding of national or regional post-secondary institutions.**

Pattern interpretations of the Alternative:

> Increasing opportunities for people to become more knowledgeable or skilled in a professional field provides a form of individual intangible wealth accrual, and is aligned with Pattern 10.
>
> This is trying to make changes well beyond the village level. It isn't clear that there has been much consideration over the broader implications of this advocacy or the results of this advocacy if it is successful. While bringing up the idea in an open-minded way would be aligned with Stage 1 of succession and Pattern 13, pursuing the idea without much more critical review skips over Stage 2 of succession and is in conflict with Pattern 14.
>
> This plan would produce more trained teachers and would facilitate the school into the future. It indicates some understanding of the importance of reaching the harmonic Stage 4, and being able to remain in that stage for as long as possible. From this perspective, such a plan is aligned with Pattern 17.
>
> If this advocacy were successful, having these institutions might make it easier to develop advanced knowledge in subjects, and thus provide a more stable occupational context for people who want to become teachers, benefiting the school

by association. Though the connection does not appear to be very strong, this plan is aligned with Pattern 18.

Our advocacy alone might be weak, but nonetheless it is aligned with Pattern 32 because we are applying pressure on other individuals to create something that indirectly supports our intended/potential school institution.

By providing academic skills to people, such institutions would constitute a form of economic and environmental support for the teaching profession. According to Pattern 39, this could incrementally increase the supported diversity of occupations, economic models, and communities that involve schools. As a form of support for school-related diversity, such institutions would essentially tend to support our design freedom by increasing the number of potentially sustainable designs. All told, this advocacy can be seen as sustainable relative to Pattern 39, because it raises the likelihood of achieving school sustainability.

If the nation were maturing economically, these institutions could be part of a sustainable economic and occupational diversification. Under this scenario, the advocacy of alternative 2c would be aligned with Pattern 40.

The contemplation of different possible solutions for how to obtain trained teachers or develop trained teachers is aligned with Pattern 41 (see alternatives 2b, 2d, 4a, 4b, and 4c).

Developing institutions that could educate future teachers would constitute action from a higher hierarchical level, in comparison to individual efforts to train teachers. Based on this, the alternative would be aligned with Pattern 43.

As interpreted, this alternative involves ten sustainability patterns. Sustainability findings for this alternative are mainly positive, but Pattern 14 points to a possible problem in the choice-making process, and alignment with Pattern 40 is dependent upon the broader economy. In all, even though advocacy for the establishment or funding of national or regional post-secondary institutions appears positive on the whole, there is a tone of weakness in its sustainability, probably owing to the fact that it is directed outward, away from the focus on the village school.

Alternative 3d: **Hire teachers from the same clan or group.**

Pattern interpretations of the Alternative:

If hiring teachers from the same clan or group helped to provide a fulcrum point for that clan or group to exert very high influence over the village education system or culture, then it would be encouraging a conflict with Pattern 3.

It is likely that people who are already grouping together for the purpose of collective survival or expression will have an easier time cooperating, and would likely view their missions as compatible. This expectation behind clan or group hiring is aligned with Pattern 21.

Teaching the children from the vantage point of one clan could have a village-wide impact over time. Unless the village has considered this through a collective decision-making process, or unless appropriate representation is planned, this plan is in conflict with Pattern 23.

If the "group" simply means villagers, then this strategy could result in village group (self-replicating group) Protection via Pattern 28, which would tend to enhance sustainability for the villagers and their unique attributes. However, if this means hiring only a subgroup from within the village, it could promote the views from one clan or group at the expense of others, disrupting subgroup Protections for many villagers and thereby diminishing subgroup sustainability via Pattern 28.

If teachers knew that people from different clans or groups were not allowed to hold teaching positions, they might be shielded from any negative feedback/consequences to their behavior toward those other groups. This would be in conflict with Pattern 30.

If teachers representing an outside clan or group are being brought in, and with little restraint espouse a set of values that are unlike those of the villagers, then they would be acting as disruptive invaders. In regard to village sustainability, this would be in conflict with Pattern 35.

The contemplation of different possible solutions that encourage teacher coordination is aligned with Pattern 41 (see alternatives 3a, 3b, 3c, and 3d).

As long as we do not simply assume that the teachers *have to* come from the same clan or group in order to reconcile differences and reach a similar coordination point, then our assumptions would not be in conflict with Pattern 42.

Setting the stage for teacher coordination through this alternative—as opposed to perhaps intervening later and trying to instill coordination—can be seen as a higher hierarchical choice, and would be aligned with Pattern 43.

The goal of this action would be to end up with better teacher coordination and better teaching. This would be aligned with Pattern 44 if we were willing to learn from the results of this strategy and adjust accordingly.

As interpreted, this alternative involves ten of the patterns. Sustainability findings for this alternative are mixed, but there is a clear negative aspect if the clan or group that the teachers are hired from is not carefully selected according to group divisions or other contextual factors.

Note that a given alternative can include separate components. Someone may, for example, view the above alternative as including first—a *decision-making process* over who to hire; second—a *strategy* over who to hire; and third—the *act of hiring* people. Separate findings of sustainability can occur with these three separate components. For example, the decision making-process to hire the teachers from the same clan or group could be handled sustainably, while the choice of who to hire could be unsustainable.

Alternative 3e: **Hire a school manager or principal with a decentralized management style, or one who is likely to align well with teacher values.**

Pattern interpretations of the Alternative:

> If there is a school manager, it is assumed that one person is having large influence upon the school. This is aligned with Pattern 3, as long as the manager decides on behalf of the school or other broad interests, and not from a lower hierarchical level, of say, the interests of just one student or one teacher.
>
> Hiring a school manager is a small boost to the labor force or economy; thus on our behalf, this action is aligned with Pattern 6.
>
> Hiring implies pay. Relative to Pattern 9, we would first need to learn about the income distribution within the village. Assuming the income distribution falls within the 2:1/5:1/20:1 ratio bounds, then as long as the manager's pay is set within these ratios, and also set within ratios relative to the pay of other school employees, then this hiring would not conflict with Pattern 9.
>
> A decentralized management style is a type of small group decision-making that aligns with Pattern 22.
>
> In order to do this in alignment with Pattern 30, the manager or principle cannot operate in such a detached manner that other staff members are not accountable for their actions.
>
> A decentralized management style would presumably be accommodating to different interpretations and behaviors displayed by different individuals. This accommodation would be aligned with Pattern 31.
>
> When compared to a micromanagement style, a decentralized management style would require managing at a higher hierarchical level and would be in line with Pattern 43.

As interpreted, this alternative involved seven patterns. Sustainability findings for this alternative are mainly positive, but negative issues are associated if pay is outside of local income ratios, and if the manager or principle doesn't require some accountability from other staff.

Sometimes, a given plan or action clearly relates to a particular sustainability pattern. For example, if an engineer is making material modifications in order to eliminate net waste production, the action clearly relates to waste and Pattern 12. However, if the engineer receives a pay raise, does that automatically involve Pattern 9, which relates to income ratios? Certainly it *relates*, but we'd need more information to know just how it relates. Or, as the engineer draws up plans for material and resource usage, does that constitute a form of Protection as described under Pattern 26? It might, or it might not. We'd need more information to determine if such a plan even pertains to Pattern 26, and if it does, we would probably need further information to determine its alignment relative to Pattern 26.

Without specific information, the relationship of a given plan or action to a sustainability pattern is sometimes ambiguous or a matter of conjecture. So, determining whether a plan or action has a significant relation to a pattern can have a subjective aspect. Because of this subjective aspect, the comparison of the forty-four patterns with alternative school designs could be done narrowly, without speculation beyond what information we were already given. Or, the comparison could be done broadly, with more speculation and assumption. The broader analysis would tend to show a much larger set of possible conflicts and alignments with the sustainability patterns. Although it is more tedious, herein the analysis is taking a slightly broader view in order to demonstrate the variety of issues and forms that can tie in to the sustainability patterns.

Alternative 3f: **Praise the teachers** for school successes whenever possible.

Pattern interpretations of the Alternative:

> The praise would be a form of positive feedback for positive behavior, and is aligned with Pattern 30.
>
> This might encourage teachers to behave in ways that benefit the school. This would be aligned with Pattern 32.
>
> Praising teachers for successes is partly an effort to get the teachers to strive for even more success. In comparison with trying to reach the same goal through intensive interactions or continual one-on-one management, praising the teachers can be seen as a higher-level choice, and thus aligned with Pattern 43.
>
> We would be aligned with Pattern 44 if we are willing to learn from the results of this strategy in terms of teacher motivation levels relative to receiving praise, and adjust accordingly.

As interpreted, this alternative involves four of the patterns. Sustainability findings for this alternative are essentially all positive.

Sometimes when an alternative is evaluated via the sustainability patterns, results from different patterns may seem nearly redundant. Above, praising the teachers was found to be aligned with Patterns 30 and 32. The reasons were similar, or at least might evoke similar images—sustainable in one case for providing positive feedback for behavior that benefits the school, and sustainable in the other case for encouraging teachers to behave in ways that benefit the school. The fact that different patterns can provide a similar prognosis may a bit redundant, but it can help lower the odds that an aspect of an alternative will slip by without notice.

Alternative 4f: If there is a government assimilation program, **encourage the villagers to abandon their native language** for the national language.

Pattern interpretations of the Alternative:

By acting in unison, the villagers should have large sway over their use of language. As long as this is the case, or is allowed to be the case, there is alignment with Pattern 4.

There isn't enough information to make a certain determination, but the village's native language and the culture it is associated with might play a valuable cultural or economic role at the national level (this is the group-level manifestation of Pattern 8). This action might cause a conflict per Pattern 8.

This choice would likely work to disconnect individuals from intangible personal wealth, including their local history, community, and culture. In this regard, there is a conflict with Pattern 10.

It isn't clear that there has been much consideration over the broader implications of abandoning the native language. As decision-making goes, bringing up the idea in an open-minded way would be aligned with Stage 1 of succession and Pattern 13, but pursuing the idea without much more critical review skips over Stage 2 and is in conflict with Pattern 14.

This could destabilize the village culture, and thereby the school (and the children). These unstable contexts would be in conflict with Pattern 18.

This would encourage the villagers to become more like another, perhaps larger, group. From the perspective of the villagers, this could be disempowering, stifling or eliminating any unique village attributes according to Patterns 27 and 28.

Unless this program had been the outfall of a cohesive group choice, an assimilation program would need to allow for individual and localized behavioral discretion in order to be aligned with Pattern 29.

If we understand that the response to this encouragement is likely to vary by individual, then our expectations would be aligned with Pattern 31.

This would encourage individuals to act against the possible interest of their local culture and with the possible interest of a national culture. If true, this would locally conflict with Pattern 32, but would nationally align with Pattern 32.

This could be part of an immunizing response (via incorporation) to a significant cultural disturbance dictated by a national government. Depending on background events, this action could be aligned with Pattern 33.

This could pull the village culture outside of its historic range of variability, which for the village presents a sustainability conflict per Pattern 34.

The village would not be losing diversity by homogenizing to a national language, but the nation would be. Lower (or lowering) diversity might be associated with smaller areas, smaller populations, more homogeneous economies or landscapes, or very new human groupings. While these are probably not the only factors impacting the ability of the nation to maintain cultural diversity, they would be influential. We would need more information to determine whether this effective reduction in national cultural diversity would be aligned or in conflict with Patterns 37, 38, and 40.

As far as choosing alternatives goes, the contemplation of different possible solutions regarding cultural destiny is aligned with Pattern 41 (see alternatives 4d and 4e).

Actions from a higher hierarchical level might be advocating for an end to the government assimilation program, or an end to the distribution of important books printed in the national language. Thus, the alternative in question is in conflict with Pattern 43.

The underlying goal of this encouragement would be to steer the villagers toward a peaceful and workable resolution of their language use. In providing this encouragement, we would be aligned with Pattern 44 if we would be willing to observe how well this action works in terms of the village culture and welfare, and to adjust accordingly.

As interpreted, this alternative involves nineteen patterns. Sustainability findings for this alternative lean mostly in a negative direction for the village.

Notice that sometimes an action taken for a particular purpose can impact sustainability somewhere else, beyond the intended target of the action (recall our consideration of compatibility back in Chapter 5, Sections 5.9, 5.10, and 5.12). If, for example, the village of Landesby plays an important tourism or spirituality role within the nation, then abandoning the local language might actually be deleterious for the nation. Similarly, eliminating the village culture could help to reduce diversity within the nation to unsustainably low levels.

Alternative 5g: **Help the villagers fight high tax rates or to find tax loopholes.**

Pattern interpretations of the Alternative:

Avoiding high tax rates or lowering taxes can be seen as a form of economic self-defense by villagers. This would be aligned with Pattern 26.

The contemplation of different possible solutions regarding school and teacher funding is aligned with Pattern 41 (see alternatives 3g, 5b, 5c, 5d, 5e, 5f, 6c, and 6d).

If high tax rates are fought by trying to change the tax code, then this effort would be acting at a relatively higher hierarchical level, and could be congruent with Pattern 43. However, if this effort occurs by appealing to the government on behalf of individual villagers be acting at a relatively lower hierarchical level, and it would not be aligned with Pattern 43.

As interpreted, this alternative involves only three of the sustainability patterns. Findings for this alternative are generally positive, but in the case of Pattern 43, it depends upon how the action is conducted.

Alternative 6a: **Pay the teachers well** to prevent them from being distracted by another job.

Pattern interpretations of the Alternative:

> Payments to teachers support teachers as well as the local economy, and can be seen as system support by those who operate the school, which is in line with Pattern 6.
>
> Assuming that the local income distribution falls within the 2:1/5:1/20:1 ratio bounds, then as long as the teacher's pay is set within these ratios and also set within ratios relative to the pay of other school employees, then this action would not conflict with Pattern 9.
>
> Assuming this would enable the teachers to avoid extreme resource competition, it aligned with Pattern 25.
>
> Paying the teachers amply as a performance incentive is a higher level of action than trying to intervene against distracted teachers. Thus, this action would be in alignment with Pattern 43.
>
> The intention of this is to sway the behavior of the teachers, and we would be aligned with Pattern 44 if we would be willing to observe how the teachers respond to their pay, and willing to make adjustments or strategy changes based on their behavior.

As interpreted, this alternative involves five patterns. Findings are generally positive, but this positive aspect can be enhanced or diminished in regard to pay ratios and a willingness to monitor teacher performance and respond accordingly.

Alternative 6g: **Establish the school to be as independent and non-reliant as possible** upon the national government.

Pattern interpretations of the Alternative:

> This goal and action would not conflict with Pattern 2 as long as it was understood that this goal might change or might need to change.
>
> The national government can be seen as a higher hierarchical level than the school and the village. If it happens that the national government is unstable, then this effort would be aligned with Pattern 18. If it happens that the national government is stable, then this effort would be in conflict with Pattern 18.
>
> This plan might be construed for financial reasons, but it would also—at least presumably—tend to ungroup the village from the national government. Unless we

clarify whether this outfall will be positive or negative for the village, this plan is potentially in conflict with, or potentially aligned with Pattern 21.

This would contribute to village Protections to a degree, and from the village self-determination perspective, this would be aligned with Patterns 27 and 28.

If the decision was made to disconnect the school from potentially unstable circumstances in response to an observation that the whole national government or nation had been pushed out of its historic range of variability, then decoupling from the nation would be aligned with Pattern 34.

Pattern 39 indicates that stronger support enables a wider array of potentially sustainable forms. If the national government is especially supportive of schools, then disengaging from it could actually reduce freedom of design, thereby reducing the number of sustainable designs via Pattern 39. On the other hand, if the nation is unsupportive or constricting toward schools in general, then disengaging from the nation could actually expand design freedom by expanding the number of sustainable designs via Pattern 39.

If this intention is carried out in blanket-fashion by minimizing all physical and logistical interactions with the national government, it would be acting from a higher hierarchical level, and would be in alignment with Pattern 43. However, accomplishing this same goal by relying upon the national government for physical and logistical support—while simultaneously trying to resist influences from that same government—would be operating from a lower hierarchical level, and would be in conflict with Pattern 43.

This strategy would be aligned with Pattern 44 if we observe its effects upon the school's success levels and revise our strategy as necessary.

As interpreted, this alternative involves nine patterns. Findings are generally mixed, with sustainability sometimes depending on the status of the national government, rather than on anything at the school or village level.

Alternative 7c: **Make sure the children have plenty of time for play and games while in school.**

Note that the original goal of this alternative was to elevate attendance, not take care of the children. Pattern interpretations of the Alternative:

If children who have ample play time are apt to behave better, this would allow such children to improve the atmosphere for their peers. If this is the case, then this plan would be aligned with Pattern 7.

This plan allows for the notion that children shift gradually during their lives—away from game playing and toward more focused learning, rather than making this shift abruptly, such as in a few weeks. In this regard, this plan is aligned with Pattern 20.

Assuming that the children will feel an inward physical and psychological demand to engage in play and games, allowing them to do so is aligned with Pattern 30.

There is likely to be a range in how much time each child "needs" to play. Accommodating this range, if any, would be aligned with Pattern 31.

If the children feel drawn to attend the school to help satisfy an inward demand for play time, this coordination between accountability and school system function would be aligned with Pattern 32.

Depending on the school hours and lifestyles prior to the school's opening, the introduction of school could constitute a significant physiological and cultural disturbance for many of the children. If allowing children to play helps them to incorporate this disturbance more easily, this plan is aligned with Pattern 33.

This plan would help to maintain the children's exposure to a diversity of social and physical challenges, broadening their learning relative to a purely sedentary focus on academic information. On behalf of the children's resilience, this could be aligned with Pattern 36.

As far as a decision-making process goes, the contemplation of different possible solutions that would help increase or improve attendance is aligned with Pattern 41 (see alternatives 7a, 7b, 7d, 8a, 8b, 8c, 8d, 9e, and 9h).

To the degree that this supports the children's needs and makes school a valuable choice for them, it is a higher hierarchical effort than one of convincing the children to attend school frequently. Action at higher hierarchical levels is aligned with Pattern 43.

The original motive for this alternative was improved attendance. This strategy could be conducted in alignment with Pattern 44 by observing the effects that it has upon the school's attendance rates and success levels, and revising the strategy as necessary.

As interpreted, this alternative involves ten patterns. Sustainability findings for this alternative are positive overall, although there is some uncertainty related to the children's needs and probable behaviors.

Notice that the underlying goal of this alternative was to increase school attendance, and yet sustainability findings in most cases have little to do with attendance. Instead, they involve things such as interdependence of a created environment, gradual events, the ability of individuals to provide self-government, resilience, decision-making strategy, etc. Indeed, this is what can be so challenging but fascinating about complex systems. A decision made for one purpose can have numerous hidden sustainable and non-sustainable elements, oftentimes involving systems or issues that weren't the target of the decision in the first place.

Alternative 8e: If the national economy holds no prospects for them, **encourage the village children to leave the country to find work** once they are educated.

Pattern interpretations of the Alternative:

Individuals require environmental support in order to survive and thrive, and if indeed the village and nation cannot offer adequate support to individuals who even expend effort, then leaving the country in search of such support is in alignment with Pattern 5, on behalf of the children themselves.

Young adults could play special roles within the village community and culture, and advocating for their departure might be in conflict with Pattern 8.

In regard to each child, leaving the nation on a permanent basis would seemingly diminish opportunities for present and future individual non-material wealth accrual, at least in regard to community and culture. This would be in conflict with Pattern 10.

It isn't clear that there has been much consideration of the implications of this emigration upon the village. While bringing up the idea in an open-minded way would be aligned with Stage 1 of succession and Pattern 13, pursuing the idea without much more critical review skips over Stage 2 as a decision-making process, and is in conflict with Pattern 14.

Depending on the place they immigrated to, children leaving the country would be deprived of their stable context, which is in conflict with Pattern 18. Also, large numbers of children leaving the village could destabilize the village, and thereby the school. On behalf of the school, this would also be in conflict with Pattern 18.

Children leaving the country can be seen as an act of ungrouping, which is appropriate when the costs of grouping truly outweigh the benefits. The weight of costs versus benefits would determine the sustainability of this alternative relative to Pattern 21.

If conditions within the village are extremely crowded or competitive relative to another location, moving could allow children to obtain Significant Expression. However the people receiving these immigrants could be faced with diminished levels of expression. This encouragement could be aligned with Pattern 25 on behalf of the village, but in conflict with Pattern 25 on behalf of the people who are receiving these children.

This encouragement to enter other people's homeland could amount to a change in restraint Protection behavior toward other people per Pattern 26, which could decrease the sustainability prospects for these other people.

Leaving the village as children without offering adequate Protections has some likelihood of eventually stripping cultural and community identities from the children, according to Pattern 28. At least in some regards, this would be seen as unsustainable on behalf of the children.

If this alternative includes an individual choice in regard to leaving the country rather than a sort of requirement, then it is more aligned with Pattern 29.

Having a capability or willingness to be influenced by physiological needs and needs of others nearby is aligned with Pattern 30, as far as our encouragement goes.

If the village loses large numbers of its children or young adults, then it could be pulled outside of its historic range of variability according to several modalities. This is in conflict with Pattern 34.

If the children leave the village to enter another region with unlike people, and if the children also lack a culture of restraint, they could have an "invasive" effect. On behalf of the receiving group, this would be in conflict with Pattern 35. On the other hand, if the children are steered to immigrate into similar groups of people and they practice restraint, this would be aligned with Pattern 35.

If the education is designed to have the children eventually enter economic sectors that the village and nation are too small, homogeneous, unsupporting, or young to contain, then having the educated children leave the country for a more economically and occupationally diverse setting would perhaps be in keeping with Patterns 37, 38, 39, and 40, on behalf of an international economy. However, this does not mean that educating the children to partake in economic sectors that do not exist within the nation or village is sustainable at the child, village, or national level.

The contemplation of different possible solutions regarding the course of graduated students is aligned with Pattern 41 (see alternative 9d).

It is unlikely that individual children could ever enter a new land and feel entirely at home; however, Pattern 42 suggests that individual migrant children could sustainably reach some of the goals or portions of goals that native individuals are able to reach.

Perhaps educated children will benefit from leaving the country, but if they are being convinced to leave the country in order to make their schooling become relevant; this would be acting from a lower hierarchical level and in conflict with Pattern 43, relative to making the schooling more intrinsically relevant.

We don't really control the children and their choices, but we might influence them through our encouragement. Alignment with Pattern 44 would entail a willingness to observe results of this encouragement upon the children and perhaps even the school, and respond accordingly.

As interpreted, this alternative involves twenty-two patterns. Findings for this alternative are mixed but with significant areas of possible or probable non-sustainability. The non-sustainability manifests itself at various levels—including each child, the village, the nation, and/or place that the children immigrate into.

Notice that sometimes the actions one group takes to improve their own outlook can degrade the prospects for another group. Sometimes sustainability even looks like a "zero-sum-game," where adding to one side subtracts from the other. Convincing educated children to leave the country could provide them with a more viable *individual* future by diminishing survival prospects for Landesby and its remaining inhabitants. Or, avoiding extreme competition in Landesby by encouraging educated children to leave may simply export the problem, and place that competition upon other groups of people. As the alternatives are evaluated against the sustainability patterns, implications for systems, structures, and institutions well beyond the village are included herein to help demonstrate these complicated relationships.

It is also important to recognize that application of the sustainability patterns is subject to opinion, to a degree. One person might perceive an alternative-pattern connection by viewing the alternative from a particular angle. Whereas, looking from another angle, a second person might not perceive a connection. By thinking *very* broadly, creatively, and by extension, practically any action or alternative *could* be argued to involve any of the patterns.

The question then becomes, "Which patterns actually provide useful sustainability insight into a given situation?" Answering this question can also be subjective, depending upon which issues are most understood or most valued by a particular individual. Pattern 2 pertains to lack of a permanently fixed outer form. Perhaps one person is evaluating alternative 8e (encourage the village children to leave the country, etc.), and they envision that a demographic shift could occur in which the village gradually transitions from keeping all of its children to sending some away to foreign lands. They might believe that this demographic shift is simply part of a sustainable gradual shift in outer form, in keeping with Pattern 2.

A second person might consider an alternative 8e and Pattern 2 connection to be less interesting from a school perspective. Or they might view the connection as too uncertain or fabricated, because we haven't been given enough additional detail to verify that this population change would in fact represent a gradual shift in the outer form of the village and culture. A third person might view the population change as an actual breakdown of the village and its culture, rather than as a shift of the village and culture. The different opinions should probably be viewed as reasonable differences in priorities or assumptions regarding what a given plan or alternative will entail.

However, just because different people may apply different sustainability patterns to a given plan or situation does not mean that interpretations should vary at random. Interpretations of the connection between a given sustainability pattern and a specific plan should not vary considerably. In other words, although application of the sustainability patterns will be somewhat viewpoint dependent, the *meaning* of each pattern as presented herein is not intended to be highly subjective and varying.

Alternative 9g: **Encourage the villagers to enter negotiations with their enemies.**

Recall that the motivation for this was to help the villagers end battles for resources in order to reduce stress upon the children and improve their learning potential. Pattern interpretations of the Alternative:

> As far as a decision-making process goes, bringing up the idea of negotiation in an open-minded way would be aligned with Stage 1 of succession and Pattern 13. Regarding the conflict itself, we don't really know what successional stage it is in. Encouraging the villagers to move beyond Stage 2 of succession (in this case, the power-struggle stage of battle) could require them to accept conditions such as a greatly diminished homeland. Such conditions could fundamentally undermine their ability to experience Stage 4 of succession (in this case, a harmony or peace stage). At the same time, if the villagers are "stuck" in Stage 2 of succession, this encouragement could help them progress. Based upon the limited information we have about the regional conflicts, this encouragement is not necessarily aligned with Patterns 14 and 16.

> This plan indicates some understanding of the importance of the village moving past the conflicts of Stage 2 and into the increasing cooperation of Stages 3 and 4. The intent of this plan is aligned with Pattern 17.

> The village is a large group. If a few people represent the village in negotiations, they should have advanced understanding of the conflict and the needs of the villagers in order to negotiate in alignment with Pattern 24.

> If they have a diversity of potential responses to conflict (including fighting and negotiation), the villagers are more resilient via Pattern 36.

> The contemplation of different possible solutions that would help maintain student attention is aligned with Pattern 41 (see alternatives 9a, 9b, 9c, and 9f).

> Partially freeing children from the logistical stress of warfare, and thereby facilitating their ability to learn, is operating at a higher level compared to struggling against their levels of distraction inside the school. Thus this effort would be aligned with Pattern 43.

As interpreted, this alternative involves eight patterns. Findings associated with this alternative are largely positive, but a key area of uncertain sustainability is the successional stage that a village conflict is in.

Alternative 10d: Depending upon the size of the building, we might need to **hire a janitor or groundskeeper.**

Pattern interpretations of the Alternative:

> The notion that all stable complex systems require internal activity (in this case, upkeep) is aligned with Pattern 1.

Assuming that the income distribution within the village falls within 2:1/5:1/20:1 ratios, and if school employee pay is referenced to these ratios, then this hiring would not be in conflict with Pattern 9.

Being concerned about the condition of the school because the children are subject to its condition is akin to being steered by the status of peers and associates, which is in line with Pattern 30.

Paying someone for this work positions them to maintain their own financial status by way of maintaining the building and grounds. In this respect, this plan would be aligned with Pattern 32.

The act of paying someone for their work can be seen as acting at a higher hierarchical level relative to trying to urge them to complete that work based upon their "good will." Also, maintaining the building can be seen as acting at a higher level compared to letting the building deteriorate and then trying to fix the problems afterwards. From both of these perspectives, this alternative is in keeping with Pattern 43.

As interpreted, this alternative involves five sustainability patterns. Findings for this alternative are all positive, though alignment with Pattern 9 is contingent upon pay scales. Notice again how a single alternative or action can implicitly include sub-issues or sub-actions. Because of this, one alternative can involve a pattern more than one time (as interpreted, Pattern 43 was involved two separate times).

Alternative/Outcome 11a: **Children will be educated** in areas of reading, history, math, science, etc.

Pattern interpretations of the Alternative:

Having these skills would allow some of the children to become teachers themselves, which might help to uphold a village school. This would be aligned with Pattern 6 and maintaining an environment. However, there are probably a limited number of teaching jobs, and unless other applications for these skills exist, this education could effectively detract from other training that is necessary for maintaining an environment, which would be in conflict with Pattern 6.

These academic subject areas may not coincide with the ways that some children are inclined to work and to contribute to their environment, and those that are inclined toward these subjects may be inclined toward different amounts of reading, history, math, etc. Depending upon the intensity with which each child will be taught these subjects relative to their inclination for them, and relative to the necessity of these subjects for maintaining the village and nation, this outfall of the school is questionable on behalf of some children, in regard to Pattern 8.

If the children's attained knowledge has some application or value, it is a form of intangible wealth accrual, which would be aligned with Pattern 10.

It isn't clear that there has been a full debate over what a school could or should teach. If a full debate has not occurred, the decision-making process behind this plan is in conflict with Pattern 13. On the other hand, generally exposing children to more subjects is congruent with Stage 1 of succession and Pattern 13.

If this education does not assist the children in an eventual livelihood, then it could slow their progress toward a successional Stage 4 in their own lives, which would put this in conflict with Pattern 17.

If this education is intended to target certain occupational sectors, alignment with Pattern 19 would be indicated by an assumption that the curriculum might have to be continually updated in its attempt to prepare the children.

The incorporation of these complicated topics by each child is likely to occur over an extended time period. The educational goal in this case is aligned with Pattern 20 if there is a plan to extend schooling for enough time (probably years) for this goal to be achieved.

This education goal assumes that by the virtue of being children in the same village, all would benefit from knowing more about the same academic subjects. Unless this assumption is verified, it is in conflict with Pattern 21.

It isn't clear that there has been any village-wide collective decision process to determine that the children should be learning these subjects. Unless this occurs or unless appropriate representation is planned, the making of this goal is in conflict with Pattern 23.

The choice of curriculum could strongly impact the conditions in the village. If this decision is to be made via representation, then the deciders have to have special information and must be in synch with a unified village. Otherwise, such representation would not be aligned with Pattern 24.

Putting the children on this educational track doesn't necessarily position them to avoid extreme competition with each other or with others in the country. Depending upon the economic opportunities available to people with this training, this curriculum may or may not be aligned with Pattern 25.

When children learn these subjects, it could effectively remove barriers and synchronize the village with other group value systems. On behalf of the village, this could be in conflict with Pattern 28. On the other hand, if these skills helped villagers to obtain local jobs that outside groups would have otherwise gotten, or if the skills could be used to help preserve and protect local interests, then they would serve as a Protection under Patterns 28.

If all choice is somehow being removed from the children and their families regarding the appropriateness of this learning path, it is likely to be in conflict with Pattern 29.

If this education provides skills that help the children to respond well to the four areas of accountability relative to alternative learning or living pathways, this school goal is aligned with Pattern 30.

An unquestioned assumption that all children will benefit from and respond to this education equally would be in conflict with Pattern 31.

If the children find these subjects boring or feel that they have little value, then they will be disinclined to attend school, creating a conflict between self-government and school system function relative to Pattern 32. However, if the children find these subjects interesting or believe that learning them will improve their lives, then they will be inclined to attend school, in alignment with Pattern 32.

Depending upon whether this number of subjects is large (diverse) or small (non-diverse) relative to the number of subjects the children *could be* exposed to, this plan could be aligned or in conflict with Pattern 36.

If the school is part of a national education system, then industries across the nation that offer jobs requiring this training would essentially exist in a more economically or environmentally supportive atmosphere. This would sustain a higher diversity of these industries via Pattern 39.

It isn't clear that there has been much exploration regarding learning alternatives for the children. The narrow assumptions of a school and the teaching of certain subjects would be out of synch with Pattern 41.

Perhaps the goals of the village school were inspired partly from similar goals in other villages, towns, or cities. Even if these situations are very different from the village, a willingness to contemplate them is in keeping with Pattern 42.

This educational outcome of the school is probably the *main goal of having a school*. How does this goal look from the perspective of the sustainability patterns? As interpreted here, this educational outcome involves twenty patterns. Findings for the outcome are quite mixed, with lots of positive potential and lots of negative potential, depending upon the way that decisions are made, what assumptions are made along with providing the education, the real economy, and other pressures the children will face after school.

In most cases, whether the weight of this goal becomes more negative or more positive depends on actions that could conceivably be taken, and on information that we might be able to obtain with a reasonable effort. For example, the issues arising with Patterns 23 and 24 have to do with proper group decision making and representation. Knowing this, perhaps a decision-making process or plan for representation could be initiated. Or, consider the outfalls relative to Pattern 28, which could diminish local culture and self-expression, or preserve local culture and potential for self-expression, depending on outside conditions. It might be possible to look at social and economic conditions and predict whether the villagers' values will be gradually enhanced or diminished by the school curriculum. Although addressing possible negative and uncertain aspects of the educational outcomes (i.e., education in reading, history, math, etc.) would require some extra effort, it seems that it could be accomplished.

In contrast, recall that when the proposed school was originally analyzed through a logical reductionist approach back in Chapter 5 (Sections 5.11 through 5.13), we eventually reached a breakdown point. We had tried to understand how our system should work by looking at all of the individual constituents and interactions. This required logical component deconstruction, analysis, and reconstruction to guide the development of school alternatives. But when we did this, there became too many ideas about how to deal with too many things (i.e., "analytical chaos"). Many of the developed alternatives seemed contradictory, and it became unclear as to how the best choices could be made. But by viewing alternatives through these sustainability patterns, the weak, strong, and questionable sustainability aspects of the alternatives can be highlighted.

If an alternative comes out of a sustainability pattern analysis with very positive prospects (such as in siting the school building away from areas undergoing environmental change, or praising the teachers for school successes), then that alternative can be retained in the plans. But when alternatives have pervasive conflicts with the sustainability patterns (such as when we encourage the villagers to abandon their native language, or encourage their children leave the country), they can be removed from further consideration.

Alternatives that generate a mixed sustainability profile can be amended in some cases, or considered against alternatives with slightly better profiles. Through this evaluation process, what was a case of "analytical chaos" can be tamed somewhat into a process of idea selection.

Having scrutinized the sixteen alternatives relative to each of the forty-four sustainability patterns, our design team determines that certain aspects of a sustainable school design have been identified, but that serious questions need to be answered and concerns need to be addressed. Depending on how those questions are answered, and how those concerns are addressed, a sustainable design might be attainable. We return to the school advocates and report our findings.

18.3 Sustainability Pattern Application Summary

Following this look at how the forty-four sustainability patterns can apply to school alternatives, here are some final points and observations on pattern application:

1. The patterns are commonly generic or abstract, pertaining to things such as time, stability, and grouping, rather than pertaining to certain prescribed scenarios. The generic quality of the patterns means their application often requires abstract thinking.

2. In some cases, the application of patterns is rather certain. For example, an attempt to make an economy more resilient by diversifying it certainly involves Pattern 36. But a similar sounding situation could be less certain: if we are making an economic change, then *if* that change makes the economy more diverse, we could infer via Pattern 36 that we are also incidentally making the economy more resilient. So intentionally diversifying the economy has a definite Pattern 36 application, but a mere change in the economy involves this pattern only "if" something else holds true, or contingent on other conditions.

3. The element of contingency, or pattern application *if* certain conditions hold true demonstrates how complicated the sustainability of a given action is. It is sustainable *if* it is carried out in a certain way or *if* certain other circumstances hold true. This points to one of the underlying difficulties with understanding sustainability—something that appears fitting and sustainable in one situation can be problematic in another. The forty-four sustainability patterns indicate that *many actions are not inherently sustainable or unsustainable*. Instead, *actions are potentially sustainable depending upon the situation*.

4. A single action can relate to, or "involve" several patterns. For example, making a change in the economy could simultaneously involve wealth, succession, and decision-making (perhaps Patterns 10, 14, 15, and 23). One reason this can occur is that the same action can be subjectively interpreted through the lens of completely different patterns (that is, one thing is being seen in different ways or from different points of view). A second reason is that many "single" actions are really composites of several smaller actions. Hiring a janitor implicitly includes the separate actions of determining what the job will entail, locating an individual for the job, and paying the person after they perform the work. Each of those sub-actions can then involve multiple patterns.

5. Sometimes, an action relates to and involves several patterns for almost redundant reasons. For example, an action may involve Patterns 21 and 23, which involve group cohesion and group decisions. While redundancy may seem to be a waste of energy, it can also be seen as beneficial because it can assure the analyst that similar results from different patterns are reasonable and pointing in the same direction, and because it helps strengthen the net that is cast—capturing issues that would otherwise be missed.

6. Analysis of an action via the forty-four patterns can sometimes indicate compartmentalized sustainability or competing sustainability. A decision-making process might be sustainable, but the actual decision might not be. Or, a decision may improve local sustainability while reducing regional sustainability, etc. For example, encouraging the villagers to enter negotiations with their enemies may be aligned with successional development. But in contrast, a decision to enter negotiations or the negotiations themselves could be conducted inappropriately relative to group representation.

7. A single action viewed through the forty-four patterns can yield very positive results (i.e., positive alignment with sustainable patterns), very negative results (i.e., lots of conflict with sustainable patterns), or a combination of positive and negative results. For example, establishing the school to be as independent and non-reliant as possible upon the national government yielded mixed results, depending on various factors. Because of the abstract nature of many patterns (point #1 above), it may not be clear at all prior to careful review whether an action will appear relatively aligned or misaligned with sustainability patterns.

Here are additional suggestions toward understanding sustainability, designing for sustainability, and the application of the sustainability patterns:

1. Even if a system has been determined to be very sustainable, there is always a possibility that an outside force will change its sustainability trajectory, making it effectively unsustainable.

2. In cases where an action is aligned with some patterns (i.e., looks sustainable), but is unaligned with other patterns (i.e., unsustainable) it might mean that the action could be sustainable if modified or if provided with additional backing or energy. Recognize, however, that any systematic subsidization required to hold a design in sustainable alignment is likely to detract from the system's functionality or sustainability anyway (this is because the system that is giving the subsidy often becomes weaker, and the weakened system may eventually have to withdraw its subsidy).

3. A single action, such as walking to the store, can be analyzed through the forty-four sustainability patterns. However, when analyzing an entire system design, multiple decisions and design aspects need to be reviewed. This review may result in multiple (evidently) sustainable features, unsustainable features, and features with mixed or contingent sustainability. The goal might be to choose all of the most sustainable features. Even so, due to a combination of internal or external constraints, it may be impossible to establish an entirely sustainable design. In such cases, one could elect to essentially embrace the most sustainable design features, reject the least sustainable features, and adopt features with mixed sustainability as needed to complete the system.

4. Analysis of an artificially isolated choice or design aspect relative to the forty-four patterns can be useful and informative, but a review of one aspect alone is not enough to determine the sustainability of a whole system to which that aspect belongs.

5. In some scenarios, alignment with or "violation" of a sustainability pattern can occur as a matter of degree. A system aspect that is just slightly out of synch with a pattern should normally be more sustainable than one that is extremely out of synch with the same pattern. For example, perhaps two projects violate Pattern 12: one generates four pounds of unrecyclable waste, and a second project generates fifty tons of unrecyclable waste. While both projects are perhaps in "complete

violation" of Pattern 12, it is likely that the violation due to the second project will have stronger negative repercussions for sustainability than the violation due to the first project.

6. Sometimes, an action will support one complex system at the expense of another complex system. Someone might even encounter situations in which it seems appropriate to support one system and purposely diminish others. However, this should be considered carefully, as destabilized systems may create unexpected conditions for the systems they interact with.

7. The forty-four patterns exist relative to each other and generally should not be thought of as variably important. In nature, these patterns interplay in a dynamic balance, without some patterns overwhelming the others. It should be assumed that an imbalanced emphasis on one pattern to the point that others are greatly undermined reduces the system's likelihood for sustainability.

8. System sustainability should be thought of as an outfall of all design aspects relative to the forty-four patterns. However in many systems, it is very likely that certain more critical or more central design aspects will have a stronger influence upon that system's sustainability. For example, the decision to teach in the native language would probably be more impactful upon a school's sustainability than the decision to have the children eat lunch inside or on the playground.

9. To some degree, different people are likely to view an action from different vantage points. No two people are expected to view an action's sustainability in exactly the same way. However, strongly different ways of viewing the sustainability of a given action is probably an indication that people are considering completely different aspects of that action, or that the meanings of one or more sustainability patterns are being misinterpreted.

The forty-four sustainability patterns don't show anyone exactly how the world *should* look. They don't show how it *will* look in the future. Rather, having the patterns is like having an owner's manual that shows what the most sustainable parts look like. It doesn't predetermine how those parts have to fit together, and it doesn't explain all the things that can be made with those parts. What will you make of them? Will you build a cabin? A mansion? A skyscraper? A playground? A bridge? Maybe. And, perhaps you will build things for which we do not yet have words.

About the Author

Steve Thomas is an ecologist living in Michigan. In his spare time he likes to watch plants. You can reach him at www.natureofsustainability.com